Industrial Intelligent Control

Industrial Intelligent Control

Fundamentals and Applications

Yong-Zai Lu
Bethlehem Steel Corporation
Chesterton, Indiana, USA

JOHN WILEY & SONS
Chichester • New York • Brisbane • Singapore • Toronto

Baffins Lane, Chichester,
West Sussex PO19 1UD, England

National 01243 779777
International (+44) 1243 779777

Other Wiley Editorial Offices

John Wiley & Sons, Inc., 605 Third Avenue, New York, NY 10158-0012, USA

Jacaranda Wiley Ltd, 33 Park Road, Milton, Queensland 4064, Australia

John Wiley & Sons (Canada) Ltd, 22 Worcester Road, Rexdale, Ontario M9W 1L1, Canada

John Wiley & Sons (SEA) Pte Ltd, 37 Jalan Pemimpin #05-04, Block B, Union Industrial Building, Singapore 2057

British Library Cataloguing in Publication Data

A catalogue record for this book is available from the British Library

ISBN 0 471 95058 0

Typeset by Thomson Press (India) Ltd, New Delhi, India
Printed and bound in Great Britain by Bookcraft (Bath) Ltd

To my wife, Feichan

Contents

Foreword

Large scale use of automatic control in industrial production has been very successful. Many industries today depend critically on automatic control which has contributed to improved quality and less of energy and raw material. The success has been achieved largely by automating control and supervision tasks. The automated systems can perform these tasks much better than humans. In spite of this success there is a common belief that current automation systems can be improved substantially by providing them with more human-like features, such as ability for learning, adaptation, reasoning, understanding and abstraction. It is, however, not clear how to design technical systems with such features. As a result there has been a significant interest in research in "intelligent" systems. Some caution is required when using this word whose everyday meaning is capacity for reasoning, understanding and similar forms of mental activity or outward manifestation of internal actions in the brain.

Professor Lu, who has a unique background from academia and industry, has written a book that summarizes some of the ideas currently explored to make better industrial control systems. One feature of the book that I appreciate very much is the emphasis that the solution lies in a combination of classical control techniques of dynamic modeling and control with new approaches. In his book he has attempted to present a unified approach. There is material on modeling, estimation, optimal control and multivariate statistics and quality control. These are combined with ideas from fuzzy systems, expert systems, neural networks and learning, fault detection and diagnosis. The key ideas are presented very clearly and they are illustrated with industrial examples. The book should be very useful to readers who would like to get an overview of different approaches to intelligent systems and their industrial use.

Professor K. J. Astrom
Lund Institute of Technology
Sweden

Preface

It is a challenge in control and computer communities to explore novel control strategies and philosophies for complex industrial processes. The application of intelligent system technologies in industrial control has been developing into an emerging technology, so-called *'Industrial intelligent control'*. This technology is highly multi-disciplinary and rooted in systems control, operations research, artificial intelligence, information and signal processing, computer software and production background. Some pioneering contributions have been made during recent decades, for example in the fields of *'learning automata and neural control'* by Tsetlin and Narendra, *'expert control'* by Astrom, *'fuzzy set theory and fuzzy control'* by Zadeh, etc. All these strategies share a common goal to design a robust and adaptive (or learning) control system under an uncertain or unknown environment with very limited (mathematical) knowledge of process principles and imprecise and/or incomplete information.

It is hard to give an exact and unique definition of an intelligent system, since 'intelligence' has different conceptual meanings and various levels. However, in this book, the term 'intelligent system' is defined as a man-made system (or machine) with some 'human-like' behavior, such as parallel information processing, associative memories, supervised and/or self-organized learning, inference and evolutionary strategies, etc. An intelligent system should be able to work effectively under a 'deterministic', 'stochastic' or 'uncertain' environment. Secondly, an intelligent system can interface with math-algorithms, data and data patterns, linguistic rule oriented knowledge base and reasoning, and extract features from the system's raw information. Finally, the most important characteristic of an intelligent system is 'self-learning', 'associative memories' and 'self-organization', which make a man-made system smarter through training and operation.

The goals of this book are to introduce fundamentals and approaches to intelligent systems and their applications in industrial control for both academic and industrial readers. The structure of this book follows the main subjects in industrial control, namely, from system modeling, estimation, dynamic control and optimization control to quality control and fault detection. In fact, the concepts, fundamentals and system architectures popularly applied in math-model based control play a key role and foundation in the design of intelligent control systems. The ideas behind intelligent systems used in this book can be summarized as follows:

- An explicit 'math-model' or 'math-control-algorithm' is extended to an 'associative memory', 'pattern mapping', or 'rule match and inference'. The knowledge representation can be converted from one form to another. For instance, the fuzzy linguistic rules representing structured knowledge can be converted into the relevant fuzzy associative memory under a numerical frame. On the other hand, the fuzzy knowledge (rules) can be extracted from the system's numerical input and output data through self-organized learning.

- The math-algorithm oriented learning and adaptation are extended to the upgrade of associative memories and feature extraction with supervised, self-organized and rule learning. The term 'robustness' or 'adaptation' commonly used in math-model based control is measured by the performance of 'generalization' in intelligent systems. It is an important and practical principle that good generalization is a necessary condition to put an intelligent system in practice. In general, the generalization depends on system architecture, parameters, learning algorithms, and particularly the difference between the training (or learning) environment and the working environment.

- Generally, analog discrete, binary and multi-valued signals are the major forms used in industrial control. The signal coding and decoding are two important measures in information processing. For example, a multi-dimensional vector can be coded into a scalar discrete number through a self-organized learning based data compression. The linguistic variables can be coded into numerical variables through fuzzy inference and defuzzification, and similarly, the analog variables can be coded into the linguistic variables with fuzzification. In comparing with math-model based control, the applications of signal coding and decoding techniques make intelligent systems more flexible and manageable.

- The strict definitions of 'optimal' solution or 'optimization' in the mathematical sense popularly used in modern control or operations research, are relaxed to an 'approximately optimal' or 'satisfactory' solution. Obviously, such changes are acceptable in the control industry.

- The system structure properties, such as 'stability', 'controllability' and 'observability' are conceptually applied in the design of intelligent control systems.

The intelligent system methodologies illustrated in this book are neural networks, fuzzy logic, rule based systems, pattern recognition, evolution (genetic) algorithms, etc., and some hybrid forms with their combinations. To make this book more practical and applicable, many industrial worked examples from the oil refineries, steel mills and others are introduced in detail. It is the author's wish that this book will help practicing control and software engineers to learn fundamentals, strategies and methodologies of industrial intelligent control, and on the other hand, will help academic researchers and graduate students to learn how to apply academic research results to real industrial environments.

This book has been written as a text book based on the author's long-time teaching and research experience from a career as a university professor and his recent research and development experiences from the industrial environment.

Some topics covered in this book were given at a number of short courses for process control and operations research engineers at the Information Technology Department, Burns Harbor Division, and Homer Research Labs, Bethlehem Steel Co., and a graduate level course of seminars at Zhejiang University, Hangzhou, China. To read this book, an undergraduate level mathematics and control background is required.

The author has made great efforts to give this book a strong applications background through avoiding too much sophisticated mathematical detail and introducing many interesting industrial worked examples. This book can be used as either a reference text book for graduate students and faculties at engineering schools, or a short course and self-learning text book for practicing engineers, researchers and consultants in the control industry (end users and control software developers). If the reader would like to learn more details of the principles and methodologies of neural networks and fuzzy systems, the books entitled *'Neural Networks: A Comprehensive Foundation'* (by S. Haykin) and *'Neural Networks: A Dynamic Systems Approach to Machine Intelligence'* (B. Kosko) can be consulted.

This book is divided into two parts. Part one includes the first three chapters and addresses the fundamentals and methodologies of intelligent systems. Chapter 1 gives a general introduction to modern control technologies, system environment and uncertainty, and intelligent systems. This chapter provides general concepts and fundamentals of industrial intelligent control. Chapter 2 introduces the concepts, fundamentals, architectures and algorithms of fuzzy logic, neural networks and rule based intelligent systems. Chapter 3 focuses on the learning strategies and algorithms used in neural network, fuzzy logic and rule based intelligent systems. Supervised learning, adaptive learning, reinforcement learning, self-organized learning, and deep reasoning and rule learning are covered in this chapter.

Part two includes Chapters 4 to 8 which cover various applications of intelligent systems in industrial control. Chapter 4 introduces math-model free techniques of process modeling and estimation. The applications of neural networks, pattern recognition and fuzzy logic in establishing an associative memory are introduced in detail. Chapter 5 describes the applications of intelligent systems in dynamic controls, namely, neural, fuzzy logic and expert controls. Expert optimization control and production scheduling are covered in Chapter 6. The major approaches to expert optimization applied in this chapter are fuzzy model based optimization, dynamic model based expert optimization, deep reasoning and rule learning, evolutionary algorithms, Hopfield network and simulated annealing. Chapter 7 introduces quality control and statistical process control (SPC) with multivariate statistics, i.e., principal component analysis (PCA) and partial least-squares (PLS), and the combination of PLS and neural networks. Chapter 8 investigates the applications of fuzzy logic and neural networks in fault detection and diagnosis. A number of industrial application examples are investigated in each chapter of part two.

Yong-Zai Lu
Chesterton, Indiana, USA
July, 1995

Acknowledgements

This book was written based on the author's long-time research and teaching experiences at Zhejiang University, Hangzhou, China, and the recent years' research and development opportunities at Information Technology Department, Burns Harbor Division, Bethlehem Steel Co. in the United States.

I wish to thank Zhejiang University, China Natural Science Foundation and many industrial sponsors, for example, China Petro-Chemical General Co., Shanghai Refinery Co., Annsan Iron and Steel Co., Coungqin Iron and Steel Co. and many others, etc., for their funding in both fundamental and application research. I am most grateful to my former graduate students: Drs. C. Xu, M. He, Y. Yang, D. Qian, N. Ye, S. Wang, S. Qin, T. Wu and Mr. Q. Yu, Mr. D. Fang and Ms. L. Zhou for their contributions in theoretical and application studies of intelligent systems.

I wish to thank Dr. W. N. Bargeron, the Chief Technical Officer and President of Bethlehem Steel Service Division, and the management team of Information Technology at Bethlehem Steel Co., T. J. Conarty, J. A. Barsic, J. A. Angstadt, A. E. Cobbs, R. L. Johns and H. R. Wetklow, for their strong support and encouragement in developing and applying intelligent systems in real industrial environments. I am grateful to my colleagues at Bethlenem Steel Co., S. Markward, S. Krass, N. Kelm, M. Barenie and many other engineers for their contributions in studying and applying neural network systems to real production lines.

In particular, I am deeply indebted to Professor K. J. Astrom, Lund Institute of Technology, Sweden, who has given freely of his time to read through the book and write a Foreword.

I am grateful to Prof. T. J. Williams, Purdue University, for providing me with great opportunities to study industrial control for large scale production systems during my sabbatical leave at Purdue Laboratory for Applied Industrial Control in early 1980s. I am also grateful to Prof. T. J. McAvoy, University of Maryland, for kindly showing me the remarkable research results of neural network applications in statistics, process modeling and dynamic control at his laboratory and for many valuable discussions when I visited his laboratory.

I am also deeply indebted to Professor S. J. Qin, University of Texas, Austin, and my colleague S. Markward for reading the manuscript, in full or in part, and making corrections and suggestions to improve the book.

I wish to thank Z. Lu, J. Wang and Dr. C. Xu for converting the Wordperfect version of the Manuscript to the Latex version.

I am grateful to A. Halligan (publishing editor), R. Hambrook (senior production editor) and A. Cresswell (editorial assistant) for their patience, understanding and efforts in publishing this book.

Finally, I would like to thank the following people and organizations for their permission to reproduce certain figures in the book:

- Figure 3.26 & Tables 3.2, 3.3 and 3.4 reprinted from Witten (1990) with kind permission of Academic Press, New York;
- Figures 3.27 and 3.28 reprinted from Greenspan *et al.* (1990), Morgan Kaufmann, San Mateo, CA;
- Figure 4.1 reprinted from Gelb *et al.* (1974) with kind permission of MIT Press, Cambridge, MA;
- Figure 4.3 reprinted from Lu and Williams (1983), ISA Publisher, Research Park, NC;
- Figures 4.11, 6.1, 6.2 and 6.4 and Table 6.1 reprinted from Lu and He (1991), with kind permission of Purdue Research Foundation, West Lafayette, IN;
- Figures 5.7, 5.8, 5.9, 5.14, 5.15, 5.17, 5.18, 5.19, 5.20, 5.21 and 5.22 reprinted from Lu (1995), Markward and Lu (1995), and Lu and Markward (1995) with kind permission from Bethlehem Steel Co., Bethlehem, PA, The USA;
- Figures 5.10, 5.11, 5.12 and 5.13 reprinted from Zhou and Kimura (1993), The IFAC World Congress, Sydney, 1993;
- Figures 6.5, 6.6, 6.7, 6.8 and 6.9 reprinted from Yang and Lu (1988) with kind permission from Elsevier Science Ltd, Amsterdam, The Netherlands;
- Figures 6.15, 6.19 and 6.20 reprinted from Choi (1994), 1st Asian Control Conference, Tokyo, Japan;
- Figures 6.17 and 6.18, and Table 6.4 reprinted from Shiba *et al.* (1994), 1st Asia Control Conference, Tokyo, Japan;
- Figure 7.2 reprinted from Kresta, *et al.* (1991) with kind permission of Canadian J. of Chem. Eng.;
- Figures 7.4, 7.6 and 7.7 reprinted from Qin and McAvoy (1992a) and Figure 8.1 reprinted from Isermann (1994) with kind permission from Elsevier Science Ltd, The Boulevard, Langford Lane, Kidlington OX5 1GB, UK;
- Figures 2.8, 2.12, 2.13, 2.14, 2.16, 2.17, 3.9, 3.13, 3.14, 3.15, 3.17, 3.18, 3.19, 3.20, 3.21, 4.7, 4.8, 4.9, 4.17, 6.13, 6.14, 6.16, 7.3, 8.7, 8.9, 8.10, 8.11 and 8.12, and Tables 4.1, 4.2, 4.3, 4.4, 4.5 and 6.3 reprinted from IEEE publications with kind permission from IEEE Operations Center. The original source and the IEEE copyright line appear prominently with each reprinted figure and table.

1 Introduction

Industrial intelligent control is a highly multi-disciplinary technology. This technology involves systems control, operation sresearch, computer technology, artificial intelligence (AI), and industrial process background and fundamentals. In general, industrial intelligent control implies applications of intelligent systems in industrial control. This chapter gives an overview of industrial control, modern control techniques, system environment and uncertainties, and intelligent systems.

1.1 INTRODUCTION TO INDUSTRIAL CONTROL

1.1.1 Historical review

For a long time, industrial control has faced three major bottlenecks: process modeling, advanced control strategies, and realistic approaches to system analysis and synthesis for complex systems. Control scientists and experts have been making great efforts to explore the future direction of industrial control during the last few decades (Benveniste and Astrom, 1993; IEEE, 1987). Seeking new technologies for advancing industrial control has become a challenge the control industry and community must face today. Recently revolutionary advances in computer technology, modern control and machine intelligence have opened a platform for the new generation of industrial control that may provide significant economic benefits through the production of quality products using an integrated computer, information and management system. Generally speaking, industrial control covers a wide variety of functions, such as dedicated dynamic control, supervisory control, fault diagnosis, production scheduling and planning, etc. They are organized in a hierarchical and/or distributed structure and intend to perform a plant- or corporation-wide optimization under desired global (production and business) performance criteria with corresponding constraints of production facilities, resources and marketing, etc.

The history of industrial control shows that the application level relied on the progress of computer, control and system engineering technologies. The control technology has been moving from conventional control to today's math-model based modern control and knowledge based intelligent control. It should be noted that several significant changes in control science and technology and related areas have been happening.

Firstly, industrial processes and facilities have been considerably changing and advancing during the past few decades. The main characteristics of today's industrial production are large scale, high speed, highly continuous, or discrete-event driven and flexible manufacturing. The serious requirements of product quality, operation safety and production optimization have been the major concerns of any industrial enterprise. The integration of strategic business plan, information, management, and production has been considered as a key in successfully managing and running an enterprise. From a business point of view, development of a new generation of industrial control with those functions has become a common goal in the control industry.

Secondly, modern control theory and techniques have been rapidly developed and successfully applied to system identification, estimation, optimization, robust and adaptive controls, particularly for linear systems. It should be stressed that these advancements have provided control engineers with more precise, quantitative and math-tools for control system design and analysis. Moreover, modern control strategies have been moving from academic research to the control industry.

Thirdly, there has been astonishing progress in computer science and hardware technology. It has been recognized by the industrial community that today's computers have large memory, high speed, low cost, and sophisticated man–machine interface environments, etc. As a result, computers and micro-processors have been widely used as basic tools in the control industry. In fact, the main challenge faced in the control industry today is not lack of reliable computers and their software environments for supporting advanced control, but shortage of realistic and feasible application software for the control industry (Lu, 1992).

All these changes have become a strong driving force in exploring novel control strategies for industrial applications. In addition to modern control technology, the applications of intelligent systems and combination of modern control with intelligent systems have been recognized as one of the most important directions in developing a new generation of industrial control. The recent research and development (R&D) made in artificial intelligence (AI) which includes knowledge engineering, pattern recognition, fuzzy logic, neural network and machine learning, etc., have provided great opportunities to solve complex systems control problems in industry.

1.1.2 General architecture of industrial control

Functionally, a hierarchical computer system for industrial control (Williams, 1984) can be generally divided into four levels, namely level one for dedicated dynamic control; level two for supervisory control; level three for coordination, production scheduling and planning, and level four for information processing and management. With the integration of business management and production, a level five for information oriented global strategic decision making and management may be added. On the other hand, the computer integrated manufacturing system (CIMS) was firstly proposed and investigated for discrete event dynamic (flexible manufacturing) systems not long ago, and recently, the applications of CIMS in process industry with continuous, batch and / or discrete event driven operations have also been explored.

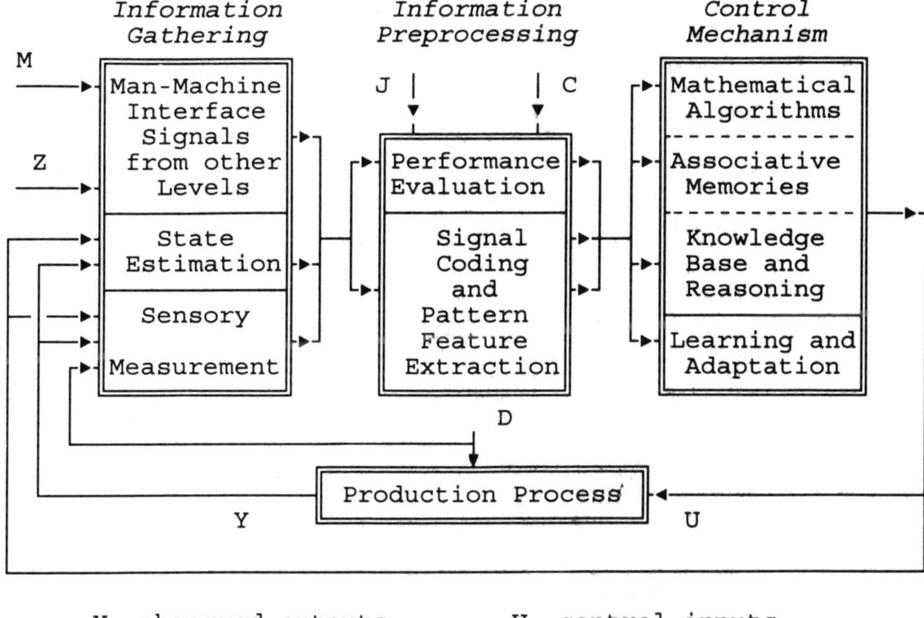

Figure 1.1 Generic architecture of industrial control systems

This section presents a functional framework of industrial control systems. This will help us to understand the problems and background that are involved in this book. In the real industrial world, there are tens of thousands of different industrial processes; however, from a system point of view, any specific physical process can be viewed as a 'system' with its input, output and state variables, which describe system external and internal information and knowledge, and consequently generic methodologies can be used in control system analysis and design. Modern control methodology illustrates the system and its components with specific mathematical representations, e.g., differential or difference equations. However, in an intelligent control system, system behavior can be described by a non-math-knowledge base, such as a rule based, fuzzy logic or neural network model. In this section, a general architecture of industrial control systems with various levels as shown in Figure 1.1 will be investigated. Generally speaking, any control or decision making system should consist of the following major functions.

Information gathering and preprocessing

Information plays a key role in systems control and decisions. In general, system information can be gathered through sensory measurements, man–machine interface and/or signal (data) communications. However, in many industrial processes, some information (or knowledge) can not be observed real-time through sensor measurement, for instance the chemical composition of a batch reactor, and the internal temperature profile and solidification progress of a steel slab in a continuous caster. As a result, this kind of internal information has to be constructed by a state estimator (or so-called 'software based sensor') in terms of system observed inputs and outputs. To ensure that the gathered information is of good quality and useful content, the signal processing, e.g. signal filtering and pattern quantization, may be used, if necessary.

Information processing (or so-called 'control mechanism')

The task of 'information processing' in a control system is to produce control actions on the basis of gathered system information. In a non-math-knowledge based system, the information processing involves signal coding and decoding, associative memory and reasoning associated with mathematical algorithms under a numerical, multivalued (fuzzy) or symbolic framework. However, in math-model based control systems, the information processing is reduced to a 'control law' or so-called 'control algorithm' with numerical computation.

Learning and adaptation

In a general sense, the function of 'learning and adaptation' in a control system is to maintain the entire system (controlled system environment plus controller) performance at a satisfactory level even when the system environment is significantly perturbed. In order to make a control system robust and adaptive, the system math-model, control law, knowledge base or any other forms of control mechanisms should be automatically adapted to the changes of system environment through on-line identification, reasoning, knowledge acquisition and learning.

Adaptive systems have been a popular multi-disciplinary topic, and several engineering and scientific disciplines study how adaptive systems respond to

Table 1.1 Different views of adaptive system studies

Discipline	Study adaptive systems as
Electrical engineer	Signal processing, nonlinear filtering, coding theory, circuit design, adaptive control
Computer scientist	Algorithm and automata theory, computer design robotics, artificial intelligence
Mathematician	Function approximation, statistical estimation, combinatorial optimization, dynamic systems
Biologist	Neuroscience, biophysics, ecology, evolution, population biology
Psychologist	Reinforcement learning, psychometrics, cognitive science

stimuli. Table 1.1 summarizes these studies (Kosko, 1992). In this book we study adaptive systems as intelligent systems. Finally, it should be emphasized that both 'feedback' and 'feedforward' control are two entities in any industrial control system.

1.2 HIGHLIGHTS OF MODERN CONTROL

Advanced industrial control can be considered as a combination of modern control, expert control, fuzzy control and neural network control (Lu, 1992). This book addresses the topics of intelligent control the concepts, strategic architecture and design methodology of modern control will be used and referred to throughout this book.

Modern control theory and techniques are based on the pioneering research results of Kalman filtering (Kalman and Bucy, 1960), maximum principle (Pontryagin *et al.*, 1962), dynamic programming (Bellman, 1957), etc. The development of modern control makes it possible to furnish industrial control with more advanced and sophisticated control strategies and algorithms. This section presents an overview of modern control techniques that will be particularly beneficial to those readers who do not have strong backgrounds in modern control.

1.2.1 System state space representation

In contrast to the system input–output (I/O) model (differential or difference equation, transfer function and frequency response) commonly used in classical control, system dynamics in modern control are governed by a state space model which describes dynamic numerical relations between system inputs, outputs and state variables in the time domain. The state space representation plays a key role in system design and analysis with modern control technology.

As we mentioned before, any real world problem can be represented as a 'system', which has an organized structure with input, output, state and other variables describing its internal and external behavior. As a result, the general methodologies of system analysis and design can be applied to a wide variety of problems in technology or non-technology areas. In general, a system's steady state behavior is governed by linear or nonlinear algebraic equations describing the static relations between system inputs and outputs. Here we give an example to describe the concept of process static modeling.

Example 1.1 (Franks, 1972)

Let a pipe heating process be a lumped-parameter system. This means that the system parameters are independent of the spatial dimension. The temperature of a pipe in a gas furnace can be derived from steady state heat transfer equations:

– Radiant heat flux from gas to pipe outer surface:

$$Q = K_1(T_g^4 - T_s^4) \tag{1.1}$$

– Heat conduction from pipe outer surface to inner surface:

$$Q = K_2(T_s - T_w) \tag{1.2}$$

– Heat conduction from pipe inner surface to process stream through a film:

$$Q = K_3(T_w - T_p) \tag{1.3}$$

where T_g, T_s, T_w, T_p represent the temperatures of gas, pipe outer surface, pipe inner surface and process stream respectively; and K_1, K_2, K_3 denote the heat transfer coefficients and can be numerically given as follows:

$$K_1 = 1.2 \times 10^{-9}$$
$$K_2 = 70 + 0.007(T_s + T_w)$$
$$K_3 = 6$$

The steady state relations between T_g and T_p can be numerically solved with equations 1.1 to 1.3.

The general form of the system steady state model can be written using the following linear and nonlinear matrix equations:

$$Y = AU \quad \text{or} \quad Y = F(U) \tag{1.4}$$

where

$U \in R^r$ is an r-dimensional input vector;
$Y \in R^n$ is an n-dimensional output vector;
$A \in R^{n \times r}$ is an $n \times r$-dimensional matrix describing the linear relations between U and Y;
$F(*)$ is an n-dimensional nonlinear matrix function describing the nonlinear relations between U and Y.

In classical control, transfer functions are used as the common language by control engineers. For instance, in the following example the transfer function shows a second order system with two poles $(-1/T_2, -1/T_3)$, one zero $(-1/T_1)$, pure time delay (τ) and gain K:

$$K(s) = \frac{Y(s)}{U(s)} = \frac{G(T_1 s + 1)e^{-\tau s}}{(T_2 s + 1)(T_3 s + 1)} \tag{1.5}$$

The transfer function 1.5 describes system I/O dynamics in the S-frequency domain for a continuous time system. In addition to the S-transfer function, system

I/O dynamics can also be described by a Z-transfer function for a time-discrete system. Both S- and Z-transfer functions can be converted to their relevant differential and difference equations respectively. However, the I/O dynamic models can only provide the dynamic relations between system observed input and output. Now we illustrate the system representation in state space form through the following example.

Example 1.2

A metal plate gas heating process as shown in Figure 1.2 is a typical distributed parameter system, which means that system parameters, e.g. plate temperatures, are not only functions of time, but also of the space. The plate unsteady-state heating process is governed by the following partial differential equation and corresponding boundary conditions:

$$\frac{\partial^2 T(x,t)}{\partial x^2} = \frac{\partial T(x,t)}{\alpha \partial t} \tag{1.6}$$

$$\left. \frac{\partial T(x,t)}{\partial t} \right|_{x=0} = K_l(T_l(t)^4 - T(0,t)^4) \tag{1.7}$$

$$\left. \frac{\partial T(x,t)}{\partial t} \right|_{x=\delta} = K_r(T_r(t)^4 - T(\delta,t)^4) \tag{1.8}$$

where

$T(x,t), x \in [0,\delta]$ is the dynamic response of the plate temperature profile in the thickness direction;
α is the thermal diffusivity;
K_l and K_r are the heat transfer coefficients for plate left and right hand sides respectively;
T_l and T_r are the gas temperatures at left and right hand sides.

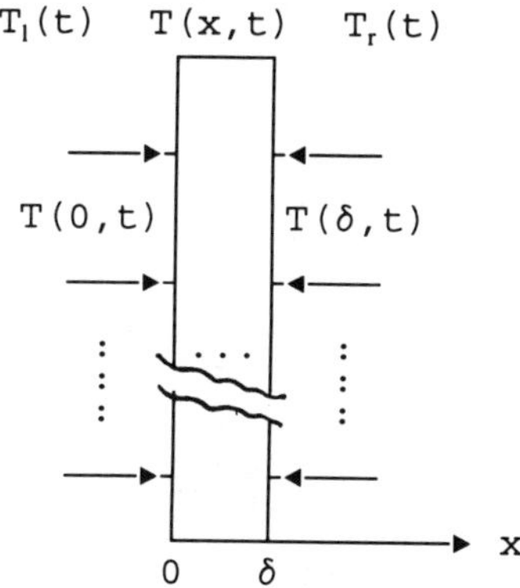

Figure 1.2 A metal plate heating process

Supposing that α, K_l and K_r are constants, equations 1.6–1.8 can be approximately converted into a linear, time-invariant, discrete state space model through discretization in both spatial and time domains (Lu *et al.*, 1983). The dynamic responses of the surface and inner temperature profile then can be described by the following state space model with state and output equations:

$$X_{k+1} = AX_k + BU_k \tag{1.9}$$

$$Y_k = CX_k \tag{1.10}$$

where

$X = [T_1, T_2, \ldots, T_n]^T$;
$Y = [T_1, T_n]^T$;
$U = [T_l, T_r]^T$;
U, Y, X are the system input, output and state vectors respectively;
A, B, C are the system matrices;
n is the number of nodes in the dimension of plate thickness;
k is the time instance.

where the elements of matrices A and B are functions of metal thermal properties, such as thermal diffusivity (density, specific heat and heat conductivity), heat transfer coefficients, and both time and spatial intervals used in discretization.

However, if the thermal diffusivity and heat transfer coefficients are considered functions of temperature, e.g. $\alpha = \alpha(T)$, $K_l = K_l(T_1, T_l)$ and $K_r = K_r(T_n, T_r)$, then a nonlinear, time-varying, state space model will be obtained, namely the system matrices A and B are functions of time, and equation 1.9 can be rewritten as

$$X_{k+1} = A(X_k, U_k)X_k + B(X_k, U_k)U_k \tag{1.11}$$

where $A(\cdot)$ and $B(\cdot)$ denote the functional relations between the elements of A, B and X_k, U_k. From this example, we can see that the system state vector X describes both external (observed) information and internal (unmeasurable) information, and the system output vector $Y (Y \in X)$ is only a subset of state vector X and represents system external information available for on-line sensory measurement. In addition, system model (A, B, C) plays a critical role in system analysis, design and synthesis, for instance in determining system stability, observability and controllability, design of state estimation and/or other model based advanced control strategies, etc. The general structure of the system state space model can be illustrated in Figure 1.3.

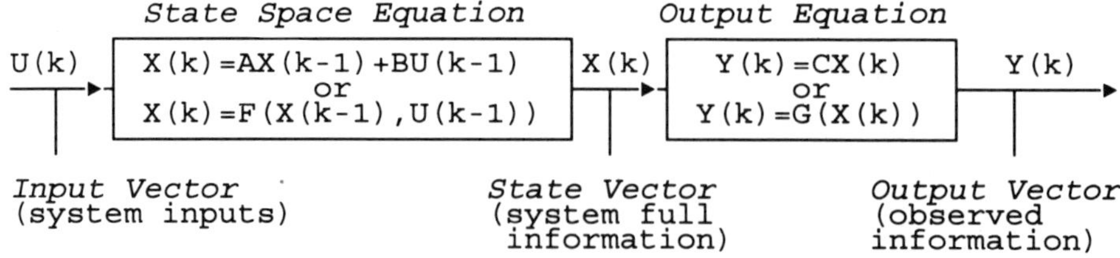

Figure 1.3 System state space representation

1.2.2 System structure properties

'Stability', 'controllability' and 'observability' are three key system structural properties being applied in math-model based control system analysis and design. However, the concepts of these properties are universal when designing any other type of control systems, e.g. expert, fuzzy logic and neural network controls, even though the relevant definitions and criteria for those math-model free systems have not been well established. In this section, we briefly review the general concepts of system stability, controllability and observability.

System stability and convergence

It should be emphasized that the concept of 'stability' is extremely important, because all workable systems must be stable. Since systematic studies on stability criteria for various control systems are beyond the scope of this book, here we only discuss the basic concepts of system stability that can be used in almost all topics (e.g., identification, estimation, control, optimization, adaptive control, and learning, etc.), involved in industrial control. The concept of 'convergence' being an independent property is also widely used in computational algorithms, heuristic searches and learning systems. Now we introduce the major definitions of stability and convergence for an autonomous nonlinear dynamic system with equilibrium state X_e as follows (Cook, 1986; Haykin, 1994):

Definition 1.1 *Suppose there exists a continuous time autonomous dynamic nonlinear system*

$$\frac{dX(t)}{dt} = F(X(t)) \tag{1.12}$$

A constant vector $X_e \in M$ is said to be an equilibrium state of the system 1.12 if the following condition is satisfied:

$$F(X_e) = 0 \tag{1.13}$$

or

$$\left.\frac{dX(t)}{dt}\right|_{X=X_e} = 0 \tag{1.14}$$

Clearly, an equilibrium state corresponds to a saddle point or so-called 'local minimum' in the trajectory of $X(t)$.

Definition 1.2 *The equilibrium state X_e is said to be uniformly stable if, for any given positive ϵ, there exists a positive δ such that the following condition is satisfied:*

$$\|X(0) - X_e\| < \delta$$

implies

$$\|X(t) - X_e\| < \epsilon, \quad \forall t > 0$$

This definition notes that the system trajectory can stay within a small neighborhood of the equilibrium X_e if the initial state $X(0)$ is close to X_e.

Definition 1.3 *The equilibrium state X_e is said to be convergent if there exists a positive ϵ such that the following condition is satisfied:*

$$\|X(0) - X_e\| < \delta$$

implies

$$X(t) \rightarrow X_e \textit{ as } t \rightarrow \infty$$

This definition means that if the initial state $X(0)$ is close enough to X_e, then the trajectory $X(t)$ will approach X_e when time t approaches infinity.

Definition 1.4 *The equilibrium state X_e is said to be asymptotically stable in the large, or globally asymptotically stable, if it is stable and all trajectories of the system converge to X_e as time approaches infinity.*

From these definitions we realize that only when both properties (stability and convergence) are satisfied, we have asymptotic stability. Both stability and asymptotic stability of an equilibrium state for a nonlinear dynamic system have been defined. The issue of determining the stability for a dynamic system is a difficult task in system design and analysis. Although the stability analysis might simply be provided through directly solving the state space model under all possible working conditions, in many cases particularly for a high dimensional system this would be time consuming and not practical. In modern control theory, Lyapunov's method has been recognized as an elegant approach to identifying the existence of stability for state space model based dynamic systems. The Lyapunov theorems can be used to identify dynamic system stability with a defined Lyapunov function. A Lyapunov function $V(X)$ is defined as a scalar, positive-definite function in state space and should satisfy the following conditions:

- The function $V(X)$ has continuous partial derivatives with respect to the elements of the state vector X;
- The function $V(X)$ should be equal to zero at the equilibrium state, i.e. $V(X_e) = 0$;
- $V(X) > 0$ if $X \neq X_e$.

The Lyapunov theorems for stability analysis based on the system state space model are given as follows (Haykin, 1994):

Theorem 1.1 *The equilibrium state X_e is stable if, in a small neighborhood of X_e, there exists a positive definite function $V(X)$ such that its derivative with respect to time is negative semidefinite in that region.*

This theorem can be simply described as that the equilibrium state X_e is stable if

$$\frac{d}{dt}V(X) \leq 0 \quad \text{for } X \in \psi - X_e \tag{1.15}$$

where ψ is a small neighborhood around X_e.

Theorem 1.2 *The equilibrium state X_e is asymptotically stable if, in a small neighborhood of X_e, there exists a positive definite function $V(X)$, such that its derivative with respect to time is negative definite in that region.*

This theorem can also be simply described as that the equilibrium state X_e is asymptotically stable if

$$\frac{d}{dt}V(X) < 0 \quad \text{for } X \in \psi - X_e \tag{1.16}$$

It should be pointed out that Lyapunov theorems can be applied to provide the global stability analysis for dynamic systems without having to solve the system state space equation. Obviously, the selection of a suitable Lyapunov function becomes a key point in applying Lyapunov stability theorems in practice. However, in many systems involved in industrial control, the energy function can be selected as a Lyapunov function.

System controllability

Conceptually, system controllability is a quantitative criterion to evaluate if the control inputs are able to steer the system output or state variables from their initial values to the desired target values. The definition of controllability for a time-discrete dynamic state space model can be expressed as follows.

Definition 1.5 *Suppose there exists a dynamic system with the following time-discrete state space model:*

$$X_{k+1} = F(X_k, U_k) \tag{1.17}$$

$$Y_k = G(X_k, U_k) \tag{1.18}$$

then the system 1.17 is said to be completely state controllable if, and only if there exists a control sequence $U_k, k = 0, 1, 2, \ldots, t_f$ that can drive the system from its initial state X_0 to a desired target state X_d within a finite time t_f.

Similarly, the system output controllability can also be defined as follows: the system 1.17–1.18 is said to be completely output controllable if and only if there exists a control sequence $U_k, k = 0, 1, 2, \ldots, t_f$ that can drive the system from its initial output Y_0 to a desired target output Y_d within a finite time t_f.

System observability

'Observability' is another important system structure property to identify if the full state variables can be constructed in terms of the observed system outputs. The definition of observability for a time-discrete dynamic state space model can be expressed as follows.

Definition 1.6 *Suppose there exists a dynamic system shown in equations 1.17–1.18 with zero control input, i.e.* $U_k \equiv 0$, *and an arbitrarily selected initial state* X_0. *The system is said to be completely observable if and only if the initial state* X_0 *can be uniquely determined by the observed output sequence* $Y_k, k = 1, 2, \ldots, t_f$ *in a finite time* t_f.

From an application point of view, system observability evaluates the feasibility and reality in designing a state estimator. For example, the chemical composition in a reactor can be estimated with the observed temperature, pressure, and other related parameters.

In fact, controllability and observability criteria provide the control engineer with an analytical tool for designing a state feedback controller or a state estimator. It should be noted that a known state space model is required in applying both criteria. The application of these two structural criteria will help the designer to check the system configuration to make a working system. However, these criteria are not widely used by practicing control engineers, since either the state space model is not available or too complicated to be used, or they do not have enough modern control knowledge. The details concerning state space representation, controllability and observability are given in many modern control text books (Ogata, 1990; Gelb *et al.*, 1974).

1.2.3 System identification and estimation

System identification and estimation are two major subjects in modern control techniques. In this section, we illustrate the basic problems of these two subjects which are also important in the design of an intelligent control system.

System identification

In modern control techniques, the goal of system identification is to identify the system mathematical model (both the model structure and its parameters) with system observed I/O data. Suppose there exists an unknown time-discrete non-linear dynamic I/O system

$$Y_{k+1} = \Phi(Y_k, Y_{k-1}, \ldots, Y_{k-n}, U_k, U_{k-1}, \ldots, U_{k-r}, \theta) \tag{1.19}$$

where θ denotes a parameter vector in system 1.19. The problem of system identification is to determine Φ and θ with the system's observed I/O data sequence, i.e., $Y_k, U_k, k = 1, 2, \ldots$. The objective of identification is to construct an identification model which produces an output Y_k^p to minimize the error between

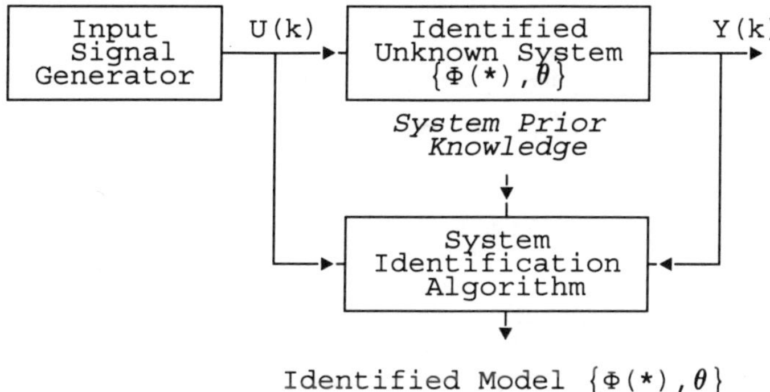

Figure 1.4 General structure of system identification

the actual output Y_k and the model predicted output Y_k^p for the same input U_k. The problem of identification can also be mathematically represented as follows:

$$\min_{\Phi,\theta} \frac{1}{2} \sum_k \|Y_k - Y_k^p\|^2 \tag{1.20}$$

However, if the math-model structure Φ is available, then the system identification is reduced to the problem of parameter estimation, which can be described as follows:

$$\min_{\theta} \frac{1}{2} \sum_k \|Y_k - Y_k^p\|^2 \tag{1.21}$$

subject to the model structure as shown in equation 1.19.

The general structure of system identification is shown in Figure 1.4. For the details of system identification the reader is referred to Ljung (1987).

State estimation

The objective of the math-model based state estimation is to estimate full state variables for a given state space model and system on-line I/O data. In general, the applications of estimation techniques involve prediction, filtering and smoothing (Gelb *et al.*, 1974). The application of state estimation techniques in industrial control provides opportunities for information and knowledge gathering, recovery and prediction. Let system 1.17–1.18 be completely observable; the goal of state estimation is to construct the system full state vector X_k^p, with system observed I/O data sequence $Y_k, U_k, k = 1, 2, \ldots$, and the system state space model. If the estimation algorithm is represented as $\mathcal{E}, \mu : Y_k, U_k \to X_k^p$, where $\mathcal{E}$ and μ denote a specific functional algorithm (Kalman filter or state observer) and the relevant parameter vector, e.g. gain matrix, respectively, then the state estimation is governed by the

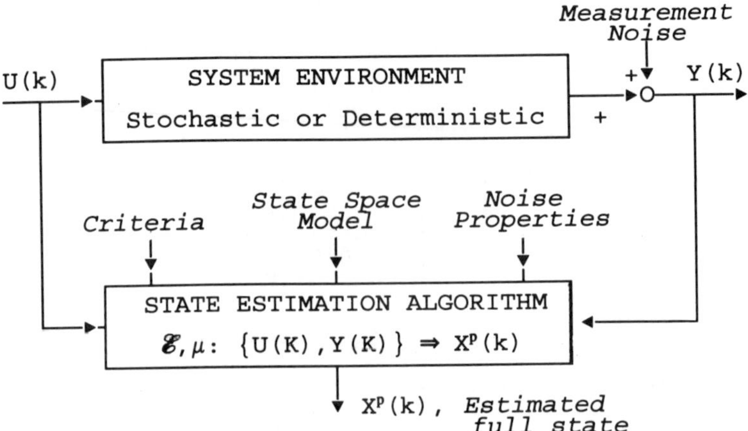

Figure 1.5 General structure of state estimation

following mathematical representation:

$$\min_{\mathcal{E},\mu} \frac{1}{2} \sum_k \|X_k - X_k^p\|^2 \tag{1.22}$$

subject to system state space model 1.17–1.18 and the observed I/O data sequence $Y_k, U_k, k = 1, 2, \ldots$.

The general structure of state estimation is illustrated in Figure 1.5.

1.2.4 Optimal and optimization control

Optimal control

An optimal control system usually yields an optimal trajectory of control variables, which optimize the given dynamic performance criteria subject to desired constraints, such as system dynamics and other constraints. Let the dynamic system 1.17–1.18 be completely controllable; the problem of optimal control can be mathematically represented as

$$\min_{U_k^*} \sum_k J(X_k, U_k, k) \tag{1.23}$$

subject to state space model 1.17–1.18 and other related system constraints.

In industrial control, the objective function J can be selected as control energy, transit time from initial state to desired target state, or control error between actual and target states. However, the most popular objective function is the linear quadratic performance index:

$$J = \frac{1}{2} \sum_k (X_k^T Q X_k + U_k^T R U_k) \tag{1.24}$$

Physically, the performance index 1.24 presents both control energy and control error with weighting matrices R and Q respectively. As a result, the optimal control law will be a linear state feedback:

$$U_k^* = KX_k \tag{1.25}$$

where the feedback gain matrix K can be solved by using the Riccati algorithm (Ogata, 1990).

For example 1.2, if we select equation 1.24 as an objective function, the resulting optimal (state feedback) control law 1.24 will minimize fuel consumption and error between actual and target temperature profile of the metal plate. However, it should be noted that, because only part of the state vector, i.e. both surface temperatures, are available for sensory measurement, to implement control law 1.25 the full state vector X must be reconstructed through introduction of a state estimator as described in section 1.2.3. In many practical applications, state feedback control usually works jointly with state estimation. In general, 'maximum principle' or 'dynamic programming' can be used to solve optimal control problems.

Optimization control

In contrast to optimal control, an optimization control or decision system (usually at level two or level three in a hierarchical computer control system) provides a steady state operating solution in a multi-dimensional control (or decision) variable space, i.e. $U^* \in \Omega_u$. The desired performance criteria are optimized subject to system constraints. The application of optimization in industrial control involves supervisory (setpoint) control, resource allocation, job-machine assignment, production scheduling and planning. In general, the approaches to linear programming, nonlinear programming, dynamic programming, heuristic searching, hill climbing, etc., can be applied to solving those optimization oriented problems. It should also be noted that many industrial control related problems, e.g. identification, estimation, adaptive control and learning, are solved by optimal control or optimization techniques.

The general form of a parameter optimization problem can be described as follows:

$$\min_{U^*} J(U) \tag{1.26}$$

subject to

$$h_i(U) = 0, \quad i = 1, 2, \ldots, m \tag{1.27}$$

$$g_j(U) \geq 0, \quad j = 1, 2, \ldots, l \tag{1.28}$$

In general, the optimization task is complicated particularly for highly nonlinear, high-dimensional systems with multiple local minima. Non-math-knowledge oriented approaches, e.g. expert system, fuzzy logic, neural network and genetic algorithm, can also be applied in solving optimization problems. Finally, it should be emphasized that for some complex industrial systems, the strict requirement of 'optimal' or 'optimization' can be relaxed to less precise but practical goals, such as 'satisfactory' or 'good'. This point is particularly important in working on intelligent optimal and optimization systems.

1.2.5 Robust and adaptive controls

Both robust and adaptive controls have an identical purpose, namely that they provide satisfactory control behavior under variable system dynamics. In this section, we introduce the basic concepts of robust and adaptive control, which are the key subjects in modern control techniques.

Robust control

The controller in a robust control system has a fixed structure with unvarying parameters. In other words, the controller in robust control is designed to be insensitive to the uncertainty of the controlled plant. However, it should be emphasized that it is impossible to design a controller which can work satisfactorily for all types of setpoint changes and disturbances and under boundless uncertainty in model structure and/or parameters. Consequently, it is necessary to provide the possible domain of the systems working environment in the design of the control algorithm. In general, the following essentials have to be given at the design stage (Morari and Zafiriou, 1989):

- process model;
- model uncertain bounds;
- type of inputs (i.e., setpoints and disturbances);
- performance criteria and relevant constraints.

The control objectives usually can be defined as closed loop stability, integral square error (ISE) and sensitivity from setpoints (or disturbances) to controlled variables. In general, the H_2- and H_∞-optimal controllers are the most popular control strategies used in design of a robust control. The H_2-optimal controller minimizes the ISE or the equivalent '2-norm' ('average') of weighted ϵ for a particular input. In contrast, the H_∞-optimal controller minimizes the worst ISE which can result from a set of inputs or the equivalent '∞-norm' ('peak') of weighted ϵ, where $\epsilon = e(d-r)^{-1}$, r, d and e denoting setpoint, disturbance and error (between system output and setpoint) respectively. The details for the design of a robust controller by using the H_2 or H_∞ method were given by Morari and Zafiriou (1989).

Adaptive control

The term 'adaptive' can be realized either in a generic or in a systems control oriented sense. Generally speaking, an adaptive system, e.g. a learning system, is capable of working from a very low level of prior information, or so-called 'incomplete information', for instance with very little knowledge of system environment and disturbance. Such systems can be found in both technical and non-technical areas. From the system control point of view, a controller or system model in an adaptive system can automatically upgrade in structure and/or in parameters to match the changes of dynamics and inputs of the controlled plant with on-line or historical information. As a result of system adaptation, satisfactory or desired reference control behavior may be provided.

Finally, it should be emphasized that the general concepts and architectures of adaptive control have been widely applied in design of an intelligent system. The details of adaptive (learning) systems methodologies will be addressed in the following chapters.

1.2.6 Large scale system methodology

In practice, some industrial processes, e.g. industrial complex or enterprise, can be considered as a large scale system with many decomposed interconnected subsystems. In a large scale system, the interconnections between subsystems are performed through mass, energy and/or information exchanges. Note that while all subsystems (local or lower level controls) work at their optimal conditions, the global system will not necessarily be optimal, due to the existing interconnections between subsystems. The main approaches to large scale systems control and optimization can be summarized as follows (Singh and Tıtli, 1978).

Decentralized control

'Decentralized control' applies when, in a large scale system, the local (subsystem) control is based only on the local information, but the resulting global system responses are stable and robust. Consequently, the application of decentralized control can significantly reduce the communication cost and provide fast response. Decentralized controls are particularly applicable to large scale systems with wide geographical distribution and the requirement for fast regulation, such as power networks, natural-gas (or crude oil) piping networks, and other industrial complexes. The concepts and strategies of decentralized controls can also be applied in designing decoupling control systems.

System decomposition and hierarchical control with coordination

A hierarchical control system with $N(n = 1, N)$ local control stations and a coordination level is shown in Figure 1.6. The optimization control for the global

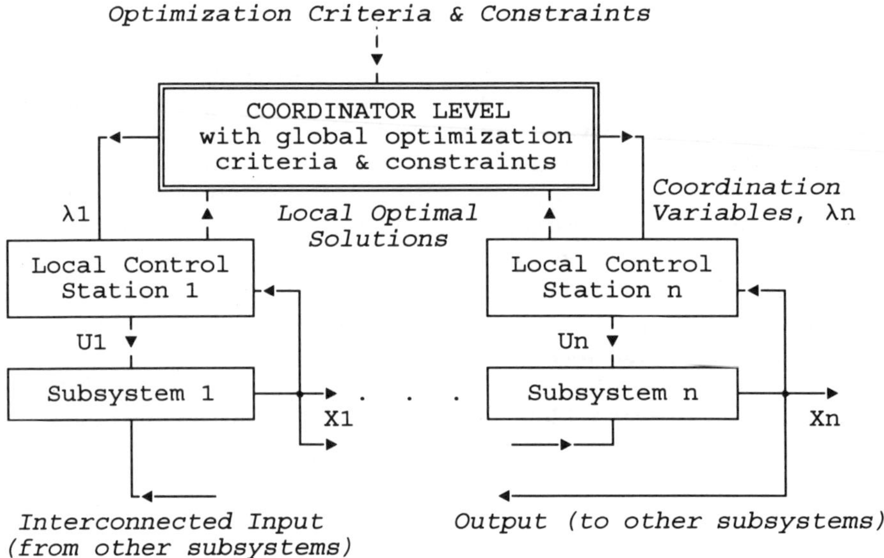

Figure 1.6 System decomposition and coordination

system is carried out in terms of the real-time information exchanges between the local level and upper coordination level. This means each subsystem finds its own optimal control solution, and then sends the local optimal solutions, X_i^*, U_i^* to the coordination level. The global criteria are used to provide the coordination variables $\lambda_i, i = 1, n$, which modify local control solutions in order to make the global system optimal or sub-optimal, while the subsystems are stable but not necessarily optimal. The methodology of system decomposition and coordination provides us with an important and practical philosophy for system design in industrial control.

The concept and methodology of large scale systems are not only applicable to math-model based control, but may also be applied in expert systems, fuzzy logic and neural network control and optimization.

1.2.7 Concluding remarks

Before closing this section, we must emphasize that the concepts, system structural properties and strategic architectures of modern control are vitally important in developing and applying intelligent systems for industrial control. In fact, many intelligent control and/or optimization systems have the same configuration as used in modern control, but use knowledge base, parallel information processing, learning and reasoning, etc., to replace mathematical descriptions and processing popularly used in math-model based control systems. The applications of hybrid systems that combine intelligent systems with modern control have great potential in developing the next generation of industrial control.

1.3 SYSTEM ENVIRONMENT

1.3.1 Concept of system environment

The term 'environment' refers to the aggregation of all external conditions and influences affecting the life and development of an organism (Narendra and Thathachar, 1989). Based on the general definition of environment, the 'system environment' in industrial control can be considered as an integration of system input, output, state variables, static and dynamic behaviors, and control objectives associated with the corresponding constraints.

In industrial control, it is difficult to find an exact stationary system with a fixed system environment; instead, many industrial systems are non-stationary in nature. In engineering practice, particularly for discrete-event driven manufacturing or batch production systems, the system criteria, constraints, structure and even system organization are heavily dependent on custom orders, status of resources, market information, etc. For instance, the system environment in the heating process shown in Figure 1.2 involves production conditions, such as metal grade and geometrical size, heating resources and heat transfer status, heating strategies and production path, etc. The key factors of the system environment can be summarized as follows:

(1) The system state is defined as an n-dimensional vector with n elements describing all system (observed and unobserved) information:

$$X = [x_1, x_2, \ldots, x_n]^T \tag{1.29}$$

(2) The system output is defined as an m-dimensional vector with m elements describing system observed information:

$$X = [y_1, y_2, \ldots, y_m]^T \tag{1.30}$$

(3) The system input is defined as a r-dimensional vector with r elements describing system observed information:

$$X = [u_1, u_2, \ldots, u_r]^T \tag{1.31}$$

(4) The system transition function $F(*)$ represents the state transition from the time instant k to $k+1$, namely

$$F : X_k, U_k \rightarrow X_{k+1} \tag{1.32}$$

(5) The system output function $G(*)$ represents the pattern mapping between state vector to output vector at the instant k, namely

$$G : X_k \rightarrow Y_k \tag{1.33}$$

1.3.2 Categories of system environment

In general, the system environment can be divided into deterministic, stochastic, structured (non-mathematical) knowledge base and adaptive categories with certainty or uncertainty in terms of natural characteristics. Here we only provide the basic concepts of various types of system environment.

Deterministic systems

A system is said to be deterministic if F and G are both deterministic mappings, in other words, the state and output are uniquely dependent on a given initial state and the specified input sequence. In industrial control, because there are many uncertain factors, it is hardly possible to find many exactly deterministic systems; however, there are still a large number of industrial processes that can be considered as approximately deterministic systems. This is the reason that many control strategies were developed for deterministic systems.

Stochastic systems

A system is said to be stochastic if at least one of F or G mapping is stochastic. For instance, if F represents a stochastic mapping, with the given current state and input X_k, U_k, the state X_{k+1} is random, in other words F gives the probabilities of state transition. Thus F can be specified in terms of the conditional probability matrices $F(\beta_i), i = 1, r$, where f_{ij}^{β} represents the probability that the system transits from state X_i to state X_j following the input, namely

$$f_{ij}^{\beta} = Pr\{X_{k+1} = X_j \mid X_k = X_i, \beta_k = \beta\}, \quad i = 1, n; \; j = 1, n; \; \beta = \beta_1, \ldots, \beta_r \quad (1.34)$$

Non-math-knowledge based systems

In non-math-knowledge based systems, the system behaviors F and G are governed by a set of (regular or fuzzy) 'if-then' rules or pattern mappings. The relationship between system input, state and output vectors can be explored by learning, reasoning and associative memory. This kind of system environment has been deeply investigated in establishing neural network, rule based expert and fuzzy systems.

Adaptive systems

In an adaptive system, F and G are automatically upgraded to adapt to the changes of a system environment, human command through a man-machine interface. In general, an adaptive system environment must have an adaptive or self-learning mechanism. Many biological systems have such adaptive abilities, for instance the human body's temperature control, and the human learning mechanism.

In industrial control, there are many systems that are non-stationary or uncertain in both structure and parameters, namely F and G are variable with time. The investigation of system uncertainty will help us to understand why robust, adaptive and intelligent controls are so demanding in industrial control. The next section will address the topic of system uncertainty.

1.3.3 System uncertainty

The main task of robust, adaptive or intelligent control is to provide specific control actions which make system behavior satisfactory while the system environment significantly changes. The system environment depends not only on the system working conditions, but also heavily on human involvement. In system controls, especially using higher levels of computer integrated production system (e.g. supervisory control, production scheduling and planning), the human factor has been part of the system. As a result, human involvement can make the system work better or worse.

In fact, many industrial processes can be represented as non-stationary environments. The design of a control system which can effectively be adapted to the significant changes of the system environment has become one of the most important issues in exploring novel control strategies and algorithms. This book focuses on this important subject.

If we assume that all control actions are made on the basis of a specified system environment, and if all facts related to control actions are known in advance, then there is no uncertainty. The solution to the problem involves a search or analysis through all alternatives to find the optimum solution. This is called control and decision making under certainty. For example, the finishing products of gasoline can be made through a gasoline blending production system with a fixed mathematical (mass balance) model, if the production environment is completely known and stationary. However, many industrial processes which have significant uncertainty factors, from varying operating conditions, are hard to describe as certain systems. Many approaches can be used to describe the uncertainty of a physical system. This section focuses on unstructured, structured, random and fuzzy uncertainties.

Unstructured uncertainty

Suppose we have a linear time-invariant system governed by the following state space equation:

$$X_{k+1} = AX_k + BU_k \tag{1.35}$$

with a specific operating condition A^*, B^*. The unstructured uncertainty can be expressed with system parametric perturbation as

$$A \Rightarrow A^* + \Delta A, \Delta A \subset \Omega_A \tag{1.36}$$

$$B \Rightarrow B^* + \Delta B, \Delta B \subset \Omega_B \tag{1.37}$$

where Ω_A and Ω_B represent the bounded perturbation for system matrix A and input matrix B respectively.

On the other hand, if the plant dynamics are governed by a transfer function, the unstructured uncertainty can then be described by the parametric perturbation bounds of system steady-state gain G, time constants $T_i, i = 1, n$ and dead-time τ as follows:

$$G \Rightarrow G^* + \Delta G, \Delta G \subset \Omega_G \tag{1.38}$$

$$T_i \Rightarrow T_i^* + \Delta T_i, \Delta T_i \subset \Omega_{T_i} \tag{1.39}$$

$$\tau \Rightarrow \tau^* + \Delta\tau, \Delta\tau \subset \Omega_\tau \tag{1.40}$$

where Ω_G, Ω_{T_i}, Ω_τ represent the bounded perturbation for G, T_i and τ respectively.

In a discrete-event driven or batch production system, the uncertainty may also be represented by a group (or family) with many labeled system operating conditions. For instance, the uncertainty in a steel rolling mill usually is governed by the combination of steel chemical composition (or grade), temperature, physical size of the slab, desired gage, width and physical properties. As a result, the huge number of combinations will cause difficulty in defining the system working bounds. However, the application of pattern clustering with supervised or unsupervised learning will significantly reduce this number.

In addition to system parametric perturbation as described above, the dynamics of system inputs (setpoint or disturbance) are also be considered as system uncertainty. The concepts of unstructured uncertainty are applicable to not only math-model based systems, but also non-math-knowledge based systems, e.g. fuzzy or neural network models. Similarly, the unstructured uncertainty in this kind of system is also only related to model parameters, such as fuzzy relational or neural network synaptic connection matrices.

Structured uncertainty

Structured uncertainty usually involves the perturbation of system model structure, for instance changes of system input, output and/or state dimensions, or their functional relations. Consider a nonlinear system represented by its state space model 1.17–1.18; the structured uncertainty means that the form of the nonlinear functions $F(*)$ and/or $G(*)$ can be disturbed within a given family, namely,

$$F(*) \in \Omega_F \tag{1.41}$$

$$G(*) \in \Omega_G \tag{1.42}$$

where Ω_F and Ω_G are defined as the sets of the families of possible functions $F(*)$ and $G(*)$ respectively.

In practice, changing production and/or control schemes may result in the system's structured uncertainty. For instance, a change of heat exchange connection in a heating network will cause a system's model structure to change. The structured uncertainty in non-math-knowledge based systems is related to the changes of knowledge base, such as a rule base, fuzzy rules or neural network architecture.

Description of uncertainty

How to describe system uncertainty has become an important topic in designing a robust, adaptive or intelligent system. Morari and Zafiriou (1989) have schematically investigated uncertainty based on frequency domain for math-model based dynamic systems. To design an intelligent system, both randomness in statistics and fuzziness in fuzzy set theory are considered as the major measures to represent system uncertainty. As noted by Kosko (1992), randomness and fuzziness differ conceptually and theoretically; however, they also share many similarities. Both systems describe uncertainty with a number in the unit interval $[0, 1]$. This ultimately means that both systems describe uncertainty numerically. Both systems combine sets and propositions associatively, commutatively, and distributively. The key distinction concerns how systems jointly treat a set A and its opposite A^c. Classical set theory demands $A \cap A^c = 0$. But fuzziness begins when $A \cap A^c \neq 0$. In practice, the uncertainty can arise from a variety of sources, such as incomplete and/or imprecise knowledge and information and inherent stochastic signals, etc. The following examples illustrate the concepts of uncertainty description under random and fuzzy senses.

Example 1.3

In steel making we have a linguistic rule describing the relations between operating conditions and finishing steel quality for the basic oxygen furnace (BOF): 'If the silicon content in the hot metal is very low, and the aim carbon and aim temperature of the steel is very high, then usually the steel turn down quality will be unsatisfactory'. In this statement, the fuzzy language 'very low' and 'very high', describing attributes of the physical variables silicon, carbon and temperature, presents the uncertainty of propositions under fuzziness. Similarly, the fuzzy term 'unsatisfactory' can also be considered as an uncertainty under fuzziness describing the unsatisfactory degree of the steel quality. Obviously, the membership of a fuzzy variable can be used to numerically measure the fuzziness. However, the fuzzy term 'usually' presents the uncertainty under randomness, namely, that probability can be used as a quantitative measure describing this uncertainty, e.g. 85%. However, this figure is dependent on the membership for the fuzzy variables in both the proposition and the conclusion, namely, 'how low?', 'how high?' and 'how unsatisfactory?'.

From this example, we find that the uncertainty under fuzziness only involves the degree of conditions or conclusions. In contrast, the uncertainty under randomness involves the probability of the occurrences of conditions, conclusions and rules. It should be noted that some event uncertainties can be described by both fuzziness and randomness.

Example 1.4

Following the previous example, we make another rule describing the relations between operating conditions and costs: 'If there are more heats with low Si in the hot metal and high aim carbon and aim temperature of the steel, then the BOF operating costs will be significantly high'. Similarly, in this rule the terms 'low', 'high' and 'significantly high' describe the uncertainty under fuzziness, and 'more' describes the uncertainty under randomness. As a result, the probability of occurrence of propositions with fuzzy variables can then present the stochastic fuzzy uncertainty. Here the rule is deterministic.

Finally, from the above discussions, it can be seen that uncertainty describes imprecise information and knowledge of system environment. The representations of uncertainty could be based on randomness, fuzziness or their combination of the events and/or their relations. The information (or so-called 'Shannon' entropy) (Shannon and Weaver, 1962) can be used to quantitatively measure uncertainty in both randomness and fuzziness. Let there be a stochastic variable X with a set of discrete values $X_i, i = 1, n$ and the relevant probabilities $\{P_i, i = 1, n\}$, $\sum_i P_i \leq 1$; Shannon entropy is defined as

$$E(X) = -\sum_i P_i \log P_i \tag{1.43}$$

On the other hand, fuzzy entropy measures fuzziness of a fuzzy set; for details refer to Kosko (1992).

1.4 INFORMATION AND KNOWLEDGE

1.4.1 System information

As we have noted above, information is the most comprehensive resource in system control and decision making. In general, system I/O data and external and internal knowledge are the major forms of information. Mathematical operations, associative memory, learning and reasoning are the key measures in information processing. To design a control system, particularly an intelligent system, it is critical to deal with system information. The following information related issues should be realized in systems design and implementation:

Data and information

The selection of a limited number of data which contain rich information is a vitally important topic in signal data processing, system identification and adaptive control. Feature extraction in pattern recognition has provided an effective measure to extract the most representative patterns which can reflect steady state behavior even while the system environment changes in a feasible domain. For dynamic systems, the specific timed signal sequences should be designed to excite system dynamics for the full operational frequency domain selected. The selected signal sequences should be the most representative of the family of system inputs in the working environment. It is difficult to construct an excitation signal for a high dimensional, time-varying, highly nonlinear and/or variable structured system. However, in general a set of pseudo-random time sequence signals is recommended for dynamic system identification.

Information recovery

The task of information recovery in system control is to extract system information from noisy I/O signals. An optimal estimator is a computational algorithm with I/O observed measurements to minimize estimation error of the system state by utilizing system knowledge, observed dynamics with specified statistics of system and measurement noises and initial conditions. In fact, the application of system estimation can not only significantly improve the information quality, but also develop and provide internal information. These functions have been changing the reliance of system information only on sensory measurements; instead, software based information gathering plays an important role in industrial control.

Self-organization and creativity

When investigating human behavior, it can be seen that the most important ability of human intelligence is not to simply accept knowledge from the environment, but to self-organize and/or create knowledge through associated memory, deep reasoning and learning. These kinds of functions make humans more clever than any man-made machines. To design an intelligent machine with some human-like capabilities, studies of self-organization and creativity are key issues in R&D of intelligent control. Finally, we should note that signal encoding, decoding and conversion can make information applications more flexible, manageable and matchable.

1.4.2 System knowledge

In a general sense, system knowledge can be viewed as understanding of system behavior. Traditionally, system knowledge is illustrated by a mathematical algorithm; as a result, a precisely numerical solution can be provided. This is so-called 'deep knowledge' which belongs to the 'unstructured knowledge' family. The deep knowledge is usually established with system first principles and/or system identification. However, as we know that since some system behavior can not be represented by deep knowledge, some other forms of knowledge representation have been proposed to describe qualitative (rule based) or fuzzy knowledge. Here we will highlight a number of major non-math-knowledge representations.

Rule base in expert system

AI systems mainly store and process propositional rules with logical implication, such as 'If A, Then B' which associates action B with condition A. Collections and organizations of rules produce a 'rule base' or 'knowledge base' representing global knowledge with forms of knowledge trees or networks. The solutions can be obtained through the inference process that is divided into the following:

- Forward-chaining inference – the effects are driven from conditions;
- Backward-chaining inference – the solutions are provided through suggesting the conditions from the observed effects;
- Combinatorial inference – the combination of both forward-chaining and backward-chaining inferences.

The tasks of the knowledge engineer are to establish the knowledge base and to develop the inference algorithms, which usually require heuristic or approximate search strategies. However, if the problem under consideration is of large scale with high dimensional variables, the 'combinatorial explosion' will cause serious problems in obtaining a solution with the desired accuracy. A knowledge base in expert systems fits a symbolic framework representing structured knowledge. In contrast to traditional mathematical representation, it is not realistic and efficient to apply a symbolic frame and processing when representing continuous func-tions, though theoretically it is feasible to convert symbols to numbers with desired accuracy and vice versa.

Fuzzy knowledge

In comparison with the rule base in expert systems, a fuzzy system can directly encode structured knowledge with fuzzy rules in a numerical framework. The relationships between system inputs and outputs can be represented either by fuzzy rules or by numerical fuzzy relational models. The conversion from fuzzy rules to the relevant numerically oriented fuzzy model can be performed through fuzzy identification or reasoning. Moreover, this characteristic makes it possible to implement an adaptive fuzzy system through fuzzy reasoning and learning. It is different from a knowledge base in AI in that a fuzzy model can be applied to system functional estimation and optimization. However, the resulting accuracy of the solution depends on the size of the reference fuzzy set. Finally, it should be noted that the application of neural network learning in fuzzy system modeling and control may extract fuzzy knowledge, e.g. fuzzy rules and/or corresponding membership with system non-fuzzy I/O data.

Knowledge representation by neural networks

The neural network has shown some interesting and powerful features in setting up associative memory. The knowledge base in a neural network is generated through network training with system training patterns which are extracted from system I/O data. Although the signal activation function of individual processing elements, or so-called 'neurons', in a network usually is very simple, an entire network organized with a large number of interconnected neurons can be used as a knowledge base to store distributed associative memory for a sophisticated system through parallel information processing and learning.

A neural network can directly encode unstructured knowledge within a numerical framework, and can be effectively applied in functional approximation for complex systems without having prior knowledge of system structure. Moreover, since neural networks have powerful learning functions, the knowledge base can be upgraded through retraining by introducing new training patterns or on-line adaptive training. Since neural network inputs and outputs could be analog or binary numbers, the networks can be used as a knowledge base representing a rule base or fuzzy rule base through corresponding encoding and decoding processes.

Finally, it should be emphasized that the mutual applications of various knowledge forms will definitely make intelligent systems more flexible and powerful.

1.5 INTRODUCTION TO INTELLIGENT CONTROL

1.5.1 What are intelligent systems?

What makes human being more clever than man-made machines? They are certainly not numerically more precise or computationally more efficient. Yet human beings are far better at pattern recognition, adaptation, learning, reasoning, associative memory, creativity and perceiving under a complicated environment with or without supervision.

'Intelligent' system or machine intelligence means the system or machine designed should have some human-like behaviors, such as parallel information processing, adaptation to environment, associative memory, learning, generalization and self-organization. On the other hand, intelligent control can be defined as the application of intelligent system methodologies in systems control, such that the control system is furnished with some intelligent behaviors as denoted above. In other words, intelligent control implies an elegant combination of modern control, system engineering and intelligent systems. From this point, the investigation of intelligent control should not conflict with math-model based control, but instead should develop new strategies and methodologies to deal with complex systems which can not be solved by existing techniques. The studies on intelligent systems in this book are based on a control and engineering background.

1.5.2 Major characteristics of intelligent systems

Parallel information processing

It is clear that a good decision usually is made through the interplay of multiple sources of information and knowledge. For example, when a lady is selecting a dress at a department store, her decision is made based on information on its fashion, color, material, price and previous knowledge, etc., through parallel processing. Similarly, when a good operator manipulates a chemical process, he receives multiple sources of information from sensors concerning reaction temperature, pressure, flow rate, chemical composition and so on; based on this information and his knowledge about the process, he adjusts the set points of various control loops in order to reach a specified performance under desired constraints.

From the examples we have just examined, we can understand that human decisions are made through parallel information processing. A number of different pieces of information must be kept in mind and processed at once. Each plays a part, constraining others and being constrained by them. In order to perform such human-like behavior, so-called parallel distributed processing (PDP) models (Rumelhart and McClelland, 1988) are implemented. The information processing takes place through sending excitatory and inhibitory signals between units in a neural network. In many cases the information received by a decision maker is imprecise or incomplete, such as fuzzy information; correspondingly, the decision making is provided through approximate inference.

Associative memory and learning

If we examine the human memory, we will find that the memory in the human brain is not stored with particular labeled addresses like the addressable memory in current computers, and also is different from some cognitive processes, in which the knowledge is stored as a static copy of a pattern, and retrieval amounts to finding the pattern in long-term memory and copying it into a buffer or working memory. Instead, in PDP models the patterns themselves are not stored; rather, what is stored is the connection strengths between units that allow these patterns to be re-created or retrieved. Consequently, we can clearly see that the knowledge representation in a neural network model is what is different from the conventional methods in AI. The advantage of this kind of knowledge representation is to provide an easy way to carry out information processing and learning.

What makes people smarter and smarter day after day? Obviously, learning is the most important factor in human intelligence. Generally speaking, learning can be divided into the following three categories (Haykin, 1994; Weiss, 1990):

- Supervised Learning – This type of learning is also called 'instruction-oriented', and the environmental learning feedback specifies the desired output pattern. In other words, the learning feedback related to the error between neural network outputs and teaching signals provides a pattern mapping between inputs and outputs; as a result, the input/output association can be established and then used as a tool for system modeling and control. If the training samples are selected to well reflect the features of the working environment, the results will satisfy the demands of many industrial systems, for instance model-free function estimation, quality control and decision making.
- Associative Reinforcement Learning – The learning feedback is a scalar signal called a 'reinforcement signal' which indicates if the actual and desired output pattern coincide, for instance 'failure' and 'success' with a binary variable; in more informative versions, this signal can also take on a continuum of values that indicate graded degree of success and failure, which can be represented by corresponding fuzzy numbers. In general, this type of learning can be called 'evaluation-oriented'. In industrial applications, the reinforcement signal could be the evaluation of product quality or status (normal or abnormal) and so on.
- Unsupervised Learning – In contrast to supervised and reinforcement learning, unsupervised learning is 'self-organization-oriented'. The objective of unsupervised learning is to capture regularities (clusters) in the stream of input patterns without receiving any learning feedback. The applications of unsupervised learning, for instance competitive learning, can extract or explore knowledge, such as pattern clusters or fuzzy associative rule generation, etc., which was not recognized before, based on the system measurable information. Obviously, this kind of learning has some human-like ability in generalization and creativity. The potential applications of unsupervised learning in industrial systems will play an important role in knowledge self-acquisition and feature extraction as well as system analysis.

In real applications, the different learning algorithms can work together to provide more flexible and powerful functions. The detailed learning strategies and algorithms for the knowledge base in expert systems, fuzzy systems and neural networks associated with their industrial applications will be described in detail in Chapters 3–8.

Finally, it must be emphasized that all learning is based on the environment; the learning results will certainly reflect the learning environment. For instance, the production data or operating conditions, and the qualification of the environment for learning in providing will be extremely critical in providing correct and desired learning sources. To perform learning systems for industrial control, the different learning sources will produce completely different learning results. In addition, the learning sources should consist of correct and comprehensive information, otherwise learning will be meaningless. These problems will be further investigated in detail in conjunction with application topics later.

Adaptation to environment

Adaptation to the changes of environment is another important human feature. In general, a control or decision system consists of environment and controller or decision maker. Adaptive control can be defined as how a control system can provide satisfactory or robust performance in terms of automatically upgrading controller parameters and/or structure when the environment under control changes with uncertainty. An adaptive control system can automatically create a best match between environment and controller, which then provides satisfactory behavior. If we examine human adaptability behavior, we find that adaptive systems can be divided into two categories based on adaptive action, namely 'environment knowledge driven' and 'control performance driven' systems.

Let us take the temperature control of a heat exchanger as an example for illustration. If the controller parameters and/or structure are adjusted directly based on the response of errors between the desired and actual outlet temperatures, this kind of adaptive system is defined as performance driven, e.g. 'model reference control', 'climbing hill optimization control', and so on. The advantage of this strategy is that only very limited prior knowledge is required and we do not need to have a real-time math-model identification or other real time system modeling. However, the design and implementation of an adaptive mechanism based optimization algorithm, fuzzy reasoning, heuristic search and neural network optimization becomes a critical task, which is difficult and costly for complex nonlinear systems.

In contrast, in an environment driven adaptive system, the controller parameters and/or structure are adjusted based on the real-time knowledge of the system environment obtained from identification or other system modeling techniques. As

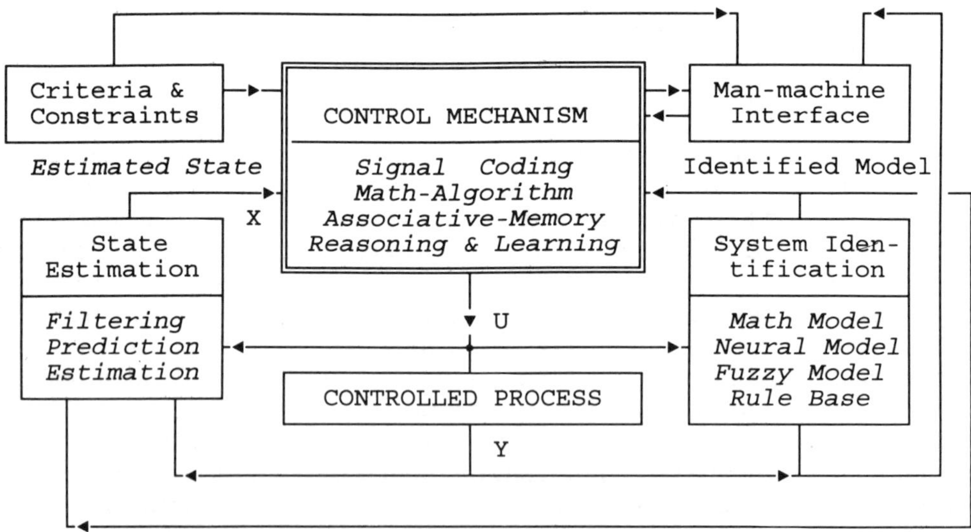

Figure 1.7 A typical intelligent system in industrial control

a result, the design of an adaptive mechanism is based on the identified knowledge of the system environment, such as math-model, rule base, fuzzy model or neural network model. Obviously, the adaptive solution can be provided directly without having to do a real-time search on the performance surface with respect to controller parameters and/or structure.

In this text both strategies (environment driven and performance driven) of math-model free adaptive controls and their industrial applications will be addressed. Finally, a diagram of a typical intelligent system in an industrial control application is given in Figure 1.7.

1.5.3 Intelligent control strategies

Expert control

Computer scientists believe expert systems with a rule base and its reasoning are similar to human-like problem solving. Consequently, applications of various expert systems have been considered as a powerful measure to solve practical engineering problems without having the knowledge of mathematical description, particularly to deal with qualitative knowledge, reasoning and human factors with symbolic operation and reasoning in a complex system. In the system control area, expert controls (Astrom *et al.*, 1986) are applicable to self-tuning, structured knowledge based adaptive control, fault diagnosis, scheduling and planning. In these subjects, human knowledge plays an important role in decision making, particularly at higher levels. Saridis (1983) has noted that in a hierarchical control system the lowest level requires more precise information and the higher levels operate with less precise information and hence need more learning and intelligent functions.

It should be emphasized that a rule based expert system is not good at processing numerical knowledge and providing precise solutions, which are usually required in systems estimation and control. Moreover, the knowledge acquisition, representation, and reasoning in developing expert control for dynamic systems have not been completely solved. The development of an expert system environment with the ability to deal with both qualitative and quantitative (such as modern control algorithms) knowledge, and to automatically explore knowledge, is a challenge of expert control R&D in the future. However, introducing fuzzy logic and neural network in systems modeling and control makes intelligent control more realistic, feasible and applicable in engineering practice.

Fuzzy control

Since fuzzy logic theory and methodology were proposed by Zadeh (1965), fuzzy control has been considered as one of the most attractive strategies in solving complex systems control and decision, particularly for nonlinear systems with imprecise and/or uncertain knowledge of system information and behavior. In contrast to a regular knowledge base in expert systems, the fuzzy rules, being structured knowledge, can be coded into relevant explicit numerical algorithms or a so-called fuzzy model with fuzzy identification or fuzzy reasoning based on system nonfuzzy I/O data. Fuzzy logic can be applied to system modeling, estimation, optimal and optimization control and adaptive control. However, only system fuzzy knowledge and I/O data are required.

Finally, it should be noted that due to the advantages of fuzzy logic technology, the applications of fuzzy logic in expert systems have become an attractive direction in establishing fuzzy expert systems with fuzzy knowledge representation and fuzzy reasoning. They are more applicable in industrial control than regular symbolic operation-oriented expert systems, since the numerical solutions can be provided by a fuzzy expert system.

Artificial neural networks

An artificial neural network can be considered as a large scale nonlinear dynamic system consisting of a large number of simple nonlinear processing elements, so-called 'nodes' or 'neurons', which are interconnected in some manner with adjustable weighted strength. The major features of neural networks are parallel processing, associative and distributed memory, learning and self-organizing which are similar to those evident in biological neural networks.

In contrast to an expert system, a well-trained neural network model can provide both qualitative and quantitative (analog, digital or logical) knowledge through information coding and decoding, and has powerful functions in learning and self-organization. These properties make neural networks more elegant in dealing with numerical data than expert systems.

In general, because the problems of system modeling, control and decision making can be cast into relevant pattern mapping in multi-dimensional hyper-space, the strength of neural networks in associative memory and learning takes advantage in dealing with these control problems.

Hybrid control

In general, the description and integration of system quantitative and qualitative behavior, and human involvement are defined as system modeling for complex systems. They consist of different levels of knowledge, such as deep knowledge, semi-deep knowledge and shallow knowledge. From the system point of view, the information resources for construction of a system model can be grouped into the categories of process principles, production data and human knowledge. Based on various types of knowledge, the following system models can be established:

- a math-model providing parametric quantitative knowledge can be developed with process principles and system identification in terms of production data;

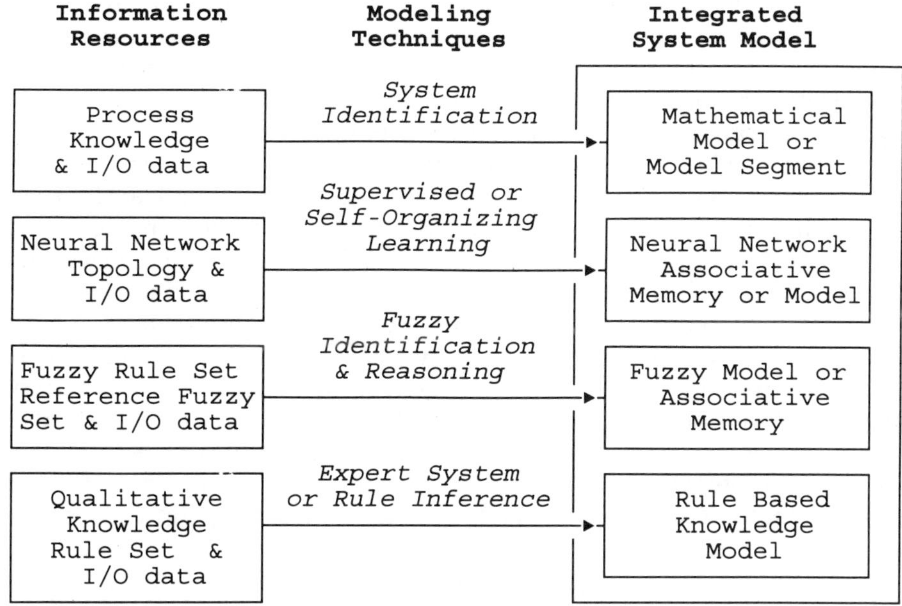

Figure 1.8 An integrated system model

- a neural network model providing non-parametric quantitative knowledge can be constructed based on process principles, human knowledge, production data, and training/learning;
- a fuzzy model providing multi-valued knowledge can be established based on process principles, fuzzy knowledge, production data and fuzzy identification/reasoning; and
- a knowledge base expert system model providing qualitative knowledge can be established based on process principles, rule base, rule inference and production data.

An integrated system modeling configuration as shown in Figure 1.8 provides us with an entire picture of a hybrid system. During the integration, a key frame should be selected based on the characteristic of the problem under consideration.

1.6 OVERALL STRUCTURE OF THIS BOOK

This text aims at two groups of readers: graduate students and academic researchers in control, engineering, and other related majors, and practicing engineers and researchers in the control industry (end user and control software manufacture). In order to make this book readable and applicable, the structure of the book follows the applications of intelligent system methodologies in the main subjects in industrial control, such as modeling, estimation, adaptive control, optimization control, quality control, and fault diagnosis. Consequently, the reader with a control and system engineering background may learn new concepts, fundamentals and methods of intelligent control much easier. Some industrial working examples are included to describe how to apply intelligent system methodologies in solving real industrial control problems.

Structurally, the book is divided into two parts. The first part consisting of Chapters 1–3 covers fundamentals and methodologies in general, and the second part consisting of Chapters 4–8 investigates applications of intelligent systems in major industrial control subjects. Chapter 1 first addressed an overview of industrial control and modern control techniques, and then the topics of system environment, uncertainty and intelligent control were introduced. Chapter 2 illustrates three major approaches, i.e. fuzzy logic, neural network and rule based systems, being used in designing an intelligent system. The focus of this chapter is on the general fundamentals, architectures and algorithms. Chapter 3 addresses the principles, strategies and algorithms of major learning approaches to neural networks, fuzzy systems and rule based systems. The description focuses on supervised, reinforcement, unsupervised, adaptive learning and deep reasoning with relevant applications.

Chapter 4 introduces the applications of intelligent systems in system modeling and estimation with fundamentals, structures and algorithms. The math-model free estimation techniques covered in this chapter are developed with fuzzy logic, neural network and pattern recognition, etc. Some worked examples are addressed in this chapter. Chapter 5 presents the applications of learning in developing adaptive control strategies and algorithms with knowledge base, fuzzy logic or neural networks. The resulting system can be effectively used for highly nonlinear and uncertain systems. Chapter 6 introduces non-math-knowledge based optimization controls, which include fuzzy, neural network and knowledge based optimization, and genetic algorithms. Some industrial worked examples are also addressed in this chapter. Chapter 7 describes applications of intelligent systems in advanced statistics and quality control. The emphasis is on the applications of multivariate statistics, e.g., principal component analysis and partial least-squares, and their combination with neural networks. Chapter 8 investigates the applications of math-model free techniques in fault detection and diagnosis in industry. The emphasis is on the applications of pattern recognition, expert systems, and neural networks.

In addition, almost all chapters begin with a brief review of math-model based methods, so that the reader can understand the problem statement more clearly, and have the opportunity to develop hybrid systems.

2

Fundamental techniques for intelligent control

In Chapter 1 we provided a general view of industrial intelligent control. In contrast to modern control techniques, intelligent control has not been formed as a systematic discipline in the theoretical aspects. However, from a practical application point of view, it has been recognized by both control and artificial intelligence communities that a variety of approaches, such as neural networks, fuzzy logic, pattern recognition and expert systems, etc., may be effectively applied in solving some complicated industrial control problems. The studies of both fundamentals and applications of intelligent systems have been progressing rapidly during the past decade. To move intelligent control from academic research to practice, it is vitally important to learn the techniques of intelligent systems under the industrial control framework. This chapter focuses on the introduction of concepts, fundamentals and methodologies of major techniques being applied in industrial intelligent control. The topics to be investigated in this chapter are fuzzy logic, neural networks and rule based knowledge engineering. The systematic theoretical subjects involved in these areas are beyond the scope of this book.

2.1 FUZZY LOGIC AND FUZZY SYSTEMS

2.1.1 Introduction

Since Zadeh (1965) proposed fuzzy logic theory and approaches, fuzzy logic has become one of the most attractive techniques in developing an intelligent control system. Generally speaking, human decision or control actions are made on the basis of imprecise and/or incomplete system environmental information, knowledge and approximation reasoning, which are different from mathematical model based modern control techniques. For instance, consider a simple example of a temperature control system for a cold–hot water mixture. The mixed water temperature can be controlled by manipulating the flow rate of any cold or hot water and its temperature. Traditionally, a math-model based control system may be designed by a feedback and feedforward control law, which can be represented by an explicit mathematical algorithm. However, the system can also be controlled by a fuzzy logic controller, which consists of many fuzzy logic rules based on human qualitative knowledge. The fuzzy control sample rules in this particular system may be:

FR1: If mixed water temperature is a little bit high, then cold water flow rate should be slightly increased;

FR2: If hot water temperature is extremely low, then cold water flow rate is significantly decreased.

From a control structure point of view, we can readily find that rule FR1 and rule FR2 correspond to 'feedback' and 'feedforward' control respectively. Both condition and conclusion in fuzzy rules are fuzzy variables described by a given fuzzy set, e.g. 'extremely low, very low, low, middle, high, very high, extremely high'. In addition, the truth of the fuzzy rules or so-called 'fuzzy proposition' can also be illustrated by a variety of fuzzy terms, such as 'possibly', 'almost impossible', 'likely' and so forth.

In general, a numerical solution is required in system functional estimation and computation, such as process modeling, estimation, control and optimization. Obviously, structured knowledge, e.g. fuzzy 'if-then' rules, are not good at dealing with this kind of numerical problem. However, it should be emphasized that fuzzy logic theory and methodology have provided us with an effective and realistic technique to code linguistic statements together with imprecise information and knowledge into a numerical framework. Consequently, the numerical solution can be produced by using a fuzzy logic based system model or so called 'fuzzy inference'. During the last two decades, fuzzy logic control has been successfully applied in solving some control problems (Mamdani and Assilian, 1974; Holmblad and Ostergaard, 1982, Dubis, 1980), particularly when the system math-model is not available, or the math-model is too complicated to be used for real-time control. In this section, the basic concepts, fundamentals and methodology of fuzzy logic and fuzzy systems will be introduced.

2.1.2 Fuzzy sets and fuzzy membership

In conventional set and logic theory, only two values, e.g. 'true and false' or '1 and 0', describe an event or a fact. Let us take a quality control system as an example to describe the basic concept of the conventional set and logic theory. For instance, the finishing product quality, x, is divided into only two categories: 'qualified' and 'not qualified'. Consider non-fuzzy subset A of quality variable space X for the qualified finishing product which has crisp or precise boundaries: $A = \{x | x_{\min} \leq x \leq x_{\max}\}$, then the product quality can be simply identified by the following bivalent membership (or characteristic indicator) function:

$$m_A(x) = \begin{cases} 1 & x_{\min} \leq x \leq x_{\max} \\ 0 & \text{otherwise} \end{cases} \tag{2.1}$$

where $m_A(x)$ denotes that x either is in A or is not. Eventually, since the membership m_A maps all real numbers x of the quality onto two numbers $\{0, 1\}$, the conventional set corresponds to two-valued logic. In addition to two-valued logic, some events can be described by multi-valued logic, for instance, products sometimes are required to be divided into a number of grades, based on given criteria. However, neither two-valued nor multi-valued logic is good at dealing with imprecise or uncertain information and its approximate reasoning, such as fuzzy linguistic statements and/or quantitative or functional relationships existing in many real-world problems.

In contrast, the application of fuzzy sets and fuzzy logic techniques makes it possible to encode structured linguistic information into a numerical framework. Thus, imprecise or uncertain information and its approximate reasoning can then be dealt with using mathematical tools. The fuzzy logic can be viewed as an extension or generalization of conventional logic. In fact, the 'truth' of an event or object, e.g. a linguistic statement, may be described by a corresponding membership function being a set of continuous values within the 'universe of discourse', $[0, 1]$. In the example of quality control as noted above, if we say 'the product quality is good', clearly this is a typical fuzzy language statement describing the quality. To represent a fuzzy statement under a numerical framework, the fuzzy membership, $m_A(x)$, can be defined and introduced to measure the 'degree of goodness' for any given product. In a general sense, the membership function $m_A(x)$ describes the degree of whatever element x belongs to subset A, namely

$$m_A(x) = \text{degree}(x \subset A) \tag{2.2}$$

In other words, the membership functions denoting a mapping from x to a unit interval, $X \subset [0, 1]$, can be defined as a fuzzy set, when the mapping is governed by some semantic description of imprecise properties of the events in X.

Let $X = [x_1, x_2, \ldots, x_n]^T$; the fuzzy subset A can be denoted by memberships $m_A(x_i), i = 1, n$ as follows:

$$A = \sum_{i=1}^{n} m_A(x_i)x_i \tag{2.3}$$

If X is an infinite set, A can be represented by

$$X = \sum_{x} m_A(x)x \tag{2.4}$$

There are many different forms describing membership functions of a set of fuzzy variables. Figure 2.1 shows a typical membership function being widely used in practice. We can clearly see that the elements in a fuzzy set can be described by corresponding numerical values, which mathematically measure the fuzzy variable within a bounded interval $[0, 1]$. We can also find that in contrast to a classic set and logic, overlapping areas exist between the boundaries. This property makes the probability $P(A \subset A^c)$ not equal to zero, whereas in a classic set, $P(A \subset A^c)$ is always equal to zero.

Similar to traditional sets, fuzzy sets also have various operations. The basic operations are comparison, containment, union, intersection and complement. Let A and B be the fuzzy subsets of X, $X = \{x_i, i = 1, n\}$, then the operations on

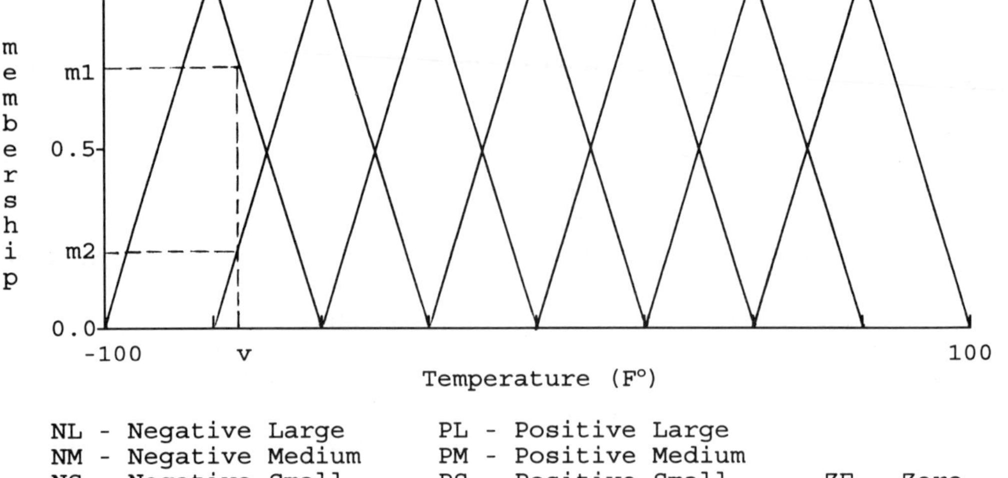

NL - Negative Large
PL - Positive Large
NM - Negative Medium
PM - Positive Medium
NS - Negative Small
PS - Positive Small
ZE - Zero
m1 - Degree of numerical value v which belongs to fuzzy set NL
m2 - Degree of numerical value v which belongs to fuzzy set NM

Figure 2.1 Sample membership function of fuzzy sets

fuzzy sets can be described as follows:

Comparison: A and B are said to be equal iff (if and only if) their membership functions are equal, i.e.

$$m_A(x) = m_B(x), x \subset X \tag{2.5}$$

Containment: B contains A, i.e. $A \subset B$, iff $m_A(x) \leq m_B(x), x \subset X$.

Union: The union (corresponds to 'or') of A and B, i.e. $A \cup B$, can be defined as

$$m_{A \cup B}(x) = \max(m_A(x), m_B(x)), x \subset X \tag{2.6}$$

Intersection: The intersection (corresponds to 'and') of A and B, i.e. $A \cap B$, can be defined as

$$m_{A \cap B}(x) = \min(m_A(x), m_B(x)), x \subset X \tag{2.7}$$

Complement: The complement of A, denoted as A^c, can be defined as

$$A^c = \{1 - m_A(x)\}, x \subset X \tag{2.8}$$

In addition, the size of the fuzzy set can be measured by the sum of all membership values, defined as sigma-count:

$$\text{count}(A) = \sum_{x_i \subset X} m_A(x_i), i = 1, n \tag{2.9}$$

The following definitions relate to fuzzy sets:

Convex fuzzy set: A fuzzy set is convex iff

$$m_A(px_1 + (1-p)x_2) \geq \min(m_A(x_1), m_A(x_2)) \quad (2.10)$$
$$x_1 \text{ and } x_2 \subset X, p \subset [0,1]$$

Extension principle: Let f be a project from $\{x_i, i = 1, r\}$ to y, i.e.

$$y = f(x_1, x_2, \ldots, x_r)$$

then

$$m_B(y) = \begin{cases} 0, & \text{if } f'(y) = 0 \\ \sup_{x_1,\ldots,x_r} \min(m_A(x_1), \ldots, m_A(x_r)), & \text{otherwise} \end{cases} \quad (2.11)$$

Cartesian product: The Cartesian product of universes $X_1, \ldots, X_n$ can be denoted as

$$X = X_1 \times X_2 \times \cdots \times X_n \quad (2.12)$$

Let $\{A_i \subset F(X_i), i = 1, n\}$, then the Cartesian product of $A_1, \ldots, A_n$ is denoted as

$$A = A_1 \times A_2 \times \cdots \times A_n \quad (2.13)$$

and define

$$m_A(x_1, \ldots, x_n) = \min(m_{A_1}(x_1), \ldots, m_{A_n}(x_n)) \quad (2.14)$$

2.1.3 Fuzzy knowledge model

A fuzzy knowledge model can be used either in describing system I/O behavior or in performing systems control and decision making. The architecture of the fuzzy knowledge model can be schematically shown in Figure 2.2.

In fuzzy systems, the numerical input values should be first converted into the corresponding fuzzy representations by using 'fuzzifiers'. The fuzzy outputs are then provided by a fuzzy model, which could be a set of fuzzy logic rules, fuzzy relations or even a simple fuzzy table, with or without deep fuzzy reasoning. Finally, the fuzzy outputs can be converted back into their relevant numerical (crisp) outputs through 'defuzzifiers'. The fuzzy model can be established (off-line or on-line) based on system fuzzy rules and/or system I/O data based fuzzy identification. The fuzzy rules usually are gathered from human knowledge, or if necessary can also be explored by using unsupervised learning in a neural network, which will be investigated in Chapter 3.

Following the general concepts of systems modeling, control and decision making as discussed in Chapter 1, we can readily find that the 'fuzzifiers' and 'defuzzifiers' perform information encoding and decoding respectively. The

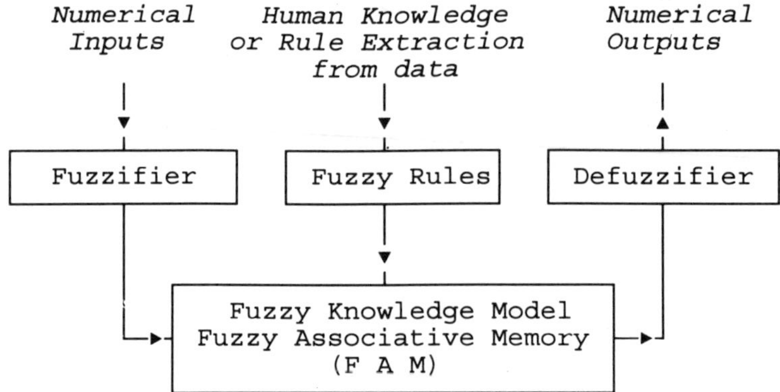

Figure 2.2 General architecture of fuzzy knowledge model

functions of a fuzzy model are quite similar to that of mathematic algorithms in non-fuzzy systems. Quantize any universe of discourse into seven trapezoidal overlapping fuzzy set values: Negative Large (NL), Negative Medium (NM), Negative Small (NS), Zero (ZE), Positive Small (PS), Positive Medium (PM) and Positive Large (PL), as shown in Figure 2.1. Suppose that the range of temperature T is within $[-100, 100]$, then based on Figure 2.1, the operation of the fuzzifier is rather simple, namely, any numerical value is converted into its relevant single fuzzy point in the given reference fuzzy set in the universe of discourse. However, the approaches to defuzzification are more complicated. Here we only introduce the following two methods, which may be used in solving fuzzy system problems in industrial control.

The simplest defuzzification method is to select the element x_{max}, which has maximum membership in the output fuzzy set A:

$$m_A(x_{\max}) = \max\{m_A(x_i)\}, i = 1, n \tag{2.15}$$

However, it was noted by Kosko (1992) that the maximum-membership defuzzification scheme has two fundamental problems. First, the mode of distribution of A is not unique. This problem affects correlation-minimum encoding more than it affects correlation product encoding. Second, the maximum membership scheme ignores the information waveform A. Again correlation-minimum encoding compounds the problem. In practice, A is often highly symmetric, even if it is unimodal. Infinitely many output distributions can share the same mode.

The alternative is the fuzzy centroid defuzzification method. The fuzzy centroid with respect to output space X, a non-fuzzy value, can be calculated as follows:

$$\bar{A} = \frac{\sum_{i=1}^{n} x_i m_A(x_i)}{\sum_{i=1}^{n} m_A(x_i)} \tag{2.16}$$

In general, a fuzzy model can be viewed as fuzzy I/O pattern mapping or a fuzzy associative memory. Before we study this issue in detail, a simple fuzzy

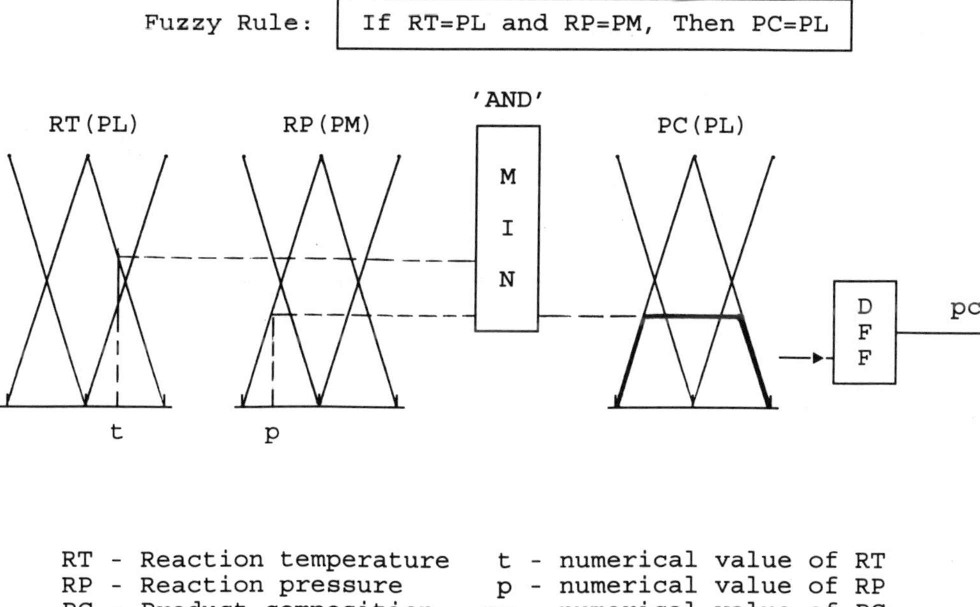

Figure 2.3 Illustration of mechanism of fuzzy rule

'if-then' rule in a chemical reactor may help us to understand how a fuzzy system works, namely how to convert system input into output in terms of fuzzy rules and reference fuzzy set in a given universe of discourse. Suppose there is a fuzzy rule describing a fuzzy relation between reaction temperature, reaction pressure and product composition: 'if reaction temperature (RT) is PL and reaction pressure (RP) is PM, then the product composition (PC) is PL'. Figure 2.3 illustrates how to produce the non-fuzzy crisp values of PC through fuzzy set mapping. It should be noted that if the fuzzy rule is combined with 'or', then 'min' should be replaced by 'max'.

2.1.4 Fuzziness and probability

It is very popular to compare the concepts and properties of 'fuzziness' and 'probability' when we investigate and apply fuzzy logic. Both 'fuzziness' and 'probability' describe uncertainty of an event with a numerical value within $[0, 1]$. However, probability measures imprecision due to the occurrence of a random event based on a large amount of statistical data, while a fuzzy set describes imprecision by using natural language. For example, an SPC (statistical process control) chart is usually provided for quality control to illustrate the probability or frequency distribution ($[0, 1]$) of quality grades of the finished product with hundreds or thousands of daily, weekly or even monthly production data. After examining an SPC chart, we might say 'only 60% of the products are ranked as grade A by using this machine'. In contrast, fuzziness describes the deterministic uncertainty of an event, for instance, 'the quality of products made by machine X is good, but the quality of products made by machine Y is very poor'. Obviously, these two fuzzy statements only describe the degree of product quality and the deterministic relations between product quality and the machine used. However, the probability can also be used to measure the frequency of a fuzzy event occurrence; then the above fuzzy linguistic statements can be modified as: '80% of products made by machine X are good', and '50% of products made by machine Y are poor'. Here we see that the probabilities of 80% and 50% measure the frequency of the occurrence of 'good' and 'poor' products respectively in a random sense. Similarly, the probabilities 80% and 50% should be based on large amounts of statistical production data.

To introduce the calculation of probability of a fuzzy event, let a fuzzy event A be a subset of universe of discourse X, $m_A(x_i), i = 1, n$ denote membership values of $x_i \subset A$, and $P(x_i)$ represent the probability of occurrences of x_i, then the probability of fuzzy event A can be calculated as

$$P(A) = \sum_{i=1}^{n} m_A(x_i)P(x_i) \tag{2.17}$$

The following numerical example will help us to understand the application of equation 2.17.

Example 2.1 (Zwick and Wallsten, 1990)

Let the fuzzy concept 'young' be represented by the following membership function:

$$m_{young}(u) = \begin{cases} 1, & \text{if } u \leq 25 \\ (1 + ((u - 25)/5)^2)^{-1}, & \text{if } 25 < u \leq 100\ . \\ 0, & \text{if } u > 100 \end{cases} \tag{2.18}$$

Table 2.1 Probability and membership distributions of numerical ages in the fuzzy set young

Probability	Age	$m_{\text{young}}(\text{age})$
0.300	50	0.038
0.100	43	0.072
0.200	37	0.148
0.200	30	0.500
0.050	28	0.735
0.025	20	1.000
0.025	19	1.000
0.100	17	1.000

Based on the above membership function, the grades of membership of the numerical ages are presented in Table 2.1. Then the probability that a randomly selected person from Table 2.1 will be young can be calculated by using equation 2.18:

$$
\begin{aligned}
P(young) &= (0.3 \times 0.038) + (0.1 \times 0.072) + (0.2 \times 0.148) + (0.2 \times 0.5) \\
&\quad + (0.05 \times 0.735) + (0.025 \times 1) + (0.025 \times 1) + (0.1 \times 1) \\
&= 0.335
\end{aligned}
$$

This method to calculate the real-valued probability of a fuzzy event was proposed by Zadeh (1968). Some other fuzzy probability models can be found in Zwick and Wallsten (1990).

2.2 FUZZY MODELING AND FUZZY ASSOCIATIVE MEMORY

In general, a mathematical algorithm, $\{y = f(x), x \in X, y \in Y\}$ can be geometrically illustrated as a pattern mapping from x to y on hyperspace X and Y, i.e. $f : x \Rightarrow y$. In a more general sense, this kind of pattern mapping can be viewed as an associative memory, which implies that the system output y can be recalled from the system input x under system environment $f(*)$. In addition to mathematical algorithms, associative memories can also be performed by pattern recognition, neural networks, rule base expert systems, fuzzy models, and even some simple tables or curve sets, etc. The associative memories behind fuzzy systems are represented by a fuzzy mapping and fuzzy models or so-called fuzzy associative memories (FAM); they map fuzzy sets (inputs) to fuzzy sets (outputs) for a fuzzy system. However, due to the powerful functions of fuzzy sets in describing system linguistic qualitative behavior and imprecise and/or uncertain information, many industrial process behavior and control laws can be modeled by fuzzy logic based approaches. In this book we focus our attention on the methodologies of fuzzy systems being used in industrial process control.

An n-dimensional fuzzy pattern can be constructed from the membership values of an n-element ordered fuzzy set. Let a fuzzy set $Y = \{y_1, \ldots, y_n\}$ and its universe, $Y \subset B$. The fuzzy pattern of fuzzy set Y can be represented by the functional relations between element $\{y_i, i = 1, n\}$ and its corresponding membership values $\{m_B(y_i), i = 1, n\}$, or $m_B(y_i) = f(y_i)$. There are a number of ways to geometrically describe fuzzy patterns. Simpson (1991) represents a fuzzy pattern as a piecewise linear curve on the two dimensional (fuzzy set element vs. membership value) plane, and Kosko (1992) describes a fuzzy pattern as a point in the n-dimensional hypercube. Eventually, the fuzzy pattern mapping or fuzzy relation between two or more fuzzy events is critical in designing a fuzzy system for industrial control. Now we will address a number of approaches to establishing fuzzy associative memory or fuzzy pattern mapping for practical applications.

2.2.1 Fuzzy relational matrix

Mathematically, vectors and matrices are effective tools to describe a set of high dimensional variables and their quantitative relationships respectively, even for a nonlinear or time varying system. The concepts and operations of both vectors and matrices have been widely applied in modern control techniques. For instance, a linear state space model $\{X_{k+1} = AX_k + BU_k, X = \{x_i, i = 1, n\}$ and $U = \{u_j, j = 1, r\}\}$ represents the numerical functional relationships, $\{A, B\} : U_k, X_k \Rightarrow X_{k+1}$. Or a nonlinear state space model $\{X_{k+1} = F(X_k, U_k)\}$ represents a nonlinear pattern mapping, $\{F : U_k, X_k \Rightarrow X_{k+1}\}$. The matrix can also represent logic relationships. Let two vectors $X = [x_1, \ldots, x_4]^T$ and $Y = [y_1, \ldots, y_4]$, then the logic of $\{x_i < y_j, i, j = 1, 4\}$, for given values of x_i and y_j can be illustrated by the following matrix:

$$\begin{array}{c} y \\ x \\ 1 \\ 2 \\ 3 \\ 4 \end{array} \quad \begin{array}{c} \begin{array}{cccc} 1 & 2 & 3 & 4 \end{array} \\ \\ \begin{bmatrix} 0 & 1 & 1 & 1 \\ 0 & 0 & 1 & 1 \\ 0 & 0 & 0 & 1 \\ 0 & 0 & 0 & 0 \end{bmatrix} \end{array} \tag{2.19}$$

From matrix 2.19, we can see that the logic relations between x_i and y_j are governed by the elements a_{ij} of matrix 2.19, namely

$$a_{ij} = \begin{cases} 1, & x_i < y_j \\ 0, & \text{otherwise} \end{cases}$$

Similarly, we can also construct a fuzzy matrix to describe the fuzzy relations between two fuzzy sets. When we define a fuzzy matrix $R = \{r_{ij}\}$ to illustrate the fuzzy relations between fuzzy sets $X = \{x_i, i = 1, n\}$ and $Y = \{y_j, j = 1, m\}$, the elements of $\{r_{ij} = [0, 1]\}$ in an $n \times m$-dimension matrix R should be the membership of $m_R(x_i, y_j)$, which measures the degree of relations between x_i and y_j. The fuzzy matrix also has various operations, such as union, intersection, etc. Let fuzzy matrices $A = \{a_{ij}\}$ and $B = \{b_{ij}\}$, and define fuzzy matrix $C = \{c_{ij}\}$ as a matrix of the union of A and B, then $c_{ij} = \max[a_{ij}, b_{ij}]$. If $C = \{c_{ij}\}$ is defined as an intersection on matrices A and B, then $c_{ij} = \min[a_{ij}, b_{ij}]$.

Example 2.2

Let fuzzy matrices A and B determine the union and intersection of A and B.

$$A = \begin{bmatrix} 0.5 & 0.3 \\ 0.4 & 0.8 \end{bmatrix} \qquad B = \begin{bmatrix} 0.8 & 0.5 \\ 0.3 & 0.7 \end{bmatrix}$$

Union:

$$C = A \cup B = \begin{bmatrix} \max(0.5, 0.8) & \max(0.3, 0.5) \\ \max(0.4, 0.3) & \max(0.8, 0.7) \end{bmatrix} = \begin{bmatrix} 0.8 & 0.5 \\ 0.4 & 0.8 \end{bmatrix}$$

Intersection:

$$C = A \cap B = \begin{bmatrix} \min(0.5, 0.8) & \min(0.3, 0.5) \\ \min(0.4, 0.3) & \min(0.8, 0.7) \end{bmatrix} = \begin{bmatrix} 0.5 & 0.3 \\ 0.3 & 0.7 \end{bmatrix}$$

2.2.2 Composition of fuzzy relation

The composition of fuzzy relations denoted by '$\circ$', a compositional operator, can be used in fuzzy reasoning. For instance, if we know two fuzzy relations $R_1 \subset F(X \times Y)$ and $R_2 \subset F(Y \times Z)$, the operation of composition of R_1 and R_2 is defined as

$$R_1 \circ R_2 = R_3 \tag{2.20}$$

where R_3 represents the fuzzy relations of X and Z; then we can further define

$$m_{R3}(x, z) = \sup_{y \subset Y} \min(m_{R1}(x, y), m_{R2}(y, z)), x \subset X, z \subset Z \tag{2.21}$$

Equation 2.21 is also defined as 'max–min composition', or if the 'min-operation' is replaced by 'product', it becomes the so-called 'max–product composition'. Now let us give examples to describe the real operations. The following laws hold in composition operation:

$$R_1 \circ (R_2 \circ R_3) = (R_1 \circ R_2) \circ R_3 \tag{2.22}$$

$$R_1 \circ (R_2 \cup R_3) = (R_1 \circ R_2) \cup (R_1 \circ R_3) \tag{2.23}$$

Example 2.3

Let fuzzy matrices

$$A = \begin{bmatrix} 0.8 & 0.7 \\ 0.5 & 0.3 \end{bmatrix} \qquad B = \begin{bmatrix} 0.2 & 0.4 \\ 0.6 & 0.9 \end{bmatrix}$$

then

$$\begin{aligned} A \circ B &= \begin{bmatrix} \max\{\min(0.8, 0.2), \min(0.7, 0.6)\} & \max\{\min(0.8, 0.4), \min(0.7, 0.9)\} \\ \max\{\min(0.5, 0.2), \min(0.3, 0.6)\} & \max\{\min(0.5, 0.4), \min(0.3, 0.9)\} \end{bmatrix} \\ &= \begin{bmatrix} 0.6 & 0.7 \\ 0.3 & 0.4 \end{bmatrix} \end{aligned}$$

It should be pointed out that in general '$A \circ B$' does not equal '$B \circ A$'.

2.2.3 Development of fuzzy associative memories

In contrast to mathematical model based associative memory, FAM may be developed from linguistic rules; then the numerical mapping between system inputs and outputs can be established through fuzzification, defuzzification and fuzzy models or inference in terms of the given discourses of fuzzy sets. Obviously, the key of FAM of a system is to develop a fuzzy pattern mapping or fuzzy model based on system knowledge and information. In general, we have discussed that the fuzzy relations between system inputs and outputs can be represented by a fuzzy relation model. For example, a first-order dynamic system, $S : u_k, x_k \Rightarrow x_{k+1}$, can be modeled by the following fuzzy relation:

$$x_{k+1} = x_k \circ u_k \circ R \tag{2.24}$$

If R is available, x_{k+1} can be readily solved from x_k, u_k and a known R through an iterative computation. Now the problem is how to find a fuzzy relation R which can provide the best pattern mapping from u_k, x_k to x_{k+1}. This problem is usually called 'fuzzy model identification'. Similar to math-model based identification, the fuzzy identification also consists of both structure identification and parameter estimation problems, which will be introduced in detail in section 4.4.

The design of a fuzzy model for process modeling, control and/or decision making can be performed in the following ways:

Human knowledge oriented method

- A set of fuzzy rules can be established based on human knowledge and/or qualitative behavior gathered from system physical principles, such that a preliminary structure of a fuzzy model can be chosen.
- Determine system inputs and outputs, and gather their off-line (or so-called training) data, then convert numerical I/O data into corresponding fuzzy sets. This procedure involves the selection of membership functions, universe of discourse and numerical data fuzzification.
- Identify both fuzzy model structure and parameters (relations) in terms of fuzzy I/O represented by relevant membership values and the preliminary fuzzy model structure.
- Recall and test the fuzzy model by using a set of working I/O data; obviously a chosen defuzzification method will be used in order to convert fuzzy outputs into corresponding non-fuzzy numerical outputs.

The most important and difficult problem is how to establish a fuzzy rule base which can well represent system behavior. This problem is similar to that of knowledge acquisition in developing an expert system and structure determination of math-model based systems identification and control. It should be emphasized that this is particularly critical for highly nonlinear and sophisticated dynamic systems; in fact, these systems need to be furnished with intelligent systems technology. However, some hybrid approaches to combining fuzzy logic, neural networks and learning with modern control may solve these problems in some degree. Now we highlight the approach to knowledge extraction through neural network self-organization learning.

Combination of fuzzy logic and neural network

In contrast to fuzzy logic, neural networks have powerful capabilities to cast system I/O data (not necessarily having rules) into an associative memory being governed by its connected weights between input–hidden-output layers in a multi-layered network. However, many researchers (Ishibuchi *et al.*, 1993; Keller *et al.*, 1992; Kosko, 1992) have been exploring hybrid technology based on combination of fuzzy logic with neural networks. In general, a set of fuzzy rules can be implemented by a trained neural network based on nonfuzzy data. The more important advantage and feature of combining these two technologies are to provide a feasible technique in performing FAM through using neural network unsupervised competitive learning based on nonfuzzy input–output learning patterns $\{x_i, y_i, i = 1, K\}$. As a result, some fuzzy knowledge can be explored by working with system regular I/O data. The details of this attractive technology and its potential applications in industrial control will be addressed in Chapter 3. The development of the FAM can be schematically illustrated in Figure 2.4. from which we can see that the FAM is established in terms of fuzzy rules and off-line data based fuzzy

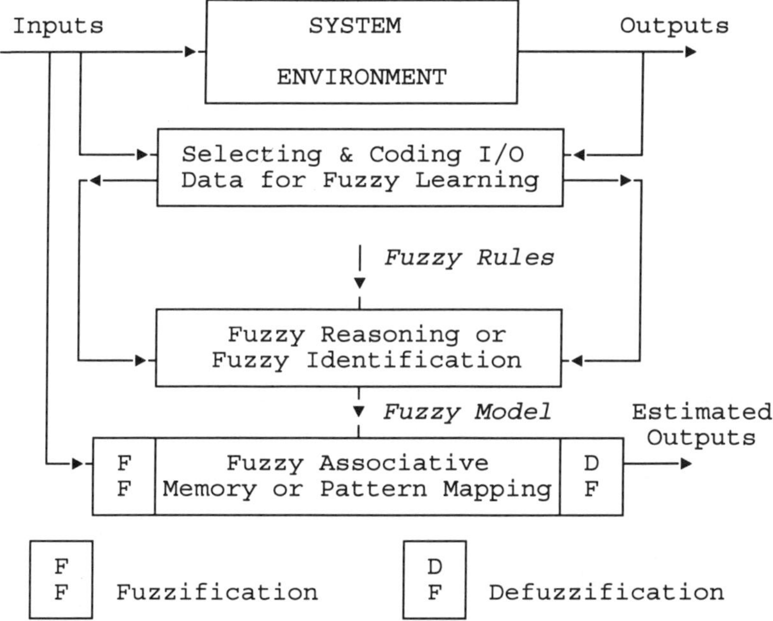

Figure 2.4 Illustration of development of fuzzy associative memory

identification. The FAM can be used to provide nonfuzzy output during recall or retrieval phase.

To make the FAM more 'robust' a so-called 'generalization' commonly used in neural network technology is another important issue in order to put fuzzy logic in practice, particularly for systems with significant uncertainty. Since this generic topic is involved not only in fuzzy modeling, but also in neural network modeling or even in math-model based identification, we will investigate this issue in Chapter 3. Finally it should be noted that fuzzy modeling has different interpretations. For instance, a fuzzy set or a fuzzy rule base can be viewed as a fuzzy model to represent human concepts. In a more general sense, fuzzy modeling can be used to deal with qualitative behavior of a system (Sugeno and Yasukawa, 1993). However, in this book, fuzzy modeling is mainly used as a tool of function estimation with structural knowledge under numerical framework. Similar to math-model based approaches, fuzzy modeling technique can also be applied in the areas of system modeling, estimation, control and optimization.

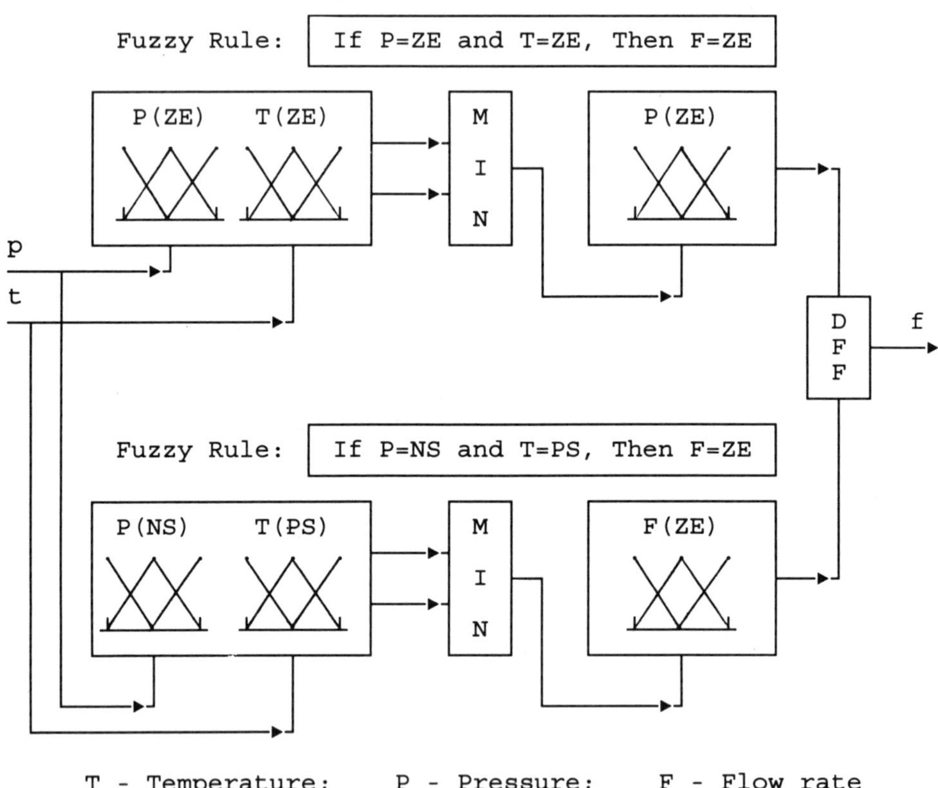

Figure 2.5 Sample inference showing how to convert the numerical inputs p, t into the numerical output f

2.2.4 FAM inference

An AI expert system makes its rule based inference through a forward or backward heuristic search on the decision tree, and as a result of the inference, the qualitative (or logical) solution to conclusions can be solved from the given preconditions. However, the FAM inference can be carried out based on the FAM rules and the relevant conditions represented by fuzzy variables. The inference procedure is illustrated in Figure 2.5. From this figure we can see that the conclusion is solved numerically by the given FAM rules associated with the numerical inputs. Obviously the numerical framework of the FAM provides us the opportunity to convert the fuzzy rules into functional relations.

2.3 INTRODUCTION TO NEURAL NETWORKS

An intelligent machine (or system) should be capable of making a reasonable (or satisfactory) decision or action under a variable or uncertain environment in terms of the abilities of parallel information processing, machine learning, associative memory, generalization and creativity. With the recent rapid advances in VLSI and supercomputer technology and the research of neural system architectures and algorithms, neural network technology has been successfully applied in pattern recognition, image and natural language processing, and robotics, etc. Neural networks have also been recognized as one of the most powerful tools in developing intelligent control systems, particularly for nonlinear system modeling, prediction, adaptive and learning control with incomplete system knowledge and imprecise system information. The most important aspects in developing a neural network are to design its architecture or so-called topology, learning algorithms and selection of training sample patterns. Now we will investigate the fundamentals of neural networks.

2.3.1 Major functions of neural networks

From an industrial control application point of view, the major functions of neural networks can be summarized as follows:

Modeling and Estimation: A neural network model can be established through off-line or adaptive learning based on system input and output data. Thus, the predicted or estimated system output or state can be recalled from a well-trained or adaptive neural network. In neural network technology, this function is related to pattern mapping, function approximation, or associative memory.

Adaptive Control: A neural network can be used as an adaptive controller to provide robust adaptive control performance under an uncertain system environment, particularly for nonlinear systems with unknown system math-model.

Optimization: A dynamic neural network with recursive feedback may provide near-optimum solution to the problem through an automatic search in the Lyapunov energy surface for some optimization problems, e.g. the combinatorial optimization.

Pattern Classification: A well-trained neural network can be used to identify a representative class from a system input pattern space. This function can be applied in data compression, feature extraction and signal coding, etc.

Pattern or Signal Filtering: When noise-corrupted patterns or signals are inputs to a neural network, noise-free or less noisy patterns or signals can be produced. Clearly, this function is similar to those math-model based filtering techniques.

Pattern Completion: A neural network may produce a complete pattern based on the input of an incomplete pattern, i.e. the missing portions of the input pattern can be filled in.

The functions of the neural networks as listed above in industrial control have been a very attractive research and development area. However, the main activities of industrial control researchers involve the applications of multilayered feedforward neural networks in system modeling, identification and adaptive control. In the following sections we will present the fundamentals, architectures and algorithms of neural networks which have great potential in applications in industrial control. Neural network training or learning will be addressed in detail in Chapter 3.

2.3.2 Neurons and neural networks

It has been estimated that there are more than 100 billion (10^{11}) neurons in a human brain (Rumelhart and McClelland, 1988). The neurons are organized as a natural network to receive information or signals from the real-world environment, and then provide the corresponding responses, e.g. decisions or actions. However, from an engineering point of view, a neural network can be viewed as a parallel information processing system with some human-like intelligent behavior, e.g. learning, associative memory and generalization, etc. In general, a neural network consists of many nonlinear processing elements, so-called 'neurons' or 'nodes'. They are organized in a single or multiple layers with or without feedback, and directly exchange information through synapse or so-called weighted connections in some particular manner. Only the neurons located in both input and output layers exchange information with a real-world environment. The neurons located in hidden layers only exchange information with the input and output layers and deal with internal information processing. An illustrative architecture of a typical three layered feedforward neural network is shown in Figure 2.6. From Figure 2.6 we can see that similar to traditional mathematical representation, a neural network is able to perform the quantitative pattern mapping from its inputs, $U = \{u_i, i = 1, r\}$ to outputs, $Y = \{y_k, k = 1, n\}$, namely

$$Y^p = \mathcal{NN}(U, W) \tag{2.25}$$

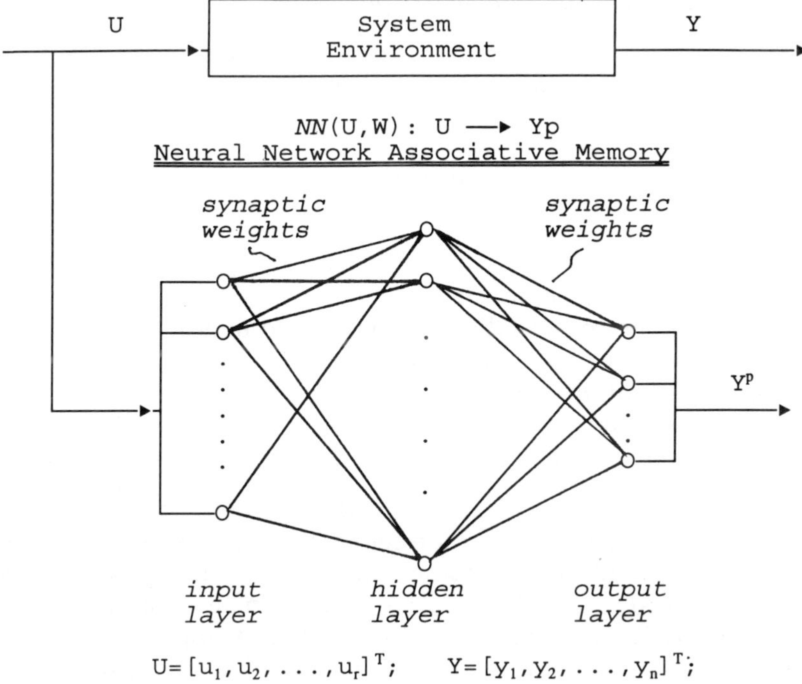

Figure 2.6 Illustrative architecture of a multi-layered feedforward neural network

or

$$\mathcal{NN} : U \Rightarrow Y^p$$

where matrix W is defined as a set of internal parameters representing the connection strength between input signals and output signals. The weight matrix is governed by the structure of the neural network, e.g. number of layers, number of neurons in each layer, connection manner between neurons and the signal (activation) function of the neurons. It is obvious that once both the neural network structure and its parameter W are determined, then an explicitly functional relation, $\mathcal{NN}(*)$, is available to calculate neural network output Y^p from its inputs U. Clearly, the $\mathcal{NN}(*)$ is similar to a nonlinear matrix function in a multi-variable system. Now we will further investigate the details of neural network functions.

2.3.3 Signal activation functions

In a neural network , the single neuron j receives the weighted input signals either from a real-world environment or from the other neurons, and then carries out the following linear combiner type of signal processing:

$$\alpha_j = w_1 u_1 + w_2 u_2 + \cdots + w_r u_r + w_0 = \sum_{i=1}^{r} w_i u_i + w_{bj} \tag{2.26}$$

$$y_j = S(\alpha_j) \tag{2.27}$$

where

$\{u_i, i = 1, r\}$ = inputs of neuron j;
$\{w_i, i = 1, r\}$ = connected weights from input i to neuron j,
α_j = activation state variable of neuron j;
$S(*)$ = neuron activation function;
y_j = output of neuron j;
w_{bj} = bias input, usually $w_{bj} = 1$.

From equation 2.26, we can see that the relationships between the inputs, u_i, and the neuron activation state, α_j, represent a linear combination. The activation function performs a nonlinear mapping from α_j to y_j. The signal processing of a single neuron denoted by equations 2.26 and 2.27 can be illustrated by Figure 2.7.

There are many different signal activation functions being used in the design of a neural network; however, here we list only the functions (Simpson, 1992; Kosko, 1992) which are the most popular in practical applications:

Linear Function The linear function can be described by the linear equation

$$S(\alpha) = c\alpha \tag{2.28}$$

where α is an activation value taking a real number, and c is a positive scalar; obviously, if $c = 1$, the neuron's function is exactly the same as a linear combiner.

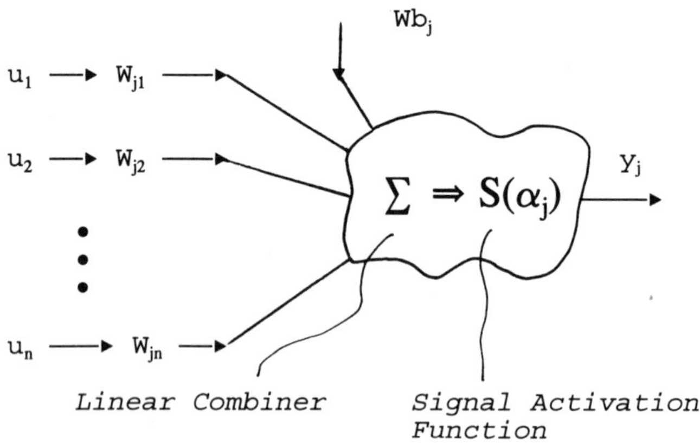

Figure 2.7 Signal processing in a single neuron

Step function The general form of step function is a bounded switch function, and can be represented as

$$S(\alpha) = \begin{cases} y_{\text{max}} & \text{if } \alpha \geq \alpha_0 \\ y_{\text{min}} & \text{if } \alpha < \alpha_0 \end{cases} \tag{2.29}$$

where y_{max} and y_{min} are the upper and lower bounds respectively. However, in real applications, equation 2.29 can be represented as a binary step function; when we assign $\alpha_0 = 0, y_{\text{max}} = 1$, and $y_{\text{min}} = 0$ or $y_{\text{min}} = -1$, then we have

$$S(\alpha) = \begin{cases} 1, & \text{if } \alpha > 0 \\ 0, & \text{otherwise} \end{cases} \tag{2.30}$$

$$S(\alpha) = \begin{cases} 1, & \text{if } \alpha > 0 \\ -1, & \text{otherwise} \end{cases} \tag{2.31}$$

Function 2.30 is used in Hopfield neural networks (Hopfield, 1982) and bidirectional associative memory (Kosko, 1988). Function 2.31 can be used in an associative reward–penalty network (Barto and Anderson, 1985).

Ramp Function The ramp function is a combination of the linear and step functions and can be defined as

$$S(\alpha) = \begin{cases} b, & \text{if } \alpha \geq r \\ c\alpha, & \text{if } |\alpha| < r \\ -b, & \text{if } \alpha \leq -r \end{cases} \tag{2.31}$$

where r is the saturation value of the function.

The family of bounded monotonic, nondecreasing nonlinear signal functions, or so-called *'sigmoid functions'*, are the most popular ones in neural network applications, since they can provide wide capability to model various nonlinear functions, and generality on system stability. Now we list some of those signal functions as follows:

Logistic Signal Functions The logistic (Sigmoidal) signal function can be represented as

$$S(\alpha) = \frac{1}{1 + e^{-c\alpha}} \tag{2.33}$$

where c is a constant. In the general case, c is assigned to be equal to 1; however, the parameter c can be used to adjust the slope of $S(\alpha)$, namely, $dS(\alpha)/d\alpha$, at $\alpha = 0$ or $S'(0)$. The activation derivative, $S' = cS(1 - S) \geq 0$, and then $S(*)$ is a monotonic non-decreasing function.

Hyperbolic Tangent Function The hyperbolic tangent function is a bipolar signal function

$$S(\alpha) = \tanh\left(\frac{\alpha}{2}\right) \tag{2.34}$$

or

$$\tanh\left(\frac{\alpha}{2}\right) = \frac{1 - e^{-\alpha}}{1 + e^{-\alpha}} \tag{2.35}$$

in which $S(\alpha)$ is bounded within the range $[-1, 1]$, and the activation derivative $S' = (1 - S^2) \geq 0$.

Threshold Exponential Function

$$S(\alpha) = \max(1, e^{c\alpha}) \tag{2.36}$$

Exponential-distribution Signal Function

$$S(\alpha) = \max(0, 1 - e^{c\alpha}) \tag{2.37}$$

Ratio-polynomial Signal Function

$$S(\alpha) = \max\left(0, \frac{\alpha^n}{c + \alpha^n}\right) \tag{2.38}$$

Some signal functions as listed above are illustrated in Figure 2.8 (Simpson, 1991), and more signal functions and corresponding derivatives $dS/d\alpha$ denoted by S' are given in Table 2.2.

Generally speaking, any neural network consists of the following three design oriented phases:

Architecture or Topology: Selection of the neuron's signal function, organization of the neurons with layers and the neuron's connection, which should match the problem under consideration;

Table 2.2 Signal functions and activation derivatives

$S(\alpha)$	$S'(\alpha)$
$\frac{1}{1 + \exp(-c\alpha)}$	$cS(1 - S)$
$\exp(-c\alpha^2)$	$-2c\alpha S$
$\tanh(c\alpha)$	$c(1 - S^2)$
$\frac{\alpha^n}{c + \alpha^n}$	$\frac{c\alpha^{n-1}}{c + \alpha^n}$

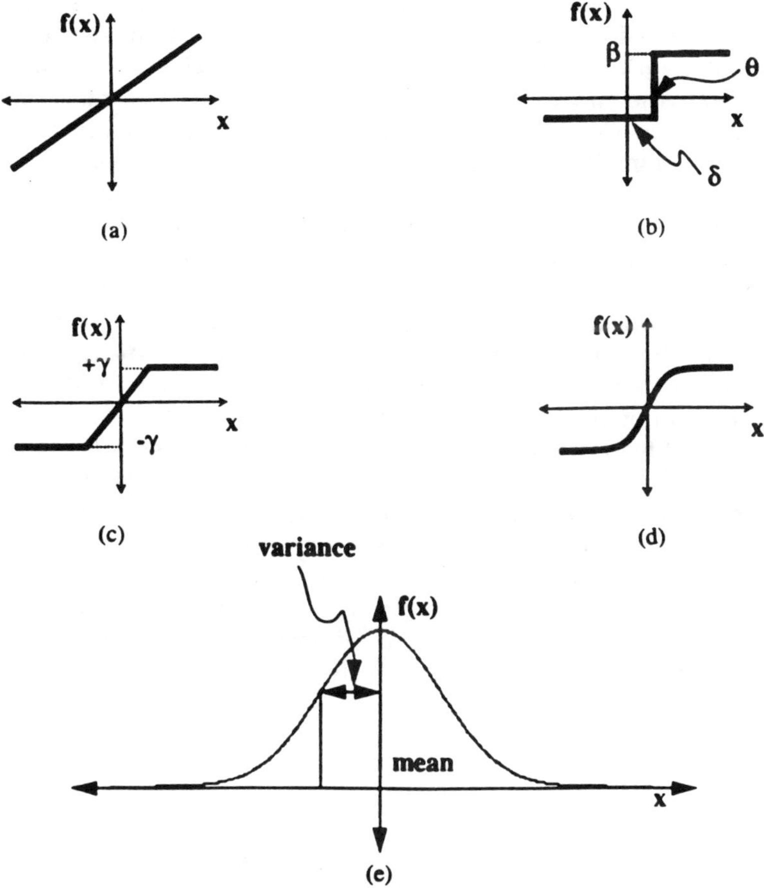

Figure 2.8 Illustration of functional relations of commonly used signal functions (Reproduced from Simpson, 1992, © 1992 IEEE)

Learning or Training: Selection of training pattern samples and learning strategy and algorithm, which should effectively cast information into a neural network and establish the relevant associative memory;

Recall or Retrieval: System outputs are retrieved from the neural network stored associative memory established during the training phase.

In the following sections, we will introduce the topologies, algorithms and functions of the neural networks which have great potential in applications of industrial control.

2.4 NEURAL NETWORKS AS FUNCTION APPROXIMATION

A multilayered feedforward neural network is the most popular one in practical applications. When individual neurons with a selected signal function are linked to each other through the weighted connections based on designed architecture (single or multilayered; feedforward or recursive feedback, etc.), a transfer function of the entire neural network, $\mathcal{NN}(*)$, can be formulated. In fact, connection weights are directly related to the information flow between the neurons; for instance, positive valued weights are excitatory connections, and negative valued weights are inhibitory connections. To create associative memory, a neural network needs to be trained by adjusting its connecting weights under a given training environment, namely the selected training pattern samples. As a result, the system information, or pattern mapping from system input vector U to output vector Y, can be cast into a neural network as an associative memory. A well trained neural network is governed by its architecture, selected signal function and trained frozen weights. In a general sense, the neuron's activation states and connection weights represent short-term and long-term memories respectively. Now we will illustrate how to derive the neural network transfer function.

A three-layered feedforward neural network with detailed information processing is shown in Figure 2.9, where the neural network consists of three layers, i.e. input, hidden and output. The neurons in the input layer receive the information from a system environment through sensory measurement, a man–machine interface and/or communication signals from another system. The function of the input layer is only to carry out signal normalization, namely, to convert various physical values $\{u(i), i = 1, N_i\}$ into standard normalized values, $\{un(i), i = 1, N_i\}$, ranged $[0.05, 0.95]$ or $[-0.95, 0.95]$, which depends on signal activation function

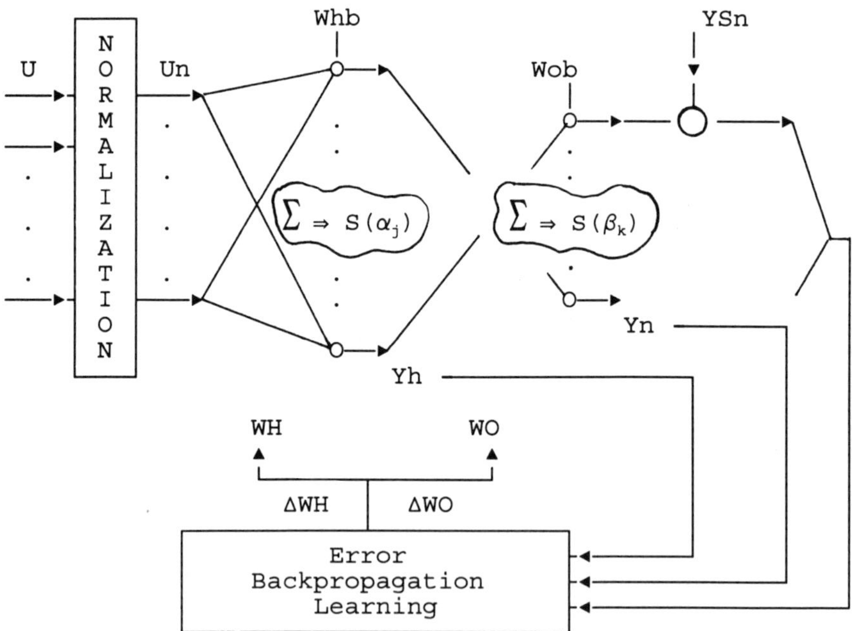

Figure 2.9 Architecture of a multi-layered feedforward neural network

selected. The normalized signals are first weighted and then sent to the neurons in the hidden layer. The outputs of the hidden layer, $\{yh(j), j = 1, N_h\}$, are further weighted and then sent to the neurons in output layers. Finally, the output layer provides the normalized outputs, $\{yn(k), k = 1, N_o\}$. Here N_i, N_h and N_o denote the numbers of neurons in input, hidden and output layers respectively. To let the neural network outputs interface with the real-world system, denormalization is required to converted the normalized outputs to the real physical values. If we define the connected weights between layers as $Wh(j, i)$ and $Wo(k, j)$, then the signal processing of the feedforward pass in the neural network can be represented as

normalization

$$u_i \Rightarrow un_i \tag{2.39}$$

hidden layer

$$\alpha_j = \sum_{i=1}^{N_i} Wh_{ji}un_i + Wh_{bj}$$

$$yh_j = S(\alpha_j),\ j = 1, N_h \tag{2.40}$$

output layer

$$\beta_k = \sum_{j=1}^{N_h} Wo_{kj}yh_j + Wo_{bk}$$

$$yn_k = S(\beta_k), k = 1, N_o \tag{2.41}$$

The mathematical representation of pattern mapping from $\{un_i, i = 1, N_i\}$ to $\{yn_k, k = 1, N_o\}$ can be expressed as follows:

$$yn_k = S(Wo_{kj}S(Wh_{ji}un_i + Wh_{bj}) + Wo_{bk}) \tag{2.42}$$

where Wh_{bj} and Wo_{bk} are defined as bias inputs of the neurons in hidden and output layers respectively; usually, the bias can be equal to one, or can also be assigned as variable to join the training.

A more compact matrix form of equation 2.42 can be directly represented as follows:

$$Yn = S(WOS(WHUn + WH_b) + WO_b) \tag{2.43}$$

where

$$Un = [un(1), \ldots, un(N_i)]^T$$
$$Yn = [yn(1), \ldots, yn(N_o)]^T$$

$$WO = \begin{bmatrix} Wo(1,1), & \dots, & Wo(1,N_h) \\ Wo(2,1), & \dots, & Wo(2,N_h) \\ \vdots & \ddots & \vdots \\ Wo(N_o,1), & \dots, & Wo(N_o,N_h) \end{bmatrix} \tag{2.44}$$

$$WH = \begin{bmatrix} Wo(1,1), & \dots, & Wo(1,N_i) \\ Wo(2,1), & \dots, & Wo(2,N_i) \\ \vdots & \ddots & \vdots \\ Wo(N_h,1), & \dots, & Wo(N_h,N_i) \end{bmatrix} \tag{2.45}$$

$$WHb = [Wh_b(1), \dots, Wh_b(N_h)]^T \tag{2.46}$$

$$WOb = [Wo_b(1), \dots, Wo_b(N_o)]^T \tag{2.47}$$

$$S(*) \quad \text{denotes a signal matrix function} \tag{2.48}$$

However, it should be pointed out that the weight matrices WH and WO in equation 2.43 are solved by using an optimization oriented learning or so-called training algorithm. For this particular architecture, the supervised learning (learning from examples, or so-called learning from teacher) can be described by the following optimization problem:

$$\min_{WH,WO} \sum_{n=1}^{N_p} \|Yn(n) - YSn(n)\|^2 \tag{2.49}$$

subject to

$$Yn = S(WOS(WHUn + WHb) + WOb)$$

where $YSn(n)$ is defined as system outputs and serves as the supervised teaching signals, and n and N_p denote the training sample instance and total number of training samples. From equation 2.49 we can see that the driving force is the error between neural network outputs Y_n and the actual system outputs YS_n. The weight modification usually can be arranged in the following two categories:

Pattern-wise training: The weights are upgraded immediately after the presentation of each training pattern;

Batch-wise training: The weights are upgraded after the completion of the presentations of all training patterns in a batch.

In general, if the neural network is well designed, after many training iterations the error between the neural network output and system output vectors approaches zero or a minimum. The optimized weights WH and WO are frozen, The associative memory or a neural network model is established. Then the estimated or predicted system outputs can be produced through retrieval or recall of the neural network model. An overall architecture of neural networks with training and retrieval is given in Figure 2.10.

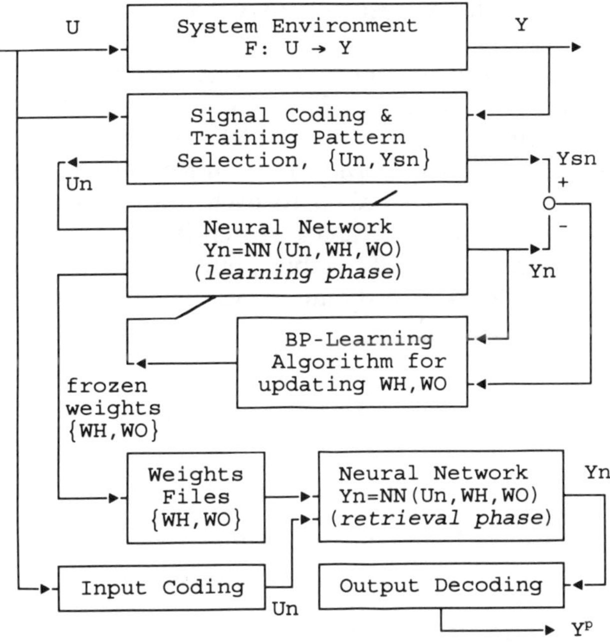

Figure 2.10 Architecture of neural network with training and retrieval

From this example, we can see that a pattern mapping from the system's inputs to its outputs in a hyperspace is established in terms of a nonlinear mathematical formula performed by a multilayered feedforward neural network which consists of many nonlinear processing elements.

It has been proved that a three layered neural network with only a single hidden layer can approximate to any nonlinear continuous function with arbitrary precision (White, 1989, Hornik, 1989). However, it should be emphasized that the design of neural network architecture, selection of the training pattern samples, and design of learning strategies and algorithms are critical to make a neural network work satisfactorily. To make the neural network capable of providing a satisfactory solution for the entire system environment which has never been trained before is a key problem of applying neural networks in the real world. This problem will be addressed in detail in Chapter 3. From the application point of view, a multilayered neural network can be used for function approximation, pattern classification and pattern mapping.

2.5 HOPFIELD NEURAL NETWORKS AND BOLTZMANN MACHINE

In this section we investigate the architectures and algorithms of the single layered neural networks, e.g. Hopfield neural networks and Boltzmann machine.

2.5.1 Hopfield associative neural networks

A Hopfield associative network is a single layered feedback network as shown in Figure 2.11 (Haykin, 1994). The major function of the Hopfield associative neural network is pattern completion, i.e. a complete pattern or information can be produced from an incomplete pattern or noisy information. Suppose there are N binary valued nodes which are linked to each other with symmetric weights ($w_{ij} = w_{ji}, w_{ii} = 0$). An iterative feedback search in a state space will drive the neural network system to a locally stable point in the state space. Obviously, the Hopfield neural network is a dynamic system. In the retrieving phase, the system dynamics are governed by

$$u_i(k+1) = \sum_{j=1}^{N} w_{ij}\alpha_j(k) \tag{2.50}$$

$$\alpha_j(k+1) = \begin{cases} 1, & \text{if } u_i(k+1) > -r_i \\ 0, & \text{if } u_i(k+1) < -r_i \\ \alpha_j(k), & \text{if } u_i(k+1) = -r_i \end{cases} \tag{2.51}$$

Define a Lyapunov energy function:

$$E(k) = -\frac{1}{2}\sum_{i=1}^{N}\sum_{j=1}^{N} w_{ij}\alpha_i(k)\alpha_j(k) - \sum_{i=1}^{N} r_i\alpha_i(k) \tag{2.52}$$

where

N = number of elements in each sample;
r_i = given threshold value;
α_i = activation values;
u_i = neural network output.

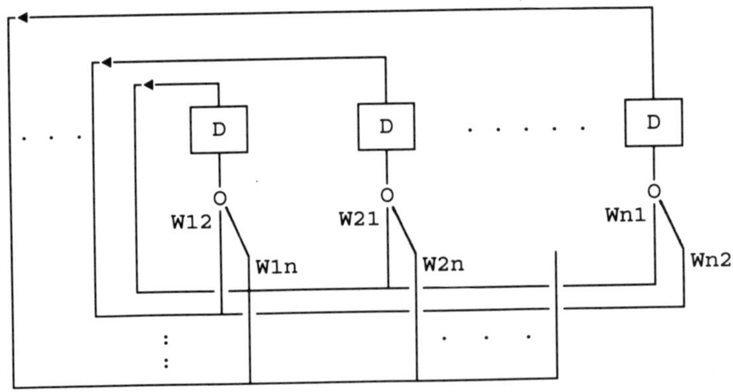

Figure 2.11 Topology of single layered feedback neural network

It can be proved (Kung and Hwang, 1989) that the energy change between two iterations is

$$E(k+1) - E(k) = -\alpha^T(k+1)(u(k+1)+r) \\ -\alpha^T(k+1)W\alpha(k+1)/2 \tag{2.53}$$

In order to guarantee that the system is stable, i.e. $E(k+1) \leq E(k)$, the weight matrix W should be a non-negative and definite one, which can be ensured by setting the weight matrix to be

$$w_{ij} = \sum_{p=1}^{P} (2b_i^{(p)} - 1)(2b_j^{(p)} - 1) \tag{2.54}$$

where P denotes the number of binary reference patterns represented by $\{b_1^{(1)}, \ldots, b_N^{(P)}\}$, which are stored in and retrieved from the associative memory.

In the learning phase, the weights are recursively upgraded as follows:

$$w_{ij}(k+1) = w_{ij}(k) + 0.5\eta b_i b_j \tag{2.55}$$

where η denotes the learning rate. In fact, this is the simplest form of Hebbian learning rule (Hebb, 1949), and the learning will be terminated when the weights reach a stable level, and the energy function becomes a local minimum. It should be emphasized that there is a limitation of memory capacity in the Hopfield network (Amit *et al.*, 1985), namely, in a Hopfield network with a total of N nodes which are all connected, only about $0.15N$ random vectors can be stored with higher retrieving probability.

2.5.2 Hopfield optimization neural networks

It is still difficult to solve an optimization problem for nonlinear systems, particularly when the system is highly nonlinear, high dimensional and/or discrete event driven. However, research results (Hopfield and Tank, 1985; Tank and Hopfield, 1986) have shown that Hopfield neural networks have great potential in solving optimization problems through search in a parameter space. In industrial control, optimization usually involves supervisory optimization control, production scheduling, planning, resource allocation and job assignment, which are furnished at higher levels in a hierarchical computer integrated production system.

The application of a Hopfield neural network in solving an optimization problem can be divided into the following phases. (1) Cast the optimization problem with the given objective function and desired constraints into the form of an energy function which represents the dynamics of a neural network. (2) Then a stable state corresponding to the minimum (local) energy function can be solved through a recursive feedback search in state space. The initial state is served as the neural network's input. (3) Finally, the neural network provides a solution while the system reaches a convergent state.

Suppose a constrained nonlinear optimization problem (Luenberger, 1984) is governed by the following formulas:

$$\min_{\Lambda} F(\Lambda) \tag{2.56}$$

subject to

$$P_i(\Lambda) = 0, i = 1, 2, \ldots, q \tag{2.57}$$

where $F(*)$ is a continuous function, Λ denotes an n-dimensional continuous or discrete valued vector, and $\{P_i(\Lambda), i = 1, q\}$ are non-negative and continuous functions. The constrained optimization problem governed by 2.56 and 2.57 can be converted to a corresponding unconstrained one with a new objective function, or so-called Lyapunov energy function

$$\min_{\Lambda} E = F(\Lambda) + \sum_{i=1}^{q} c_i P_i(\Lambda) \tag{2.58}$$

where $\{c_i, i = 1, q\}$ are large positive constants and related to the penalty of the constraints. In order to solve the optimization problems with continuous inputs and outputs, the Hopfield associative network should be modified by using a sigmoid function:

$$S(\alpha) = \frac{1}{1 + e^{-\alpha/T}} \tag{2.59}$$

and the dynamic response in recall phase then becomes:

$$u_i(k+1) = \sum_{j=1}^{N} w_{ij} \alpha_j(k) \tag{2.60}$$

$$\alpha_j(k+1) = S(u_i(k+1)) = \frac{1}{1 + e^{-u_i'(k+1)/T}} \tag{2.61}$$

where

$$u_i'(k+1) = k_1(u_i(k) + r_i) + k_2(u_i(k+1) + r_i) \tag{2.62}$$

where k_1 and k_2 are constants which determine the convergent rate and system stability. From 2.61 we can see that if T approaches zero, the sigmoid function turns to a binary threshold function.

It should be pointed out that a Hopfield neural network can not guarantee to provide a global optimal solution, when the system has multiple minimum points. To move the solution from a local minimum to the global minimum, a Boltzmann machine with a simulated annealing algorithm can be introduced.

2.5.3 Boltzmann machine

A Boltzmann machine (Hinton and Sejnowski, 1986) can be established by modifying a Hopfield model by introducing a thermodynamic stochastic property. If we define the coefficient T in 2.61 as a variable parameter, the so-called *'annealing temperature'*, and let $k_1 = 0$ (no time delay), then equation 2.61 becomes

$$\alpha_i(k+1) = S(u_i(k+1), r_i) = \frac{1}{1 + e^{-(u_i(k+1)+r_i)/T}} \tag{2.63}$$

Further, instead of the deterministic Hopfield neural network we addressed, if now we define the neural network as a stochastic system and the neuron response to be a discrete value (0 or 1), then the neurons' state can be represented by the probability of the activation values:

$$Pr(\alpha_i(k+1) = 1) = \frac{1}{1 + e^{-(u_i(k+1)+r_i)/T}} \tag{2.64}$$

If we schedule the annealing temperature T as a deterministic function of time, $T(k)$, namely, gradually reduce T during neural network iterations, which is similar eventually to the physical annealing process, the perturbation produced by decreasing annealing temperature will help the neural network state to gradually release from its local minimum and finally to move to a global minimum, when the weights are frozen while the T becomes a very small positive number.

The learning algorithm of a Boltzmann machine is quite different from a Hopfield network. The Boltzmann machine learning can be viewed as a two step-learning, i.e. phase+ and phase−. In the first step (phase+, called the environment-driven phase), the vectors (input or input/output) to be learned are clamped on the relevant nodes, and the neural network iteratively run under a given annealing schedule until convergence, which corresponds to thermal equilibrium at a low temperature. In step two (phase−, called the free-running phase), we remove the clamping of the nodes, i.e. now the network freely runs under the same annealing schedule until convergence. After running these two phases, the weights can be upgraded as follows:

$$w_{ij} = w_{ij} + \eta(Pr_{ij}^{+} - Pr_{ij}^{-}) \tag{2.65}$$

It can be seen that the learning will not be terminated until a set of synaptic weights, w_{ij}, are found that provides a close match between Pr^{+} and Pr^{-}, in other words, the activation outputs in the phase− closely match the desired outputs in the phase+.

Finally, it should be pointed out that both the Hopfield neural network and the Boltzmann machine have only shown potential application in industrial control; further R & D is required to investigate their real applications.

2.6 SELF-ORGANIZING NETWORKS

In general, the system input pattern, $X = \{x_1, x_2, \ldots, x_n\}$, represents the working status of the system environment. If a system has high dimensional inputs, there may be a huge number of possible patterns governed by combinations of system input pattern values. For example, in a force prediction system for a steel rolling mill, the predicted force can be represented as a complicated nonlinear function of its input pattern, i.e. steel chemical composition C, Mn, P, S, Si, ..., gauge, temperature, speed, etc. if each pattern is described by a point in the hyperspace of the input variables, there are a huge number of points with a non-uniform distribution. It is important to classify the patterns into a number of clusters, in order to select a limited number of the represented patterns for neural network training, fuzzy model identification and even for math-model based identification. Moreover, the pattern clustering can also be used in constructing a system model (neural network, fuzzy logic and math-model) with a hierarchical and/or a decomposed structure. This section presents the concept, structure and algorithms of self-organizing neural networks.

The major function of self-organizing networks is to automatically classify input patterns into a number of disjoint clusters. The patterns located in the same cluster have similar features. The self-organizing network is formed in terms of unsupervised learning, i.e. learning without a teacher, for instance winner-take-all competitive learning. Here we introduce the algorithm of competitive learning self-organizing networks (Kohonen, 1988). In a self-organizing network, a vector quantizer can be performed by adjusting weights from N input nodes to M output nodes. When the input patterns have been presented sequentially to the network without specifying the desired output, the input patterns can be automatically classified into M clusters. In other words, after enough input pattern presentations, the created weights can provide a feature map, namely, when any input pattern is presented, the network will indicate what particular group the pattern should belong to. The detailed algorithm can be summarized as follows (Lippmann, 1987):

Step 1: Randomly initialize small valued weights $W_{ij}(0), i = 1, N; j = 1, M$.

Step 2: Present new pattern $x_i, i = 1, N$.

Step 3: Calculate the distance between the input vector and the weight vector for all individual output nodes:

$$d_j = \sum_{i=1}^{N} (x_i(k) - W_{ij}(k))^2 \tag{2.66}$$

where $x_i(k)$ is the input to the node i at time k, and $w_{ij}(k)$ is the weight from input node i to output node j at time k.

Step 4: Select the most active output node j^*, or the so-called *'winner'* which has the least distance, i.e.

$$d_{\min} = \min\{d_j, j = 1, M\} \tag{2.67}$$

If $d_j = d_{\min}$, then $j = j^*$ and $y_j^* = 1$, otherwise $y_j = 0$ (j is not equal to j^*).

Step 5: Upgrade the weights for the 'winner' node

$$w_{ij}(k+1) = w_{ij}(k) + \eta(k)y_i(x_i - w_{ij}(k)) \tag{2.68}$$

where $\eta(k)$ denotes learning rate and is defined as a time decreasing function within the range $[0 < \eta(k) < 1]$.

Step 6: Repeat by going to Step 2.

From equation 2.68 we can see that eventually the weights are upgraded only for the winner node j^*. However, in practice, the weights can also be upgraded not only for the winner j^*, but also for all nodes in the neighborhood of the winner. The size of the neighborhood $NE_j(k)$ can be predefined and can start large and slowly decrease in size with time. The weights upgrading may follow a modified version:

$$w_{ij}(k+1) = w_{ij}(k) + \eta(k)(x_i - w_{ij}(k)) \tag{2.69}$$

for all j which are located in the neighborhood $NE_j(k)$. The structure of self-organizing neural networks and the geometrical explanation of competitive learning are schematically illustrated in Figures 2.12 and 2.13 (Haykin, 1994) respectively.

It should be pointed out that the self-organizing network has shown powerful functions in signal processing, pattern recognition, data compression and imaging/vision recognition. However, since the concepts and methodologies of competitive learning and self-organizing are new to control engineers, these technologies are still at a very early stage of application in the industrial control area. Based on the features of self-organizing networks, we will find that there are many potential applications in system identification, data analysis and compression, pattern feature extraction, production scheduling and resource allocation, etc.

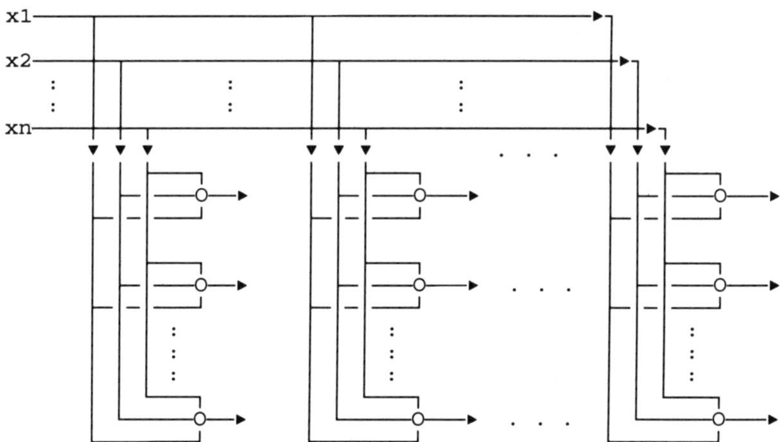

Figure 2.12 Topology of Kohonen self-organizing network (Reproduced from Lippmann, 1987, © 1987 IEEE)

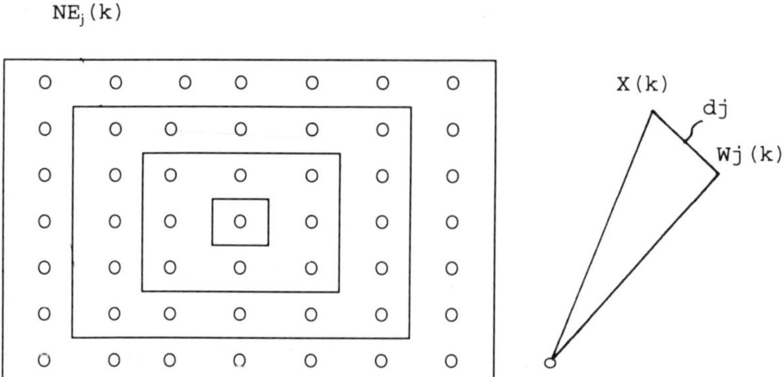

Figure 2.13 Geometrical illustration of competitive learning

2.7 DUAL (MEAN–VARIANCE) CONNECTIONS

If we examine the structures of the neural networks we have addressed, we will find there is only one synaptic connection between any two nodes, and the connecting weights have no special physical meanings. Figure 2.14 (Simpson, 1991) shows the path from input to output for a single neuron with dual connections. This section presents only the structure of neural networks with 'mean–variance' dual connections. There are other forms of dual connections, e.g. 'min–max': see Simpson (1991).

In a neural network with 'mean' and 'variance' connections, one set of connections represents mean of a class and the other represents the variance of the other class (Lee and Kil, 1989; Simpson, 1991). In this case, the output of the neuron can be represented as a nonlinear function of its inputs, and both connecting sets, W_j

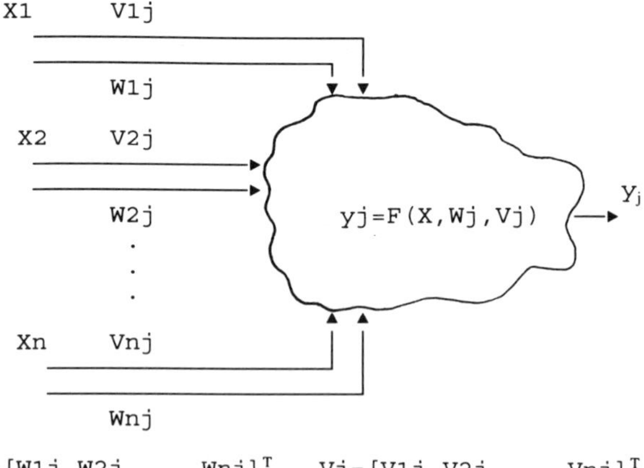

Figure 2.14 The dual connections between neurons (Reproduced from Simpson, 1992, © 1992 IEEE)

and V_j:

$$y_j = f(X, W_j, V_j) \tag{2.70}$$

where $X = [x_1, \ldots, x_n]^T$ denotes the input pattern, and the connecting weighting vectors, $W_j = [w_{1j}, \ldots, w_{nj}]^T$ and $V_j = [v_{1j}, \ldots, v_{nj}]^T$ represent mean and variance connections respectively. The neuron's output can be calculated as follows (Simpson, 1991):

$$y_j = f\left(\sum_{i=1}^{N}\left(\frac{w_{ij} - x_i}{v_{ij}}\right)^2\right) \tag{2.71}$$

where the signal function $f(*)$ is the nonlinear Gaussian function shown as follows:

$$f(\alpha) = \exp\left(-\frac{\alpha^2}{2}\right) \tag{2.72}$$

The Gaussian signal function is a symmetric radial function about the origin. If the variance is known as a stationary value, then the dual connection can be simplified to a single connection network.

2.8 SIGNAL CODING METHODS

In order to let neural networks interface with the real-world environment, neural networks must have normalized inputs and denormalized outputs. The approaches to signal normalization and denormalization mainly rely on the selected signal function and desired problem solution. There are many signal coding methods; however, here we introduce only the following two commonly used methods.

2.8.1 Min–max linear approach

For a continuous valued network with sigmoid function, the min–max linear coding methods are the most applicable in system modeling, optimization and control. The conversion from the real physical value, u, to the corresponding normalized value, u_n, can be represented by the following linear functional relation:

$$u_n = \frac{(u_n^{\max} - u_n^{\min})(u - u^{\min})}{(u^{\max} - u^{\min})} + u_n^{\min} \tag{2.73}$$

Similarly the denormalization is the conversion from neural network output, y_n, to its relevant real physical value, y

$$y = \frac{(y_n - y_n^{\min})(y^{\max} - y^{\min})}{y_n^{\max} - y_n^{\min}} + y^{\min} \tag{2.74}$$

where $u_{\min}, u_{\max}$ and $y_{\min}, y_{\max}$ denote the minimum and maximum values of the physical system's input and output respectively, and $u_n^{\min}, u_n^{\max}$ and $y_n^{\min}, y_n^{\max}$ represent the minimum and maximum values of the neural network normalized input and output respectively, which should match the threshold values of the selected signal function, in the range [0.05, 0.95] or [−0.95, 0.95]. In practice, min–max values of physical variables can be obtained from statistical analysis of the production data.

2.8.2 Mean–variance approach

We can see that the min–max linear coding method is easy to use. However, if the production data cover a wide range of possible working areas, but high probability of the data is only located in a much narrower area, then neural networks can not provide accurate results, since the frequent working area is small, while the normalization (min–max) area is large. In contrast, the mean–variance method can provide coding results with statistical sense. The normalization and denormalization algorithms can be represented as follows:

$$y_n = \frac{y - y_m}{y_{\text{var}}} \tag{2.75}$$

$$y = y_n y_{\text{var}} + y_m \tag{2.76}$$

where the mean value and variance can be calculated from the system data as follows:

$$y_m = \sum_{k=1}^{K} y_k / K \tag{2.77}$$

$$y_{\text{var}} = \sqrt{\frac{1}{K-1} \sum_{k=1}^{K} (y_k - y_m)(y_k - y_m)} \tag{2.78}$$

where K is the window length of the data to be used for signal coding which should be selected based on statistical properties of the data, namely, the window size is suitable to represent the data statistics. Moreover, in stead of a fixed window, a moving window can be used to upgrade the signals for coding; then the coding results will follow the changes of system environment.

It should be pointed out that the mean–variance coding method does not provide min–max bounded values, consequently, instead of sigmoid functions, with fixed boundaries, only a linear signal function or other unbounded function can be used in the output layer. However, if the signals are first coded using the min–max method, then further using the mean–variance method, the results may match the desired threshold values.

In addition, the analog signal can also be converted to the relevant binary signal, e.g. a sequence of 0 and 1, with an A/D converter; then a binary neural network can be used.

2.9 STRUCTURED KNOWLEDGE REPRESENTATION

2.9.1 Introduction

The rule based expert system has been recognized as one of the most powerful tools for production scheduling, job-machine assignment, fault diagnosis and expert adaptive control, etc., in industrial process control. Many expert system tools are available for practical applications. However, the applications of expert systems in system dynamic control are not as good as we expected, since there are some major difficulties in applying an expert system to solve dynamic control problems, such as process modeling, estimation, knowledge acquisition and representation, reasoning and learning under an uncertain dynamic environment.

Table 2.3 Comparison between traditional and fuzzy/neural expert systems

	Traditional expert system	Fuzzy/neural expert system
Knowledge acquisition	Human expertise: certain rules Math. model and simulation	Human expertise: fuzzy rules Math. model and simulation Fuzzy model identification Fuzzy associative memory Neural network associative memory
Reasoning and learning	Rule based reasoning under certain system environment Rule learning and deep reasoning	Fuzzy model and reasoning under fuzzy system environment Neural network supervised learning, and generalization and recall Neural network unsupervised learning and creativity
Knowledge structure and framework	Structured knowledge Symbolic framework	Structured knowledge for fuzzy rules Unstructured knowledge for neural model Numerical framework for both fuzzy and neural models

In this section, we will focus on the analysis of fuzzy and neural representation of structured knowledge, and combination of expert system with fuzzy logic and/or neural networks which will significantly enlarge the development, functions and applications of expert or intelligent systems in solving industrial control problems. With the introduction of fuzzy logic and neural network technologies in expert systems, in comparison with the traditional concepts and methodologies, some fundamental changes have been made as shown in Table 2.3.

Now we will further investigate the functions of fuzzy association memories (FAM) and neural network models in the development of an expert system in terms of problem background and knowledge/information resources available.

A regular rule in an expert system can be described as

$$\text{If } OP(A_1, A_2, \ldots, A_n), \text{ then } B$$

where the operation (OP) represents 'and' or 'or' of pre-conditions $\{A_i, i = 1, n\}$. Both the states of events $\{A_i, i = 1, n\}$ and conclusion B are certain symbolic values, $\{0 \text{ or } 1\}$, or $\{-1, \text{ or } +1\}$, and then the event-driven (or object-driven) inference or reasoning is made on a basis of given certain knowledge, which consists of pre-conditions and conclusions as well as a set of rules. Obviously, rule based knowledge representation and inference engine in an expert system are quite different from the behavior of human-like decision making. The major characteristics of human decision making can be summarized as follows:

- Human decision making is usually made under an uncertain or fuzzy environment; this means incomplete and imprecise information and knowledge, and approximate reasoning;
- The process of human decision making is an information distributed parallel process. For instance, when a human operator sets up the set point values of the control loops for a chemical reactor, the decision is made from parallel processing of the production information and knowledge, such as observations of sensory measurements, knowledge of control criteria (or objectives) and any production constraints, and his past experience. Moreover, instead of the addressable memories of current computers, sophisticated knowledge is stored in the memory with a huge number of synaptic connections. Consequently, the memories are stored in a distributed way, and the decision making is made through 'recall' or 'retrieval' in a parallel manner.
- Learning, generalization and creativity are the most important behaviors in human decision making; they can be described as follows:
  - Learning from examples or teachers, and/or deep reasoning;
  - Feature extraction or knowledge acquisition through 'unsupervised' (or so-called 'self-organized') learning, namely, some new knowledge or higher level meta-knowledge can be explored;
  - The learning results can be effectively extended from the learning environment to working conditions which have not been learned before; so-called 'generalization'.

The following sections will address how to introduce fuzzy and neural technologies into knowledge representation and reasoning in expert systems.

2.9.2 Applications of fuzzy knowledge in expert system

The application of fuzzy logic in rule based expert systems has been an attractive research and development area in the past decade (Zimmermann, 1987; Buckley, 1990; Qian and Lu, 1989), and some fuzzy expert system shells, such as FLOPS (Siler *et al.*, 1987), have been developed. In contrast to a non-fuzzy rule, a fuzzy rule usually can be described as *'If OP* $(A_1, \ldots, A_n)$*, then B with cf (certainty, or confidence factor)'*, in which both $\{A_i, i = 1, n\}$ and B are fuzzy variables, and the degree of truth of the rule is measured by a given cf (certainty factor).

Suppose a fuzzy proposition P, *'A is X'*, where A is linguistic variable defined on the universe of discourse U and X is a fuzzy subset defined on the same universe of discourse U. The probability of fuzzy proposition P may be represented by the possibility distribution function, $p_A(u) = m_X(u),\ u \subset U$, where $m_X(u)$ is the membership function of X.

As for a dynamic system, the time oriented temporal descriptor, $T(P)$, can be defined to describe the time properties of fuzzy proposition $T(P)$ as follows:

- time point descriptor: $AT(P, \int_T m(t)/t,\ unit)$
- time interval descriptor: $AT(P, \int_T m_1(t)/t, \int_T m_2(t)/t,\ unit)$

where '$\int_T m(t)/t$' denotes the time point of the P (fuzzy proposition) occurrence with membership $m_1(t)$ on the universe of discourse T, and '$\int_T m_1(t)/t, \int_T m_2(t)/t$' describes the time interval from P appearing to P disappearing with the membership $m_1(t)$ and $m_2(t)$ respectively. The 'unit' can be any time unit, e.g. second, minute, hour, etc.

In practice, the temporal descriptor T can be physically described by the following different time coordinates:

- Absolute or so-called 'real clock' time, for instance, 'the temperature increases at around 5 o'clock' can be represented as:

  $$AT\ (P,\ 0/4.50',\ 0.5/4.55',\ 1.0/5.00',\ 0.5/5.05',\ 0/5.10'),\ \text{o'clock})$$

  where $0, 0.5, 1, 0.5, 0$ denote the corresponding membership values of P's occurrence time of 4.50′, 4.55′, 5.00′, 5.05′, 5.10′ respectively.

- Relative time description, for instance

$$AT(P1, 1/0,\ \text{min}) \wedge AT(P2, (0/1.5', 0.5/1.55', 1.0/2.00', 0.5/2.05', 0/2.10'),\ \text{min})$$

  means that the proposition $P1$ occurs about 2 minutes earlier then that of $P2$.

- Before/after chain description, for instance

$$AT(P_1, 1/0, \text{min}) \wedge AT(P_2, (1/t > 0), \text{min})$$

denotes that the occurrence of $P1$ is before that of $P2$.

In addition to the time descriptor given above, there are more descriptors, such as after, before, overlapping, during, contain, etc., representing the corresponding temporal rules. Moreover, a temporal fuzzy rule can be generally described as

$$\text{If } (T_1(P_1) \wedge \cdots \wedge T_m(P_m)), \ \textit{then } T_{m+1}(P_{m+1}) \ \textit{with confidence } W$$

where $\{P_i, i = 1, 1+m\}$ are fuzzy propositions and $\{T_i, i = 1, m+1\}$ are the corresponding temporal descriptors. Then, a set of fuzzy rules can be organized as a knowledge chunk S_k with the following form:

$$\textit{If} \begin{bmatrix} T_1^{(k)}(P_{1(1)}^{(k)}) & \wedge & \cdots & \wedge & T_{mk}^{(k)}(P_{mk(1)}^{(k)}) \\ \vdots & & \ddots & & \vdots \\ T_1^{(k)}(P_{1(dk)}^{(k)}) & \wedge & \cdots & \wedge & T_{mk}^{(k)}(P_{mk(dk)}^{(k)}) \end{bmatrix}$$

$$\textit{then} \begin{bmatrix} T_{mk+1}^{(k)}(P_{mk+1(1)}^{(k)}) \\ \vdots \\ T_{mk+1}^{(k)}(P_{mk+1(dk)}^{(k)}) \end{bmatrix} \textit{with confidence}: \begin{bmatrix} W_1^{(k)} \\ \vdots \\ W_{dk}^{(k)} \end{bmatrix} \tag{2.79}$$

A fuzzy reasoning network can consist of a number of knowledge chunks $S_i, i = 1, n$ organized with either hierarchical or decomposed structure. In the network, the conclusions inferred from each knowledge chunk might serve as the

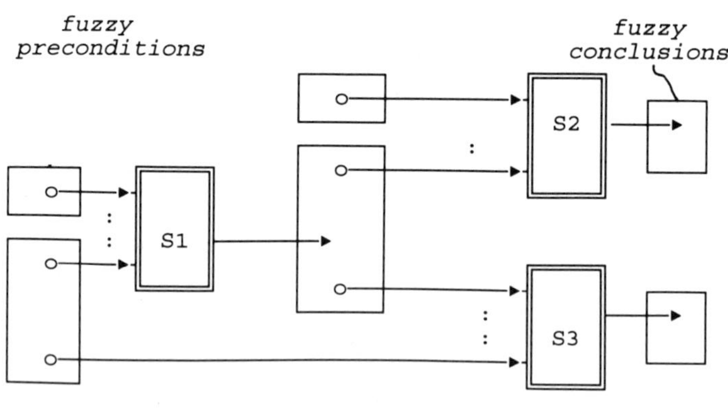

Figure 2.15 Sample fuzzy reasoning network

fact in the preconditions of the other knowledge chunks. A simple example of fuzzy reasoning network is shown in Figure 2.15 (Qian and Lu, 1989).

In practice, the preconditions and conclusions can have either an identical or a different universe of discourse and temporal descriptor, and they can be divided into different groups, each using forward or backward search strategies in problem solving (refer to Qian and Lu, 1989 for details).

Since Zadeh (1988) proposed a fuzzy logic, the applications of fuzzy reasoning have been playing an important role in developing fuzzy expert systems. Suppose a fuzzy rule R ($A \rightarrow B$ with cf), and if both the fuzzy value of A and cf in rule R are known, the problem of fuzzy reasoning is to infer the fuzzy value B and the degree of validity associated with the inference. In order to solve this kind of simple fuzzy reasoning problem, we can assume A and B are the fuzzy sets on the universes of discourse, X and Y respectively, and the membership functions of A and B are represented as $m_A(x)$ and $m_B(y)$ respectively. The membership of $R(A \Rightarrow B)$ can be described as

$$m_{A\rightarrow B}(x,y) = [m_A(x) \wedge m_B(y)] \vee [1 - m_A(x)] \tag{2.80}$$

In approximate reasoning, if we have $(A \Rightarrow B)$, then from the condition A_1 being the subset of A, the conclusion can be solved as

$$B_1 = A_1 \circ (A \Rightarrow B) \tag{2.81}$$

However, in contrast to rule based systems, the structured knowledge in fuzzy systems can be encoded into a numerical framework. As a result, the fuzzy knowledge and fuzzy reasoning can be more effectively managed by a fuzzy relational model or fuzzy associative memory as fuzzy pattern mapping in hyperspace from unit cubes to unit cubes. Now let us give a traffic control association as an example to illustrate the approaches to fuzzy association.

Example 2.4 (Kosko, 1992)

For traffic control association, the linguistic rule 'if the traffic is heavy in this direction, then keep the light green longer'. To describe the numerical relationship between the traffic density and the duration of green light, let the fuzzy set pair (A, B) encode the traffic associate control (heavy, light). Then quantize the traffic density and green light duration n and p with numerical variables, namely, $X = [x_i, i = 1, n], Y = [y_j, j = 1, p]$. As a result, A defines a point in the n-dimensional hypercube $I_n = [0, 1]_n$, and B defines a point in the p-dimensional fuzzy cube I_p.

In fact, A and B define membership functions m_A and m_B, which denote how much x_i belongs to A and how much y_j belongs to B respectively. The following fuzzy 'max–min' composition can be used to represent the numerical relationship between A and B:

$$A \circ M = B \tag{2.82}$$

The component b_j can be solved by the following 'inner product' of vector A with the jth column of M:

$$b_j = \max\min(a_i, m_{ij}), \quad i = 1, n \tag{2.83}$$

Let vector $A = [0.3, 0.4, 0.8, 1]$ and fuzzy matrix

$$M = \begin{bmatrix} 0.2 & 0.8 & 0.7 \\ 0.7 & 0.6 & 0.6 \\ 0.8 & 0.1 & 0.5 \\ 0 & 0.2 & 0.3 \end{bmatrix}$$

then $b_j, j = 1, 3$ can be calculated by using equation 2.83:

$$\begin{aligned} b_1 &= \max\{\min(0.3, 0.2),\ \min(0.4, 0.7),\ \min(0.8, 0.8),\ \min(1, 0)\} \\ &= \max(0.2, 0.4, 0.8, 0) = 0.8 \\ b_2 &= \max(0.3, 0.4, 0.1, 0.2) = 0.4 \\ b_3 &= \max(0.3, 0.4, 0.5, 0.3) = 0.5 \end{aligned}$$

More approaches to fuzzy association memories, e.g. fuzzy Hebb matrix (Mamdani, 1977; Togai and Watanabe, 1986), can be used to solve this kind of fuzzy reasoning problem.

It should be emphasized that directly encoding structured knowledge into a numerical framework through introducing FAM in an expert system makes fuzzy expert systems much more attractive and applicable to industrial control, since the introduction of FAM makes knowledge representation, knowledge acquisition, inference and learning more realistic, feasible and simple. The abilities of knowledge representation and reasoning of fuzzy reasoning networks (Qian, 1987; Qian and Lu, 1989) involves temporal knowledge representation with hierarchical or network structure in a fuzzy set, and fuzzy network based reasoning.

2.9.3 Rule-based inference by neural network

Neural network technology has had an enormous influence in development of new approaches to system modeling, estimation and control. However, how to enlarge the application domain of neural networks has been one of the major concerns in engineering societies. On the basis of the characteristics and functions of neural networks and the difficulties in developing expert systems, the combination of neural network and rule based expert system is definitely the great potential for intelligent control. Knowledge acquisition, real-time reasoning and learning have been recognized as three important issues in developing an expert system which can work well in an uncertain environment. This section will investigate the strategic resolution and advantage of such a combination.

Control engineers consider the neural network as a tool for function approximation, pattern mapping or adaptive algorithms. Computer and information scientists view the neural network as a knowledge base, an associative memory or an information processing tool with parallel and distributed structure in nature. Nevertheless, a neural network can describe the behavior of a system (or simply an event) from its inputs (or conditions) to the relevant outputs (or conclusions), and can automatically classify a system environment into a variety of families based on their similarity with or without supervised learning.

The combination of neural network and rule based expert system has been studied in the past decade (Narazaki and Ralescu, 1992; Lacher *et al.*, 1992; Fu and Fu, 1990). A recent research result is the application of neural networks like connectionist mechanisms in rule-based inference (Narazaki and Ralescu, 1992). This work consists of the following points: (1) representation of knowledge in a network, (2) the mechanism of a node in a network, and (3) analysis of the behavior of the mechanism. The details of the problem formulation can be summarized as follows:

Rules A rule of 'If p, then q with certainty factor $r \in [0, 1]$' can be interpreted as the combination of the two inequality constraints

$$1 - V(p) + V(q) \geq r \tag{2.84}$$

$$V(q) \leq r \tag{2.85}$$

where $V(p)$ and $V(q)$ denote the truth values of p and q respectively. In other words, $V(p)$ must be constrained as

$$r + V(p) - 1 \leq V(q) \leq r \tag{2.86}$$

The inequality 2.86 is equivalent to the equality constraint

$$g = g_1 + g_2 = 0 \tag{2.87}$$

where

$$g_1 = \max\{0, r + V(p) - 1 - V(q)\} \tag{2.88}$$

$$g_2 = \max\{0, V(q) - r\} \tag{2.89}$$

Note that g_1 and $g_2 \geq 0$. If $V(p)$ and $V(q)$ do not satisfy 2.84 and 2.85, then g_1 or g_2 become positive respectively.

General Constraints The general constraint with equality form can be defined as

$$h(V(p_1), \ldots, V(p_n)) = 0 \tag{2.90}$$

Facts Facts give the truth values of propositions. Facts are treated as given data and fixed throughout the inference process.

Preferences The preferences are expressed by the least-square error minimization index:

$$J = \frac{1}{2}\sum_{i=1}^{n} \mu(V(p_i) - P(p_i))^2 \tag{2.91}$$

where $P(p)$ is defined as a preferred truth value for a proposition p and $\mu \geq 0$ is a weight given by the user.

Formulation of the Inference Problem The inference problem can be defined to decide the truth value distribution $V(p_i)$, so that

$$E' = \frac{1}{2}\left(\sum_{i=1}^{n_1} \rho_i(g_i) + \sum_{i=1}^{n_2} (\mu_i h_i^2)\right) \to 0 \tag{2.92}$$

where n_1 and n_2 are the number of rules and the number of general constraints in the knowledge base respectively, and ρ_i and μ_i are the weights of the rule and equality constraints respectively. E' is the total amount of constraint violation.

The energy function E to be minimized can be constructed by combining J in 2.91 with E':

$$E = E' + J \tag{2.93}$$

If E' is reduced to 0, a feasible solution has been found. If the minimum E does not reduce E' to 0, then we call it a local minimum. If we can not find a feasible solution, the knowledge base is inconsistent. The detailed method was given by Narazaki and Ralescu (1992).

Finally, we should realize that it has been an attractive research area to explore the combination of neural networks and fuzzy systems through either introducing learning functions and nonlinearity of neural networks into fuzzy systems or incorporating the knowledge structure of fuzzy logic into the resulting neural networks.

2.9.4 Fuzzy rule-based neural network

When we design an industrial control system, two different kinds of information, namely sensor measurement oriented numerical data and linguistic (fuzzy) information, are gathered to be processed. For instance, an expert control system may consist of both math algorithm based deep knowledge and 'if-then' fuzzy rules. To coordinate numerical and qualitative knowledge in a hybrid system has been one of the important subjects in the design of an intelligent system with machine learning. The neural network has remarkable learning ability for establishing associative memory in terms of the training samples. In this section, we will further investigate combination of neural network with fuzzy logic. The emphasis will be the strategic approaches to constructing fuzzy-neural networks.

Some recent publications (Ishibuchi *et al.*, 1993; Hirota and Pedrycz, 1994) have addressed the applications of neural networks in learning and constructing fuzzy rules and structured knowledge. Here we introduce the design of a neural network utilizing expert knowledge represented by fuzzy 'if-then' rules (Ishibuchi *et al.*, 1993). In other words, we establish a fuzzy pattern mapping from fuzzy input vector, $X = [x_i, i = 1, r]$ to fuzzy output vector, $Y = [y_i, i = 1, n]$ in terms of neural network learning, namely, $\mathcal{NN} : X \to Y$. Both fuzzy inputs and output can be linguistic values defined by their membership functions. In order to simplify the neural network computation, the linguistic values usually are restricted to convex and normal fuzzy sets on a real line. Obviously, the key point in performing neural network based fuzzy pattern mapping is to design the architecture and learning algorithm which should be able to deal with fuzzy variables. Here we provide a number of simple fuzzy rules to describe the fuzzy relations between production conditions and finishing product quality in a distillation column:

Rule 1: *If reflux rate (R) is large and and reboiler steam flow rate (S) is middle, then the column's top distillate of column (P_t) is qualified.*

Rule 2: *If reflux rate (R) is small and reboiler steam flow rate (S) is high, then the column's top distillate of column (P_t) is non-qualified.*

It can be seen that the sample rules given above represent a classification problem in a multi-dimensional input pattern space. Without loss of generality, if we consider the classification problem with only two classes, than the general form of the fuzzy rule set can be given as follows:

If x_{p1} is A_{p1}, and, ..., and x_{pn} is A_{pn},
then $X_p = (x_{p1}, \ldots, x_{pn})$ belongs to $G_p, p = 1, 2, \ldots, P$

where $\{A_{pi}, i = 1, n, p = 1, P\}$ denotes the fuzzy number (linguistic value) of x_i for pattern p, and G_p represents the class of pattern p. This rule can be further simplified as the following compact form:

$$A_p = (A_{p1}, \ldots, A_{pn}) \text{ belongs to } G_p$$

where A_p is a fuzzy vector. The membership function of a fuzzy vector $A = (A_1, \ldots, A_n)$ is defined as

$$m_A(x) = \min\{m_{A1}(x_1), \ldots, m_{An}(x_n)\} \tag{2.94}$$

Now we introduce an architecture and the relevant learning algorithm of neural networks which can perform fuzzy pattern mapping. The fuzzy pattern mapping can be geometrically represented as mapping from a cube in input space to another cube in output space as we noted before. The fuzzy activation functional mapping can be illustrated in Figure 2.16 (Ishibuchi, *et al.*, 1993). A multi-layered feedforward neural network with BP learning can be designed for fuzzy pattern mapping. The

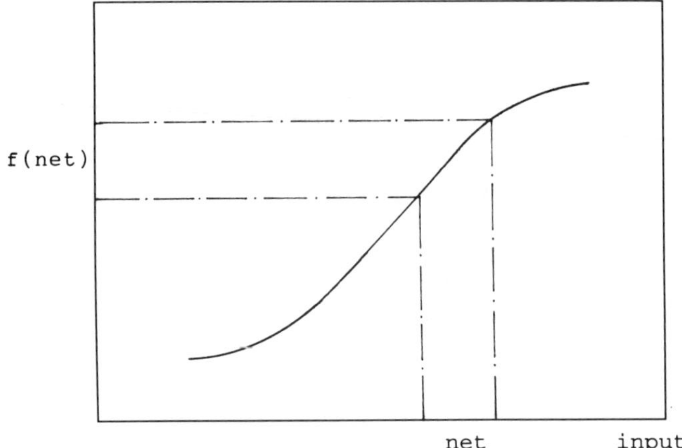

Figure 2.16 Fuzzy pattern mapping through fuzzy activation function (Reproduced from Ishibuchi *et al.*, 1993, © 1993 IEEE)

fuzzy input pattern $A_p = (A_{p1}, \ldots, A_{pn})$ is presented to the input layer, and then the forward-path can be calculated as

(1) Place the neural network inputs at the input layer:

$$u_i = A_i, \quad i = 1, 2, \ldots, N_i \tag{2.95}$$

(2) Calculate the outputs of the hidden layer:

$$\alpha_j = \sum_i WH_{ji} u_i + WH_{bj}, \quad j = 1, \ldots, N_h \tag{2.96}$$

$$y_{hj} = f(\alpha_j) \tag{2.97}$$

(3) Calculate the outputs of the hidden layer:

$$\beta_k = \sum_j WO_{kj} y_{hj} + WO_{bk}, \quad k = 1, \ldots, N_0 \tag{2.98}$$

$$y_k = f(\beta_k) \tag{2.99}$$

where f is defined as fuzzy activation function; WH, WO and WH_b, WO_b are real-valued connecting weights and bias respectively, and all other variables, such as, u, α, y_h, β and y, denote fuzzy numbers with relevant fuzzy memberships.

When completing the presentation of each pattern (i.e. pattern-wise) or a batch of patterns (i.e. batch-wise), similar to non-fuzzy network, the connection weights can be adjusted to minimize the cost function. Suppose that the output is a scalar, then the cost function can be defined in terms of both supervised fuzzy value, r, and fuzzy output, y, provided by fuzzy network. A fuzzy set A can be represented by its level sets: $[A]_{h1}, [A]_{h2}, \ldots, [A]_{hm}$ where $0 < h \leq 1$, and m denotes the number of level sets.

For instance, we can select five levels: $h = 0.2, 0.4, 0.6, 0.8, 1.0$, then the cost function for the pth pattern can be represented as

$$e_p = \sum_h e_{ph} = \sum_h h \max\{(r_p - y_p)^2,\ y_p \in Y_p \mid h\} \tag{2.100}$$

The weights are modified with the following learning algorithms:

$$\Delta WO_j(t+1) = \eta\left(-\frac{\partial e_{ph}}{\partial WO_j}\right) + \rho \Delta WO_j(t) \tag{2.101}$$

$$\Delta WH_{ji}(t+1) = \eta\left(-\frac{\partial e_{ph}}{\partial WH_{ji}}\right) + \rho \Delta WH_{ji}(t) \tag{2.102}$$

where η and ρ denote the learning rate and momentum constant respectively.

The fuzzy neural network can be applied to fuzzy pattern classification or fuzzy control. The following example shows the application results.

Example 2.5 (Ishibuchi *et al.*, 1993)

Let the following fuzzy 'if-then' rules apply for two-dimensional input space $[0, 1] \times [0, 1]$:

If x_1 is small and x_2 is small then y is small;
If x_1 is small and x_2 is large then y is small;
If x_1 is large and x_2 is small then y is medium;
If x_1 is large and x_2 is large then y is large;

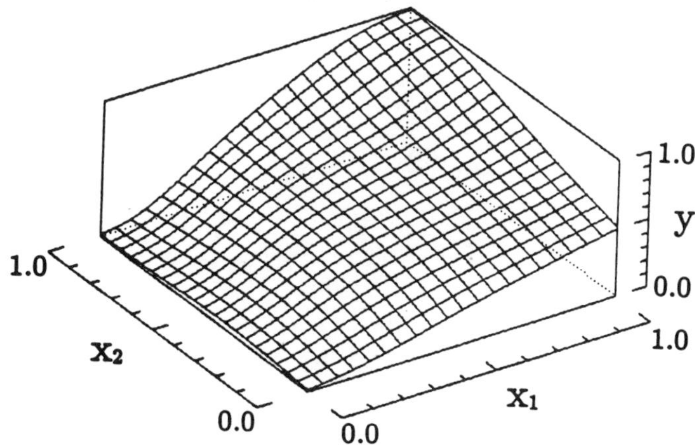

Figure 2.17 Relationship between inputs and output for example 2.5 (Reproduced from Ishibuchi *et al.*, 1993, © 1993 IEEE)

These fuzzy rules can be represented as four fuzzy input–output pairs: $[(S, S), S], [(S, L), S], [(L, S), M], [(L, L), L]$. A multilayered neural network with two input neurons, five hidden neurons and one output neuron is trained with the above four pairs of inputs and outputs. As a result, the nonlinear fuzzy pattern mapping is established as shown in Figure 2.17 (Ishibuchi, *et al.*, 1993).

The applications of fuzzy min–max neural networks with supervised learning and self-organizing learning in pattern classification and clustering have also been investigated. For details see Simpson (1992, 1993).

Finally, it should be pointed out that the techniques for designing an intelligent system being addressed in this book are not restricted to neural network, fuzzy logic and rule based knowledge engineering; some other methods, for instance the genetic algorithm and pattern recognition, will be introduced in conjunction with the specific application areas, and learning as a key body in any intelligent system will be introduced in Chapter 3.

3

Learning strategies and algorithms

The subject of learning has been studied extensively in many different areas over past decades, such as learning a natural language to use a machine, learning to recognize a pattern and learning mathematical concepts and operations, etc. We realize that 'learning' makes a man-made machine 'intelligent' and 'learning' is one of the most important features of an intelligent system. In this chapter, we introduce the learning concepts, strategies and algorithms to be applied in industrial control. This chapter focuses on 'supervised learning', 'unsupervised learning' and 'reinforcement learning' in neural networks, fuzzy logic and rule based conceptual learning. The learning methodologies to be illustrated in this chapter may be applied to system modeling, estimation, adaptive control and optimization, etc.

3.1 POSITION OF LEARNING IN INDUSTRIAL CONTROL

'Robustness' and 'adaptation' are considered as the most important properties for any industrial control system. With the rapid progress of machine learning technologies, the applications of learning in neural networks, fuzzy logic and expert systems have provided great opportunities to develop *non-math-knowledge* based (or so-called 'math-model free') control systems with good robust and adaptive behavior.

In a general sense, any control or decision making system should at least consist of a controlled environment and a decision maker (or so-called 'controller'). Generally speaking, the major function of learning in system controls is to automatically upgrade the structure and/or parameters of a system model or a controller, or the relevant knowledge base through learning and inference from historical data and past information and knowledge. As a result of learning, the control law and its parameters may be adapted to the system environment changes to provide satisfactory or robust behavior. In addition, it should also be emphasized that learning is not only derived from past experience, e.g. examples, patterns or historical data, but may also reveal hidden knowledge which was not originally explored by the designer. The latter is performed by 'self-organization learning' or 'rule learning and deep reasoning'. From an application point of view, the main functions of learning in industrial control can be summarized as follows.

3.1.1 Learning in system modeling

The learning task in system modeling for a static system is to establish and/or upgrade a system model, $\Phi_k : U_k \to Y_k$, with supervised learning. Suppose that a static system is governed by an unknown nonlinear functional relation:

$$Y_k = \Phi_k(U_k, \theta_k) \tag{3.1}$$

where

$Y \in \Re^m$ is an m-dimensional output vector;
$U \in \Re^r$ is an r-dimensional input vector;
$\theta \in \Re^p$ is a p-dimensional parameter vector;
$\Phi(*)$ is a non-linear pattern mapping, which can be a neural network, fuzzy logic or pattern recognition model, etc.

Then learning for model 3.1 is to determine or upgrade either θ_k or both Φ_k and θ_k with the system's historical data, $\{U_k, Y_k, k = k, k - K\}$ associated with a fuzzy rule set, $\mathcal{F}$, or a neural network topology $\mathcal{N}$, describing the relations between the system inputs U and outputs Y.

For a dynamic system, instead of 3.1, an unknown state space model can be represented as follows:

$$X_{k+1} = \Phi_k(X_k, U_k, \theta_k) \tag{3.2}$$

$$Y_k = \Psi(X_k) \tag{3.3}$$

where

$X \in \Re^n$ is an n-dimensional state vector;
$\Psi(*)$ is a pattern mapping from system state vector to output vector.

Similarly, the learning task in dynamic system modeling is to determine or upgrade the system dynamic model, Φ_k: $U_k, X_k \to X_{k+1}$, and can be further represented by the following optimization problem:

$$\min_{\Phi_k, \theta_k} \sum_k \|Y_k^p - Y_k\|^2 \tag{3.4}$$

where Y_k^p denotes the predicted output vector. Here we give an example to describe how to apply learning in system modeling.

Example 3.1

Suppose a batch chemical reactor with the chemical reaction, $A + B \Rightarrow C$, and its input and output vectors can be denoted as follows:

$$U = [C_A, C_B, W_A, W_B, P, T_0]^T$$
$$Y = [C_C, W_C, T_f]^T$$

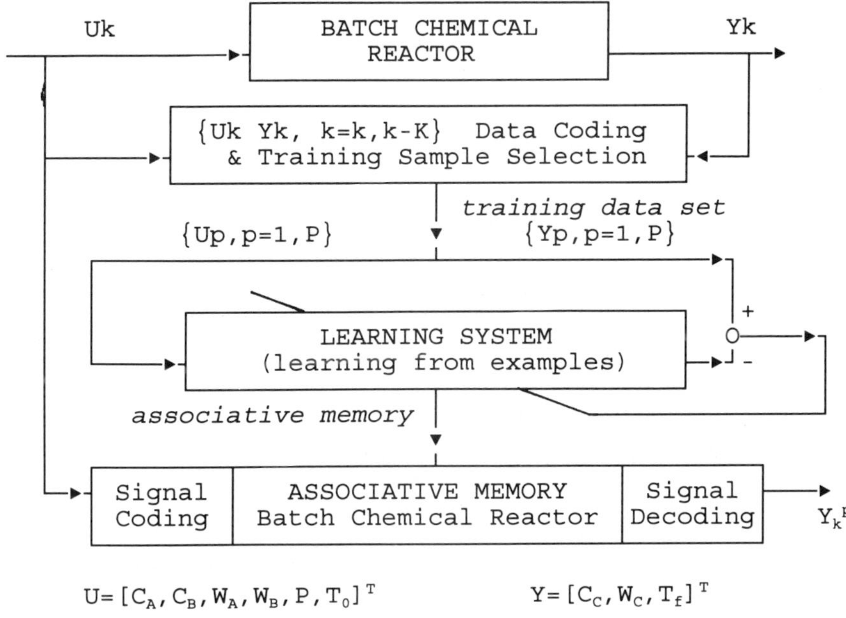

Figure 3.1 Learning system for modeling a batch chemical reactor

where

C_A, C_B, C_C are the chemical compositions of the reactants *A, B* and the finishing product *C*;
W_A, W_B, W_C are the weights of reactants *A, B* and the finishing product *C*;
P, T_0, T_f are the reaction pressure, the initial and final temperatures respectively.

The model of the chemical reactor can be established using either a neural network model:

$$\mathcal{N} : U_k \Rightarrow Y_k \tag{3.5}$$

or a fuzzy model:

$$\mathcal{F} : U_k \Rightarrow Y_k \tag{3.6}$$

with the production data $\{U_k, Y_k, k = k, k - K\}$ and other process knowledge through supervised learning. To follow the changes of system environment, the reactor model, $\mathcal{N}$ or $\mathcal{F}$, must be modified through learning from the updated production data. The learning system in example 3.1 is illustrated in Figure 3.1.

3.1.2 Learning in state estimation

The task of state estimation in industrial control is to estimate (or predict) a system's full state vector with the system observed I/O data. The state observer and Kalman filtering are the major state estimation tools for deterministic and stochastic systems respectively. If considering a system with state space model (3.2), the task of state estimation is to estimate the system state X_k with the system model (3.2), system observed outputs Y_k and inputs U_k, i.e. $\mathcal{E}$: Y_k, $U_k \Rightarrow X_k^p$. The general form of state estimation can also be represented by the following optimization problem:

$$\min_{\mathcal{E}, \lambda_k} \sum_k \|X_k^p - X_k\|^2 \tag{3.7}$$

$$X_{k+1}^p = \mathcal{E}_k(X_k^p, Y_k, U_k, \xi_k) \tag{3.8}$$

where

X_k^p is the estimated state vector;
$\mathcal{E}_k, \xi_k$ are the functional pattern mapping and parameter vector of the state estimator.

In a math-model free estimation system, the functional pattern mapping, $\mathcal{E}_k$ and its parameter vector ξ_k can be established and further modified through a neural network, fuzzy model or any hybrid knowledge based learning. If we keep the math-model based estimation frame, i.e. $\mathcal{E}$ denotes an explicit mathematical algorithm, then the function of learning will be reduced to only determining the parameter vector, ξ_k.

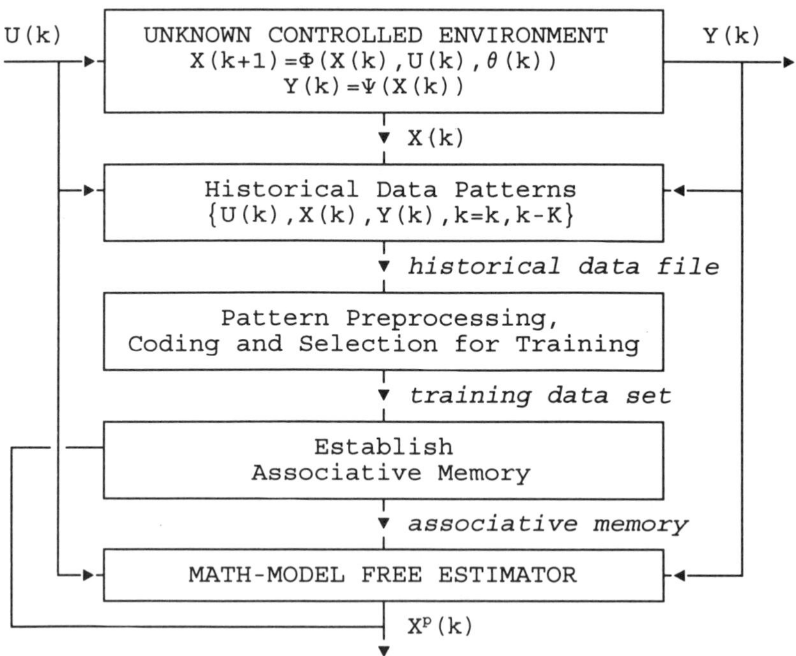

Figure 3.2 Architecture of a math-model free estimation system

The general architecture of an estimation system with learning is given in Figure 3.2. Now an example will be given to describe the concepts of math-model free state estimation in detail.

Example 3.2

Consider a continuous chemical reactor with the chemical reaction: $A + B \Rightarrow C + D$. The system dynamic behavior of the reactor can be represented by the following pattern mapping representation:

$$X_{k+1} = \Phi_k(X_k, U_k, \theta_k) \tag{3.9}$$

$$Y_k = \Psi(X_k) \tag{3.10}$$

where

$X = [C_C, C_D, T_r, P_r]^T$, state vector;
$U = [F_A, F_B, T_j]^T$, input vector;
$Y = [T_r, P_r]^T \in X$, output vector;
C_A, C_D are the chemical compositions of the reaction products C and D respectively;
F_A, F_B are the flow rates of the reactants A and B respectively;
T_r, P_r, T_j are the reaction temperature, pressure, and jacket temperature.

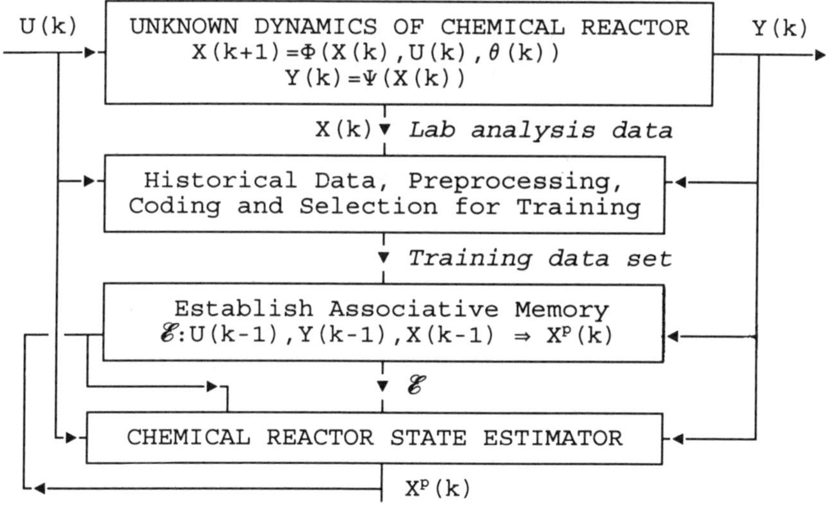

Figure 3.3 A math-mode free estimation system for the chemical reactor shown in example 3.2

The task of the state estimation in this example is to estimate the system internal information C_C and C_D with the system observed outputs $[T_r, P]$ and inputs $[F_A, F_B, T_j]$. Obviously, the function of the learning for this problem is to establish a corresponding pattern mapping from system input and output vectors to state vector, i.e., $\mathcal{E} : U_k, Y_k, X_k^p \Rightarrow X_{k+1}^p$, based on the system information and knowledge. In addition, $\mathcal{E}$ will be upgraded when the reactor's operating conditions significantly change. The architecture of state estimation for example 3.2 is illustrated in Figure 3.3.

3.1.3 Learning in adaptive control

The function of learning in an adaptive control system is to automatically adjust the parameters and/or structure of the controller to provide a satisfactory or desired control response. The problem of adaptive control with self-learning can be mathematically described as follows:

$$\min_{\mathcal{C}, \beta_k} \sum_k \|X_k^d - X_k\|^2 \tag{3.11}$$

subject to

$$X_{k+1} = \Phi_k(X_k, U_k, \theta_k) \tag{3.12}$$

$$U_k = \mathcal{C}(X_k, \beta_k) \tag{3.13}$$

where

X_k^d is the desired trajectory of the state vector;
$\mathcal{C}_k$, β_k are the control law and parameter vector, respectively.

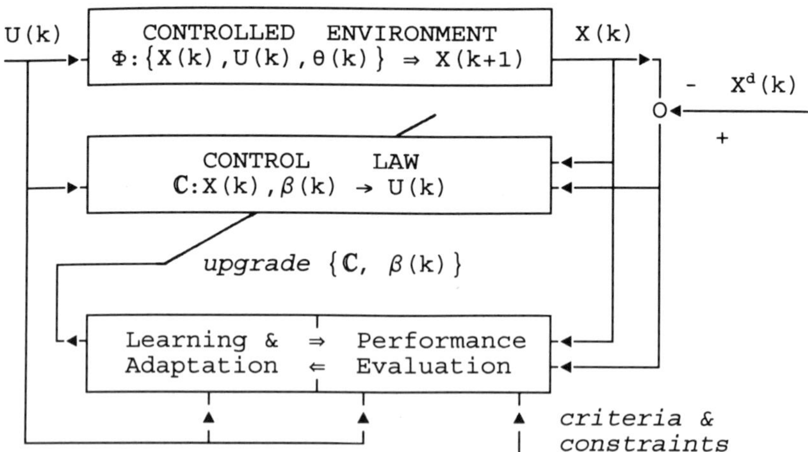

Figure 3.4 Conceptual illustration of learning in adaptive control

Learning in an adaptive control system is to establish and/or upgrade the control law, $\mathcal{C}: X_k \Rightarrow U_k$, which is designed as a pattern mapping from X_k to U_k and may be implemented by a neural network, fuzzy logic, rule base or any hybrid form.

It should be pointed out that in addition to the applications of learning in local control loops, learning can also be applied at a high-coordination level which makes a number of local controllers cooperatively work together to reach a global criterion. The general function of learning in adaptive control can be illustrated in Figure 3.4.

3.1.4 Learning in optimization control and decision

The math-model, heuristic search and expert system based optimization techniques have been widely used in the supervisory (set point) control, production scheduling, job assignment and production planning in an integrated production system. In general, these optimization problems can be grouped into categories: static and dynamic, or continuous and discrete-event driven, etc. However, here we only investigate the optimization techniques being applied in solving static problems. In a supervisory control, the optimization problem can be described as follows:

$$\min_{U_k} J = J(Y_k, U_k) \tag{3.14}$$

subject to

$$Y_k = \Psi(U_k, \Theta_k) \tag{3.15}$$

and other related constraints, where $J(*)$ is defined as an objective function. The goal of the optimization is to find a specific control action U_k, e.g., the set points of the control loops or any other optimization oriented decisions, to minimize a given objective function J. It should be emphasized that in an intelligent optimization system, the solution to the optimization is given under a non-stationary environment, which may include any changes in the objective function, system model and/or production constraints. Obviously, the problem becomes more difficult if we use a math-model based optimization approach. This is why the applications of learning in solving an optimization problem under a non-stationary environment are critical.

3.1.5 Learning for self-organization and knowledge acquisition

In comparison with the learning functions described above, learning for 'self-organization' and 'knowledge acquisition' has a different concept and philosophy in both fundamental and practical aspects. Generally speaking, the applications of self-organization learning are able to automatically extract the knowledge, and to classify the system input patterns into a number of families based on their similarity.

Most self-organized learning strategies are unsupervised learning, namely, learning without an instructor. Self-organized learning has been applied in pattern recognition, imagine and signal processing for dealing with incomplete information, such as pattern completion or recovery, etc. However, in industrial control, this technology is still quite new. Based on the features of self-organized learning, introducing this technology to control researchers and engineers will result in finding great opportunities for developing novel industrial control strategies. Generally speaking, the application areas of self-organized learning may involve knowledge acquisition, information extraction, data compression, and pattern classification and recovery.

3.2 INTRODUCTION TO LEARNING MECHANISM

3.2.1 Introduction

In information and system control technologies, the learning process is to design computer software (or with hardware) which is able to learn and to infer from the system's historical information (patterns, data or any other information resources) under an uncertain or variable environment. As a result of learning, the system responses or behaviors can be consistently improved even when the system environment is significantly disturbed. In addition, learning may also result in discovering more knowledge, that makes a system more robust and adaptive. In general, the learning process involves information processing, associative memory, vector quantization, deep reasoning, etc.

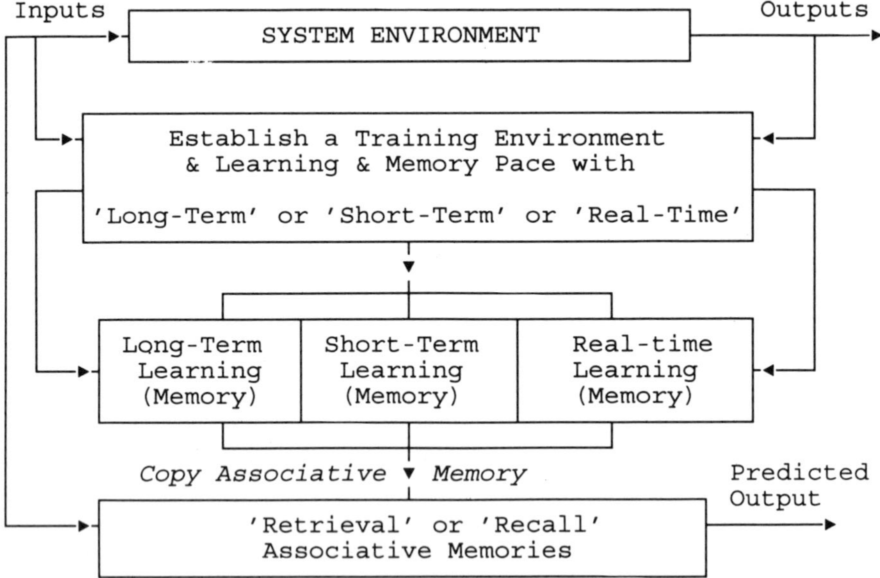

Figure 3.5 Learning and retrieval process in a learning system

From an application point of view, many learning systems consist of two general phases: the 'learning' and the 'retrieval' or so-called 'recall'. The learning phase can be viewed as a process of establishing a knowledge base, e.g. distributed associative memory, through information processing and reasoning. On the other hand, the retrieval or recall process can be viewed as an application of the learned knowledge from an associative memory. Obviously, learning is much more time consuming and difficult than recall, because the task of learning is to cast the system's sophisticated information and knowledge into an associative memory. The process of learning and recall is given in Figure 3.5. It should be pointed out that based on the pace of the memory upgrading, the learning can be performed in either an 'adaptive' or a 'short term' and 'long term' manner, depending on how frequently the new learning information is introduced and existing memory is upgraded.

3.2.2 Learning, generalization and adaptation

In general, a learning system can provide good performance while the 'working environment' is close to the 'learning environment'. However, in practice, there is not always a good match between the working environment and the learning environment; as a result, the 'generalization' becomes a critical measure to examine the learning performance. In a practical sense, the generalization means that an associative memory or system model formed by learning not only provides a good performance under the learning environment, but also provides a satisfactory performance even for a working environment never learned before. Obviously, if it lacks generalization, the resulting associative memory or model can not be applied in practice. The level of generalization mainly depends on the statistical distribution of the selected training patterns, the structure of the intelligent model and the learning approach used. Finally, it should be noted that the generalization commonly used in intelligent systems is related to the robustness or the adaptation popularly used in modern control techniques. How to design a learning system with good generalization is a key topic in any intelligent control system.

3.2.3 Classification of learning approaches

If using the environmental learning feedback as a classification criterion for learning systems, learning can be divided into the categories of supervised, reinforcement, and unsupervised learning (Williams, 1989).

Supervised learning

'Supervised learning' is 'instruction-oriented'; this means the purpose of learning is to establish a pattern mapping or an associative memory which approximates the functional relations between the environmental system inputs and outputs in hyperspaces.

Either a neural network or a fuzzy model can be formed by using supervised learning. The 'learning feedback' or driving force in supervised learning is the error between the model's output and the system teaching patterns. An optimization approach, e.g. the gradient descent algorithm, can be used to minimize the error through iteratively upgrading the structure and parameters of the associative memory (or model), e.g. the synaptic weights in a neural network or the fuzzy relational matrix in a fuzzy model.

The associative memory formed by supervised learning can be either 'hetero-association' or 'self-association'. The hetero-associative paradigm is to establish an associative memory from one set of patterns to the other. In contrast, the self-associative paradigm is to establish a memory in which the pattern is associated with itself; the latter is commonly used in signal processing. To perform supervised learning, the system informative knowledge is highly demanding, namely, it is necessary to have rich information concerning system inputs and outputs. However, in some production systems, e.g., quality control, detailed information on the system outputs might not be available; consequently it will be difficult to use supervised learning to form a quality prediction or analysis model; instead, evaluation-oriented reinforcement learning may be applied to those systems only with evaluation information.

Reinforcement learning

'Reinforcement' learning is 'evaluation-oriented'; the learning feedback (or so-called 'reinforcement signal') is evaluative. The objective of reinforcement learning is to maximize a desired function of the reinforcement signal. In the simplest case, the reinforcement signal can be assigned a binary number, $[0, 1]$, or $[-1, +1]$, which represents 'success' or 'failure'. In a more informative version, the reinforcement signal can also be denoted by a number of values, which illustrate the degree of 'success' or 'failure'; then a 'reward–penalty' algorithm can be used to upgrade the associative memory during the learning phase.

In fact, a sensible definition of reinforcement learning (Sutton *et al.*, 1991) can be described as: *If an action taken by a learning system is followed by a satisfactory state of affairs, then the tendency of the system to produce that particular action is strengthened or reinforced. Otherwise, the tendency of the system to produce that action is weakened.* On the other hand, reinforcement learning can also be divided into 'non-associative' and 'associative', corresponding to a single reinforcement signal and additional information from an environment respectively.

Unsupervised learning

There might be some confusion in grasping the concept of 'unsupervised learning' or so-called 'self-organized learning' for control people, because in contrast to control systems, in unsupervised learning there is no learning feedback signal. In fact, unsupervised learning is 'self-organization-oriented', or learning without a teacher, and the learning process is based only on system input patterns. The major application of unsupervised learning is to capture regularities or clusters from a set of input patterns.

The general conceptual illustration of three learning strategies as addressed above is shown in Figures 3.6–3.8. Finally, this chapter focuses on the introduction to those learning techniques that may be applied in industrial control.

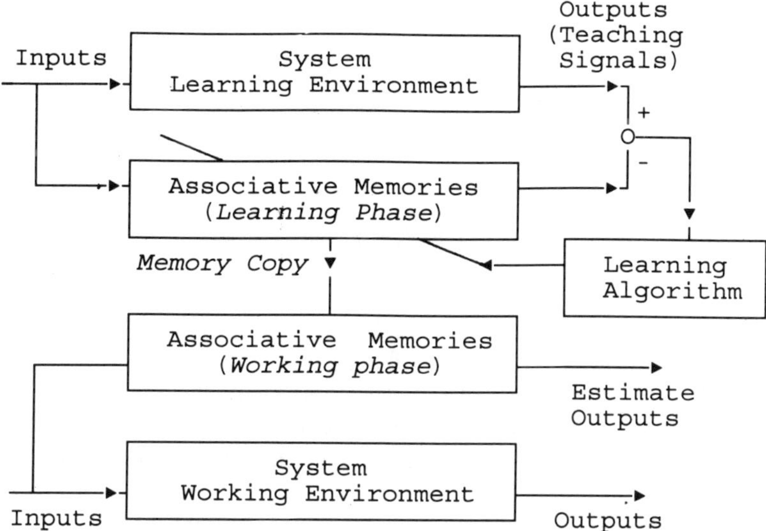

Figure 3.6 Conceptual diagram of supervised learning

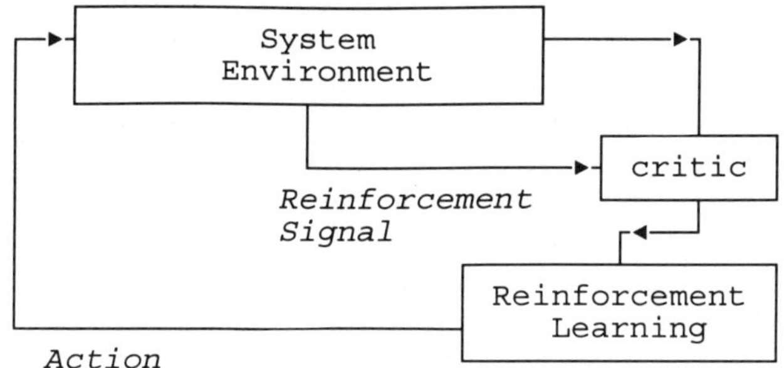

Figure 3.7 Conceptual diagram of reinforcement learning

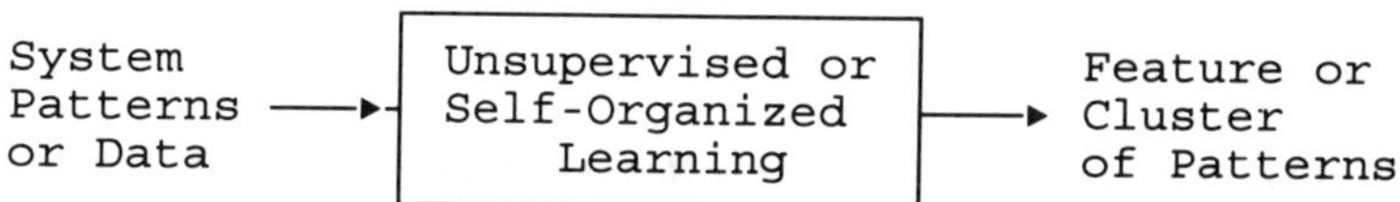

Figure 3.8 Conceptual diagram of unsupervised learning

3.3 SUPERVISED LEARNING I: NEURAL NETWORK BP LEARNING

3.3.1 Problem formulation

Consider a multilayer feedforward neural network; the problem of back propagation (BP) learning can be described as the following optimization problem:

$$\min_W J(W) = \sum_p \|Y_p - Y_p^d\|^2 \tag{3.16}$$

subject to

$$Y_p = \mathcal{N}(U_p, W) \tag{3.17}$$

where

$\mathcal{N}(*)$ is the neural network transfer function;
W is the neural network synaptic connection matrix;
$Y_p \in R^m$ is the neural network output vector;
$\{U_p \in R^r, Y_p^d \in R^m, p = 1, P\}$ is the set of training I/O patterns presented to the neural network,
Y_p^d serves as a teaching signal.

Geometrically, the learning process is to find a local or global optimization solution W through the search on the performance (error) function surface. Physically, the function of BP learning can be viewed as a process of casting system information and knowledge contained in the training patterns into the neural network synaptic connection matrices. In fact, neural network associative memory is constructed through minimizing the error between the teaching signals and the neural network outputs by upgrading the synaptic connection matrices, or so-called weights. The learning action will not stop until the error gradient is small enough, in other words, the state of the neural network has been almost at its local minimum. Moreover, it should be pointed out that instead of upgrading the weights, the topology of a neural network, e.g. the numbers of hidden layers and/or neurons, and the type of neuron activation functions, can also be upgraded during the learning process. In this section, we present the BP learning algorithms and the corresponding behaviors that are important in designing and performing an intelligent system for industrial control.

3.3.2 The generalized delta rule

The weights upgrading in BP learning follows a 'generalized delta rule', that is a fundamental rule in developing a neural network BP-learning algorithm. The delta rule, or so-called 'Widrow–Hoff rule' (Widrow and Hoff, 1960) can be derived from an optimization problem with the following objective function:

$$E_p = \frac{1}{2}\sum_j (y_{pj} - y^d_{pj})^2 \tag{3.18}$$

where E_p is the measure of the error for I/O pattern p, and $E = \sum E_p$ is the overall measure of the error. The generalized delta rule can be derived with a *gradient descent approach* to minimizing E_p. Supposing a linear signal processing element, the delta rule can be simply represented as follows:

$$-\frac{\partial E_p}{\partial W_{ji}} = \delta_j u_j \tag{3.19}$$

or

$$\Delta W_{ji} \propto \frac{\partial E_p}{\partial W_{ji}} \tag{3.20}$$

$$\Delta W_{ji} = \eta \delta_j u_i \tag{3.21}$$

where η denotes the *learning rate*, or so-called 'plasticity coefficient'; $\{\delta_j = (y_j - y_j^d), j = 1, n\}$ are the error signals used as learning feedback; $\{u_i, i = 1, r\}$ are neuron input signals; and r and n denote the neuron numbers in the input and output layers respectively.

It can be seen from 3.21 that the upgraded weights are dependent on the error signals δ_j, input signals u_i and the learning rate η. The delta rule performs an error correction function in BP learning and the learning rate is an adjustable parameter to manipulate the convergent property of the learning process. In general, BP learning can be divided into two categories: 'pattern-wise' and 'batch-wise'. Now we present BP algorithms in detail.

3.3.3 Pattern-wise BP learning

The technique of BP learning was originally developed by Werbos (1974), and then many other researchers (Rumelhart *et al.*, 1986; Parker, 1982) applied BP learning in neural networks. In fact, the BP learning algorithms for a multi-layer neural network with nonlinear signal processing elements can also be developed by using the gradient descent method; then the δ_j depends on the signal activation function of the neurons, namely,

$$\delta_j = -\frac{\partial E_p}{\partial \alpha_{pj}} = -\frac{\partial E_p}{\partial y_{pj}}\frac{\partial y_{pj}}{\partial \alpha_{pj}} \tag{3.22}$$

where

$$\frac{\partial y_{pj}}{\partial \alpha_{pj}} = S'(\alpha_{pj}) \tag{3.23}$$

where

α_{pj}-signal activation value of neuron j for pattern p;
$S(*)$-signal activation function.

The applications of the delta rule in both hidden and output layers in a multi-layer feedforward network can be described as follows:

(1) *Output layer*:

$$\Delta Wo_{kj} = \eta\delta_k yh_j \tag{3.24}$$

$$\delta_k = (y_k - y_k^d)S'(\beta_k), \quad k = 1, No; \ j = 1, Nh \tag{3.25}$$

where

No, Nh are the numbers of the neurons at the output and hidden layers respectively;
Wo is the synaptic weight matrix between the hidden and output layers;
yh_j is the output of the neuron j at the hidden layer;
β_k is the activation value of the neuron k at the output layer;
$S'(*)$ is the derivative of the signal activation function;
y_k, y_k^d are the output of the neuron k at the output layer and its relevant teaching signal respectively.

(2) *Hidden layer*:

$$\Delta Wh_{ji} = \eta\delta_j u_i, \quad i = 1, Ni \tag{3.26}$$

$$\delta_j = S'(\alpha_j)\sum_k \delta_k Wo_{kj} \tag{3.27}$$

where

Ni is the number of neurons at the input layer;
u_i is the input of the neuron i;
Wo is the synaptic weight matrix between the input and hidden layers;
α_j is the activation value of neuron j in the hidden layer.

The calculation of the derivative for the sigmoid activation functions, $S'(*)$, has been given in section 2.3.3. The pattern-wise BP learning means that after presenting each pattern, the learning action (i.e. upgrading weights) will be immediately taken in terms of the relevant error performance. The algorithm of pattern-wise BP learning for a multi-layer feedforward network can be summarized as follows:

1. Set the initialized iteration labels: $ii = 1, p = 1$, and randomly set the initial weights, $\{Wh_{ji}(p)$ and $Wo_{kj}(p)$, $i = 1, Ni$; $j = 1, Nh; k = 1, No\}$ with small real values.

2. Place the system's normalized inputs $u_i(p)$ in the training set to the neurons at the input layer, and then calculate the neural network outputs $y_k(p)$ through a feedforward path, i.e. $N : \{u_i, i = Ni\} \Rightarrow (y_k, k = 1, No\}$:

$$yh_j(p) = S\left(\sum_i Wh_{ji}(p)u_i(p) + Wh_{bj}\right) \tag{3.28}$$

$$y_k(p) = S\left(\sum_j Wo_{kj}(p)yh_j(p) + Wo_{bk}\right) \tag{3.29}$$

where Wh_{bj} and Wo_{bk} are the bias for both hidden and output layers respectively.

3. Calculate the error, $e_k(p) = (y_k(p) - y_k^d(p))$.

4. Upgrade the synaptic weights for both output and hidden layers:

$$Wo_{kj}(p+1) = Wo_{kj}(p) + \eta\delta_k(p)yh_j(p) \tag{3.30}$$

$$Wh_{ji}(p+1) = Wh_{ji}(p) + \eta\delta_j(p)u_i(p) \tag{3.31}$$

where

$$\delta_k(p) = (y_k(p) - y_k^d(p))S'(\beta_k(p))$$

$$\delta_j(p) = S'(\alpha_j(p))\sum_k \delta_k(p)Wo_{kj}(p)$$

5. If $p < P$ (the total number of patterns in the training set), then $p = p + 1$ and return to step 2; If $p = P$ and

$$\sum_p |\delta_k(p)|/P \Rightarrow 0 \text{ and } \sum_p |\delta_j(p)|/P \Rightarrow 0 \tag{3.32}$$

then end the learning iterations and write the neural network's frozen weights, $Wo_{kj}(P)$ and $Wh_{ji}(P)$, as a weight data file; otherwise, set $p = 1$, $ii = ii + 1$ and return to step 2.

In designing the learning algorithms, some other learning criteria can also be selected, for instance:

- *Average absolute error*:

$$\sum_p \sum_k |e_k(p)|/(\text{No P}) \le \varepsilon \tag{3.33}$$

where ε denotes a given small scalar value.

- *Error mean and standard deviation*:

$$e_k^m = \sum_p e_k(p)/P \le \varepsilon_k^m$$

$$e_k^{sd} = \left(\sum_p (e_k(p) - e_k^m)^2/(P-1) \right)^{1/2} \le \varepsilon_k^{sd}$$

where e_k^m and e_k^{sd} denote the error mean and the error standard deviation vectors respectively; ε_k^m and ε_k^{sd} are the given criteria vectors of error mean and error standard deviation respectively.

From the pattern-wise BP learning algorithm, we can see that the synaptic weights in both the output and the hidden layers are upgraded by each pattern presentation with gradient descent search results. The pattern presentation order can be randomly selected. In addition, to make the weight change smoother, a 'momentum term' ρ can be introduced in both the hidden and the output layers, then the weight increment can be represented as:

$$\Delta W(p+1) = \eta m + \rho \Delta W(p) \tag{3.34}$$

where m represents the gradient search direction on the error surface. In general, ρ can be selected as $\{0.9 \le \rho \le 0.99\}$.

The learning process is of a search on the error surface with respect to the weights. Obviously, the optimization search made in the pattern-wise algorithm is not the direction of steepest descent with respect to the global criterion, because the search action taken is only based on the local (individual training pattern) information. However, the pattern-wise learning algorithm can approach the gradient descent search in a global sense, only if the search step length (equivalent to the learning rate η) is small enough. On the other hand, the pattern-wise learning algorithm can not provide the search with the orthogonal direction that has been recognized as an effective way to get a solution. To overcome these disadvantages of pattern-wise BP learning, batch-wise BP learning with *conjugate gradient approach* has been proposed (Leonard and Kramer, 1990).

3.3.4 Batch-wise conjugate gradient BP learning

In contrast to the pattern-wise learning, *batch-wise conjugate gradient learning* has the following major features:

- The neural network weights are upgraded at the completion of all pattern presentations, i.e. $\{u_i(p), y_k^d(p), p = 1, P\}$ based on global error performance and an overall direction of descent;
- The combination of pattern batch presentation with the conjugate gradient approach can be viewed as batch-wise BP learning with dynamic and adaptive

adjustment of both momentum term and learning rate. As a result, the descent search direction will be controlled under a global sense, and the learning convergent property can also be under control through optimization of the learning rate.

The batch-wise BP learning algorithm with conjugate gradient method can be summarized as follows:

1. Set the initialized iteration labels: $ii = 1, p = 1$, and randomly set the initial weights, $\{Wh_{ji}(p)$ and $Wo_{kj}(p)\}$ with small values.
2. Place system normalized inputs $u_i(p)$ in the training set to the neurons at the input layer, and then calculate the neural network outputs $y_k(p)$, and the error $e_k(p) = (y_k(p) - y_k^d(p))$.
3. Calculate the local gradients for the individual pattern p, i.e. $\{m_0(p) = \delta_k yh_j\}$ and $\{m_h(p) = \delta_j u_i\}$.
4. If $p < P$ (the total number of patterns in the training set), then $p = p + 1$ and return to step 2; if $p = P$, then
   - calculate the overall gradient for the output and the hidden layers respectively:

$$M_{ii} = \sum_p m(p) \tag{3.35}$$

   - upgrade the weights for both Wh and Wo with the overall gradients:

$$\Delta W_{ii+1} = \Delta W_{ii} + \eta M'_{ii} \tag{3.36}$$

   where

$$M'_{ii} = M_{ii} + \rho M'_{ii-1} \tag{3.37}$$

$$\rho = \|M_{ii}\|^2 / \|M_{ii-1}\|^2 \tag{3.38}$$

   - use a line search method to optimize the learning rate:

$$\eta(ii + 1) = \eta_{opt}$$

   - if the overall gradient approaches zero or the performance for test set has no further improvement, then end learning, otherwise $ii = ii + 1$ and go to step 2.

The learning rate can be either a scalar value η or a k-dimensional vector $\{\eta_k, k = 1, No\}$. The value of each element η_k is optimized on the basis of the relevant output learning performance, i.e.,

$$\min_{\eta_k} \sum_k e_k e_k^T \tag{3.39}$$

The learning rate in the hidden layer can be the average of η_k,

$$\eta_h = \sum_k \eta_k / No \tag{3.40}$$

Finally, it should be noted that the 'cross-validation' with the comparison between train set and test set is commonly used to test the learning performance.

3.3.5 Structure learning of neural networks

It is still an open problem how to design a neural network architecture which may provide good generalization. The design of a neural network topology mainly includes the following items:

- select the number of hidden layers;
- select the neurons in each hidden layer;
- select the signal activation function;
- design the connection layout among all the neurons.

Similar to the determination of a math-model structure in system identification and curve fitting in nonlinear regression, the neural network architecture can be determined through dynamic structure learning. Azimi-Sadjadi *et al.* (1993) developed an algorithm for recursive dynamic node creation for multilayer neural networks. This algorithm performs recursive weight adaptation and node creation in BP learning. The derivations of the methodology are based upon the application of the orthogonal projection theorem (Alexander, 1986). When starting to train a neural network, we can select a small number of neurons in the hidden layer. During the training process, if the error reduction is stabilizing, then an additional hidden neuron can be added. The initialized weights connected with this new member should be randomly set as a small number. In addition, the number of hidden layers can also be dynamically adjusted based on the learning response through dynamic recursive training, such that an optimal topology of a neural network can be formed.

The structure learning can significantly improve the convergent property of the error reduction. In general, when a neural network approximates the neighborhood of an optimal solution, the speed of error deduction will be extremely slow. However, the introduction of a new neuron in the hidden layer will disturb the over-balance status, and activate the neural network to move to the optimal direction faster.

3.4 GENERALIZATION AND ROBUSTNESS

3.4.1 Problem statement

It should be stressed that in general, the learning strategies can be divided into three categories: learning from examples, learning through comparison, and learning through deep reasoning. In this section, the focus is on the first category, since this is most directly related to the applications of learning in system controls. Instead, the second and third categories are more related to learning in signal processing and expert systems respectively, which will be addressed in the later sections.

In system modeling and identification, all system models are established and/or upgraded with system I/O data and other related information and knowledge. In other words, the behavior of the resulting model is heavily dependent on the system learning environment. To investigate the issues of generalization and robustness, let us first consider learning in system modeling. Suppose we have a nonlinear static model with an unknown functional relationship between the system inputs and outputs:

$$Y = F(X, \Theta) + E \tag{3.41}$$

where

$X = [x_1, x_2, \ldots, x_r]^T \in R^r$, input vector;
$Y = [y_1, y_2, \ldots, y_n]^T \in R^n$, output vector;
$E = [e_1, e_2, \ldots, e_n]^T \in R^n$, residual vector describing model noises;
$\Theta = [\theta_1, \theta_2, \ldots, \theta_p]^T \in R^p$, model parameter vector.

We select K pairs of the I/O training samples, $\{X^k, Y^k, k = [k, k-K+1]\}$, then the training data set can be described as the following data matrix:

$$\begin{aligned}
X_s &= \begin{bmatrix} x_1^{k-K+1} & x_1^{k-K+2} & \cdots & x_1^k \\ \vdots & & & \vdots \\ x_r^{k-K+1} & x_r^{k-K+2} & \cdots & x_r^k \end{bmatrix}_{r\times k} \\
Y_s &= \begin{bmatrix} y_1^{k-K+1} & y_1^{k-K+2} & \cdots & y_1^k \\ \vdots & & & \vdots \\ y_n^{k-K+1} & y_n^{k-K+2} & \cdots & y_n^k \end{bmatrix}_{n\times k}
\end{aligned} \tag{3.42}$$

The task of learning in system modeling is to construct or update the system model $F(*)$ and its parameter vector Θ in terms of the selected learning data matrix $\{X_s, Y_s\}$, namely, $\{X_s, Y_s \Rightarrow F_k(*) \& \Theta_k\}$. The learning results should optimize the desired learning criterion.

Control engineers are more familiar with 'robustness' and 'adaptation' than 'generalization'. We know that even though conceptually the robustness is different from the adaptation, in fact they share the same purpose, namely, both robust and adaptive systems should work satisfactorily under an uncertain environment. On the other hand, the generalization used in an intelligent system means that a learning system can provide good results even under an environment that has never been learned before. It is apparent that the generalization has very similar features to the robustness and adaptation. In other words, good generalization can result in

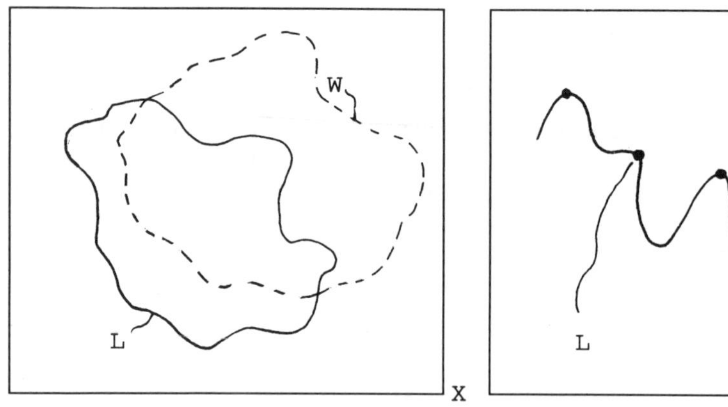

(a) poor generalization

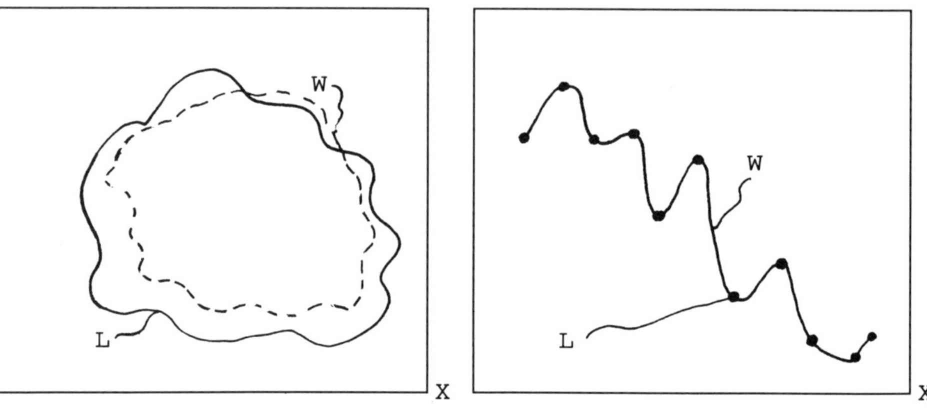

(a) good generalization

L - learning environment W - working environment

Figure 3.9 Geometrical illustration of the generalization (Reproduced from Haykin, 1994, © 1994 IEEE)

better robustness and adaptation. In general, the generalization depends on the following key factors:

- system model structure, e.g. math-function, fuzzy rule set or neural network topology, should match system behaviors, e.g. dynamics and nonlinearity, etc.;
- model parameters, such as a math-model parameter vector, fuzzy relations or synaptic connection matrices, should be identified to minimize the model errors;
- difference between the learning environment and the working environment. For instance, in industrial control, to have better generalization, the operating

conditions (or system dynamics) at the learning phase and working phase should be as close as possible.

Consider a multi-dimensional nonlinear pattern mapping, $F : X \in R^r \Rightarrow Y \in R^n$. We select a set of examples, $\{X^k, Y^k, k = 1, K\}$, as training patterns and establish a system associative memory. Geometrically, the degree of generalization can be measured by the statistics of the error between the actual and predicted values of the system output for those samples which are not included in the training set, the so-called the working set of the data. The geometrical illustration of the generalization is given in Figure 3.9 (Haykin, 1994).

3.4.2 *Neural network topology and generalization*

Determination of neural network inputs and outputs

To construct a neural network model or an associative memory, we should first select neural network I/O which match the system's I/O. If some important system inputs are not included in the neural network's inputs, or if some inputs not being correlated to the system outputs are included, then the resulting neural network will lose its generalization. On the other hand, sometimes it is difficult to determine what inputs really contribute to the system outputs in the design stage. However, the sensitivity analysis of the neural network model will help us to select the most sensitive system inputs as neural network inputs. Qin and McAvoy (1992b) proposed a sensitivity analysis approach to selecting input variables in a neural network with a partial least squares (PLS) scheme. Here we introduce the physical concept of sensitivity analysis and its applications in selecting neural network model inputs. Suppose that a physical system has the following inputs and outputs:

$$X = [x_1, x_2, \ldots, x_r]^T$$

$$Y = [y_1, y_2, \ldots, y_n]^T$$

A neural network model can be established through sample pattern training, namely

$$N : X \Rightarrow Y$$

then the sensitivity matrix can be solved with the resulting neural network model $N : \{X \Rightarrow Y\}$:

$$\Psi = \frac{\Delta Y}{\Delta X} = \begin{bmatrix} \psi_{11} & \psi_{12} & \cdots & \psi_{1r} \\ \psi_{21} & \psi_{22} & \cdots & \psi_{2r} \\ \vdots & & & \vdots \\ \psi_{n1} & \psi_{n2} & \cdots & \psi_{nr} \end{bmatrix} \tag{3.43}$$

where $\psi_{ji} = \Delta y_j / \Delta x_i$ denotes the sensitivity from x_i to y_j.

Now we can select the system inputs based on its sensitivity to the outputs, i.e. the input x_i can be eliminated while all values of its sensitivity vector are less than or equal to a given small value ϵ. The selection criteria can be described as follows: if $\psi_{ji} \leq \epsilon$, $\forall j$, then x_i can be eliminated from the input vector.

Selecting the signal activation function

As we have noted before that a multilayer feedforward neural network with at least one hidden layer can perform a universal approximation for any continuous function, how to select a suitable activation function is an important issue in neural network topological design.

In addition to the sigmoid functions introduced in section 2.3.3, there are some other applicable signal functions which can be used in practice. The following signal functions (Moody and Yarvin, 1992) can also provide good generalization behavior:

– *Polynomials*:

$$S(\alpha) = a_1\alpha + a_2\alpha^2 + \cdots + a_n\alpha^n \tag{3.44}$$

– *Rationals*:

$$S(\alpha) = \frac{a_1\alpha + a_2\alpha^2 + \cdots + a_n\alpha^n}{1 + (b_0 + b_1\alpha)^2 + (b_2\alpha + b_3\alpha^2)^2 + \cdots} \tag{3.45}$$

– *Flexible Fourier Series*:

$$S(\alpha) = \sum_{i=0}^{n} a_i \cos(b_i\alpha + c_i) \tag{3.46}$$

The relevant learning algorithms for the signal functions as listed above can be developed based on the gradient-descent algorithms. On the other hand, the signal normalization has to match the selected signal function, for instance, if the 'mean-variance' normalization method is used for signal preprocessing, then the normalized values will not be bounded within the ranges of either [0, 1] or [−1, 1]. As a result, the activation function in the output layer can be selected as a linear function. From the principle of functional approximation in a mathematical sense, the activation function plays an important role in performing a desired pattern mapping. However, since the learning in a neural network mainly works with data without having other structured knowledge, this makes it difficult to develop an analytical approach to design of the activation function. In practice, the computer simulation may help us to find what activation function is the best for the problem under study.

Determination of hidden layer and its neuron numbers

From the analytical algorithm of the forward path, we can clearly see that the number of hidden layers and neurons will also significantly affect to the functional approximation. However, similar to the structure identification for a nonlinear system, it is hardly possible to develop a universally analytical approach to designing a neural network topology based only on the training patterns.

It should be noted that it is not always true that the more hidden layers and/or hidden neurons, the better the learning quality. Suppose a system is not highly nonlinear or in other words can be described by a lower order polynomial, then obviously, too many hidden layers or hidden neurons will over-complicate the implementation of the information storage. Usually, a single hidden layer is good enough to construct an associative memory for many practical applications in industrial control. The required number of hidden neurons is related to the properties and desired capacity of information storage, which mainly depends on the dimensions of the input and output spaces, and their relations. However, quantitatively, these problems have not yet been completely solved. In a practical engineering application, the method of structure learning introduced in section 3.3.5 can help the designer to determine the neural network structure parameters case by case.

Decomposed structure of neural networks

Many industrial systems usually work under an uncertain environment, particularly for discrete event-driven production systems in manufacturing, and batch or semi-batch production systems in process industry. Consequently, an intelligent system with a global structure can hardly provide good learning behavior, or generalization, in all working environments.

It is an effective measure of improving the generalization for an intelligent system to decompose a global system into a number of subsystems. As a result, these subsystems have exactly the same topology but with different parameters, e.g., the neural network weights or fuzzy relations, corresponding to the different system environment, such as different raw material, products or operating conditions. We take the neural network model as an example to describe how to establish a neural network model with decomposed structure. The technique to be introduced can also be used in other learning systems, such as fuzzy modeling, pattern recognition and rule based expert systems.

Suppose that a system has the input vector $X \in \Omega_X$ and the output vector $Y \in \Omega_Y$; Ω_X and Ω_Y denote the global domains in X and Y hyperspace respectively. The neural network model with a global structure can be established through pattern training as follows:

$$N(W) : X \in \Omega_X \Rightarrow Y \in \Omega_Y \tag{3.47}$$

where W is a weight matrix in the neural network N.

We decompose the global domain of the system input Ω_X into M sub-domains $\{\Omega_X^1, \Omega_X^2, \ldots, \Omega_X^M\}$, and thus $\{\Omega_X^m, m = 1, M\}$ are the subsets of Ω_Y, namely

$$\{\Omega_X^m, m = 1, M\} \in \Omega_X \tag{3.48}$$

The neural network model with a decomposed structure can be established through sub-domain based input/output pattern learning:

$$\begin{aligned} N(W^1) &: X \in \Omega_X^1 \Rightarrow Y \in \Omega_Y \\ N(W^2) &: X \in \Omega_X^2 \Rightarrow Y \in \Omega_Y \\ &\vdots \\ N(W^M) &: X \in \Omega_X^M \Rightarrow Y \in \Omega_Y \end{aligned} \tag{3.49}$$

From equation 3.49 we can see that the decomposition does not change the neural network topology, but only divides the global input domain Ω_X into M sub-domains, and creates a number of weight matrices $\{W^m, m = 1, M\}$ which correspond to the relevant sub-domain $\{\Omega_X^m, m = 1, M\}$. In other words, if system

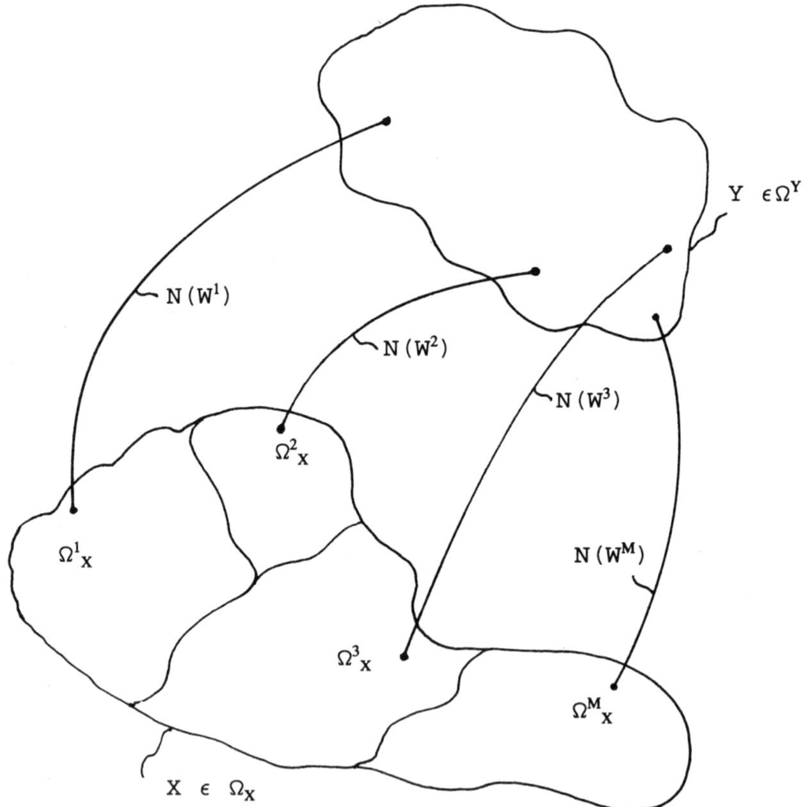

Figure 3.10 Geometrical illustration of decomposed pattern mapping

input is located in a sub-domain Ω_X^p, then the corresponding weight matrix W^p will be used to produce the network's output. The geometric illustration of decomposed pattern mapping is given in Figure 3.10.

The procedures of both learning and retrieval for a decomposed neural network can be summarized as follows:

- *Learning Phase*:
  a. Decompose the system input space Ω_X into M sub-spaces $\{\Omega_X^m, m = 1, M\} \in \Omega_X$ in terms of the system environment.

  b. Select the system I/O data patterns $\{X_k, k = 1, K\} \in \Omega_X$ for training, which can be further classified into M sub-datasets, $\{X_k^m, m = 1, M, k = 1, K_m\}$ through pattern classification based on the criteria given in step **a**. M and K_m denote the total number of groups and the number of patterns in the group m respectively, and we have $\sum_m K_m = K$.

  c. Establish the neural network model with its corresponding weight matrix data files, $\{W^m, m = 1, M\}$ through the pattern learning, namely $N(W^m)$: $X^m \in \Omega_X^m \Rightarrow Y \in \Omega_Y$.

- *Retrieval Phase*:
  a. Receive system input data pattern $X \in \Omega_X$ and classify it into a particular class, for example class j $(1 \leq j \leq M)$, in terms of the same pattern classification algorithm and criteria used in the learning phase.

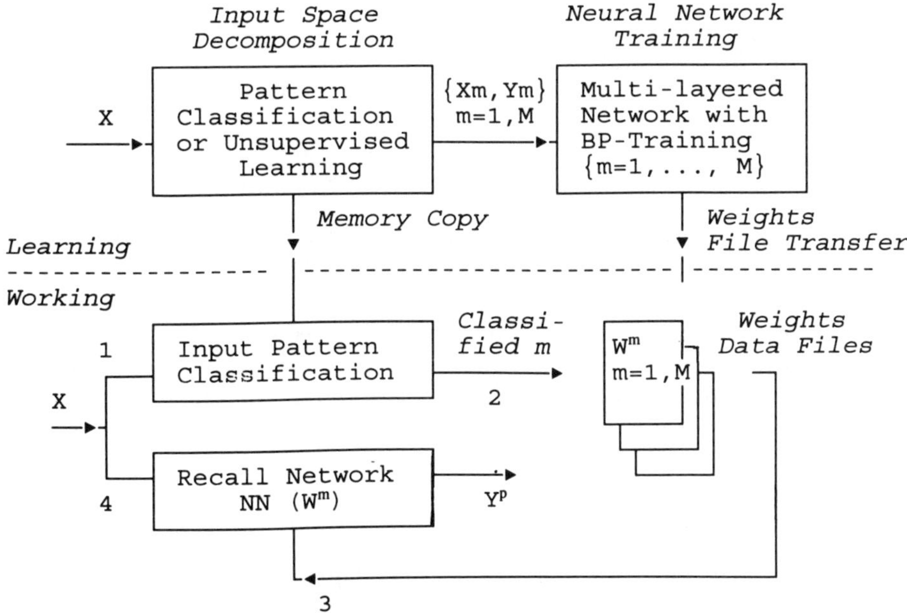

Figure 3.11 A neural network with decomposed structure

b. Call the corresponding data file of weight matrix W^j.

c. Run the forward path of the neural network and provide the neural network output, namely, $Y^p = N(W^j, X)$.

d. Return to step **a**.

The procedure described above is shown in Figure 3.11. The neural network model with a decomposed structure can provide much higher performance in both learning and generalization, since each subsystem only works under a subdomain of the entire system environment.

The environment classification can be based either on the system operating conditions or through pattern classification and competitive learning, which will be discussed later.

3.4.3 Selection of the training samples

Since an associative memory or knowledge base of a neural network is established through learning from examples, the selection of the training samples becomes a critical factor in making the trained neural network with good robustness or generalization. This section addresses the major concerns and approaches to the training sample selection. The general concerns in selection of the training samples can be described as follows:

1. The training samples should cover the entire operating environment. For instance, if the temperature set-point T^{sp} of a chemical reactor is within the range [400–800°F], the training samples should cover the same range. However, if the training samples are only located in the range [500–700°F], then the trained neural network will provide poor performance while the reactor works under the conditions $[400 \leq T^{sp} < 500°F]$ and $[700 < T^{sp} \leq 800°F]$.

   If a neural network is used to construct a dynamic model for a chemical reactor through pattern learning, e.g. $N : U(t), Y(t-1) \Rightarrow Y(t)$, then the training signals should be designed as a time-sequential function. In general, a PRBS (pseudo random block signal) function can be used as a test signal. To improve the generalization performance of a trained neural network, both static and dynamic features of the designed PRBS should match the system dynamics in the working environment. For instance, if the system under study has a larger time constant, then a larger time interval Δ in PRBS should be applied. Moreover, the frequency distribution analysis may help us to select a suitable PRBS function for neural network or fuzzy model training.

2. The generalization performance is also significantly dependent on the information content in the training patterns. When investigating raw production data, we will find in many cases that even though there are a huge number of data patterns, the system information content is still poor; this is so-called 'data rich–information poor'. In practice, we wish to establish the training data

set which should contain rich information but has a limited number of patterns.

Now we introduce a pattern classification oriented algorithm for the selection of the training patterns in the input hyperspace. As we know that similar input patterns will provide similar output patterns, pattern selection can be made with analysis of pattern similarity. In fact, pattern similarity can be quantitatively measured by the Euclidean distance between the patterns in a hyperspace.

Suppose we have the system I/O data pairs $\{X^k, Y^k, k = 1, K\}$, or $\{[x_1^k, x_2^k, \ldots, x_r^k]$ and $[y_1^k, y_2^k, \ldots, y_n^k], k = 1, K\}$, and a given least distance ϵ_d, the Euclidean distance between any two valued patterns X^i and X^j may be defined as

$$d_{ij} = \|X^i - X^j\| \tag{3.50}$$

or

$$d_{ij} = \left(\sum_{l=1}^{r} (x_l^i - x_l^j)^2 \right)^{1/2} \tag{3.51}$$

The algorithm of selecting the training patterns can be described as follows:

(1) Set $k = 1, m = 1$ and an initialized given criterion of the least distance ϵ_d being the minimum distance between any two patterns in the training data set. Read the system I/O data pair $\{X^k, Y^k, k = 1\}$, and write it into a defined training data file, e.g. 'train.dat' with the label m, i.e. $\{X^m, Y^m, m = 1, M(= 1)\}$ as a member of 'train.dat'.

(2) Set $k = k + 1$ and read system I/O data pair $\{X^k, Y^k, k = k\}$, and calculate the m-dimensional vector of the Euclidean distance $\{d_{mk}, m = 1, M\}$ with equation 3.51. The $\{d_{mk}, m = 1, M\}$ represents a set of distances between X^k and $\{X^m, m = 1, M\}$.

(3) If $\{d_{mk}, \forall m(= 1, M) > \epsilon_d\}$, then write the data pair $\{X^k, Y^k\}$ into the data file 'train.dat' with the label $m + 1$, i.e., $\{X^m, Y^m, m = M + 1\}$, otherwise go to step (2).

(4) If $k \geq K$, then end the algorithm.

From this algorithm, we can see that the I/O data pairs $\{X^m, Y^m, m = 1, M\}$ in the training data file 'train.dat' are selected through eliminating those patterns near others. In other words, the Euclidean distance between any two patterns in the training data file must be greater than the given least distance ϵ_d. The density of the training patterns can be adjusted through upgrading the given ϵ_d. This algorithm can also be readily applied in on-line training pattern selection with a moving window for sampling system I/O patterns. The geometrical illustration of the least distance based training pattern selection is shown in Figure 3.12.

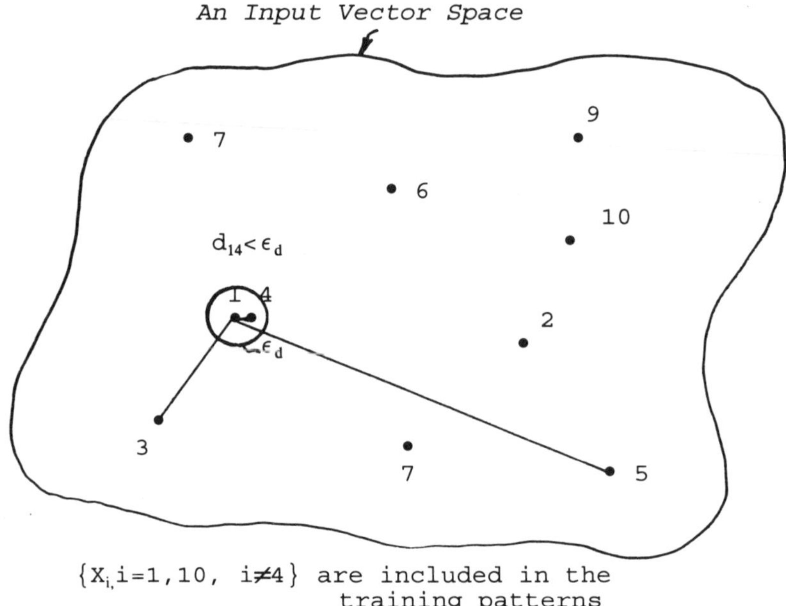

Figure 3.12 Geometrical illustration of the training pattern selection with desired uniform distribution

Finally, from a statistical viewpoint, instead of selecting a training data set with a uniform pattern distribution, the desired statistical distribution of the training patterns can also be constructed in terms of defining ϵ_d as a function of the system working environment. As a result, a set of training patterns with a non-uniform distribution can be established. In general, the statistical distribution of the training patterns can be designed to approximately match the system working patterns.

3.4.4 Quality of training patterns

The other important issue involving learning performance and generalization is the quality of the training patterns. The following problems should be dealt with while selecting and/or preprocessing the training samples.

Missing data

It is common to have missing data in a historical data file due to the dummy measurements, e.g. chemical analysis and other operation or quality related sensor measurements. Obviously, the missing data will cause loss of information, particularly for those systems having very limited or infrequent (long-period) measurement data. A number of methods have been developed to deal with the missing data problem.

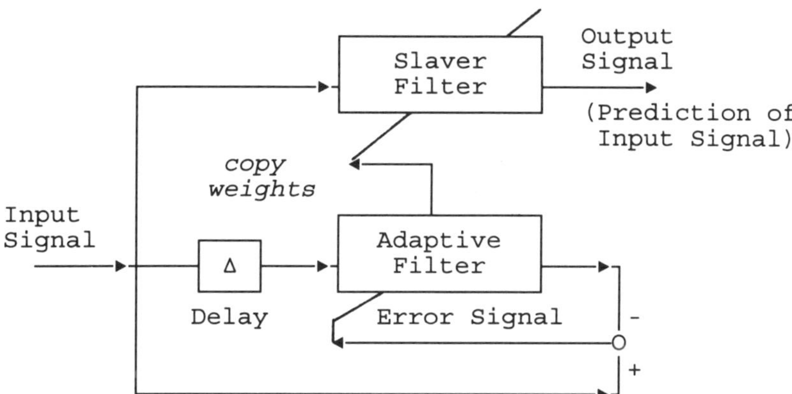

Figure 3.13 An adaptive statistical predictor (Reproduced from Widrow and Winter, 1988, © 1988 IEEE)

In statistical prediction, one can estimate the future values of the time corrected digital signals from present and past input samples. If the auto-correlation function is known, Winner's theory can be used in development of an optimal linear least squares filter for signal prediction. However, in many real-world systems the auto-correlation function is unknown or time variant; then an adaptive filtering may be applied in optimal prediction as shown in Figure 3.13 (Widrow and Winter, 1988).

From Figure 3.13 we can see that the pattern mapping is established from the system input, a delayed signal $x(t - \Delta)$, and output, its non-delayed signal $x(t)$. As a result, the slaver filter (retrieval phase in a neural network) can then provide an optimal prediction of the input Δ time units into the future.

In addition, if the missing data are evenly distributed through the data set, some other statistical methods can be used, e.g. partial least squares (PLS), principal component regression (PCR) and the integration of PLS with neural networks (Geladi and Kowalski, 1986; Wold *et al.*, 1987; Qin and McAvoy, 1992a).

Noise canceling

In signal processing, both the optimal Kalman filter and the Wiener filter are the most popular classical tools in separating a signal from additive noise. However, either a Kalman filter or a Wiener filter can do the perfect job in removing the noise without distorting the signal, only when a precise signal dynamic model with known stochastic properties of the noise is available. In many industrial systems, it is rather difficult to meet this requirement.

The other approach to this problem is adaptive filtering as shown in Figure 3.14 (Widrow and Winter, 1988). This approach is feasible only if the 'reference input' is available containing noise n_1, which is correlated with the original corrupting noise n_0. From Figure 3.14 (Widrow and Winter, 1988), we can see that the noisy 'primary input', $s + n_0$, serves as a desired response, and the system output ϵ acts as an error signal for the adaptive filter. As noted by Widrow and Winter (1988), *adaptive noise canceling generally performs much better than the classical approach since the noise is subtracted out rather than filtered out.*

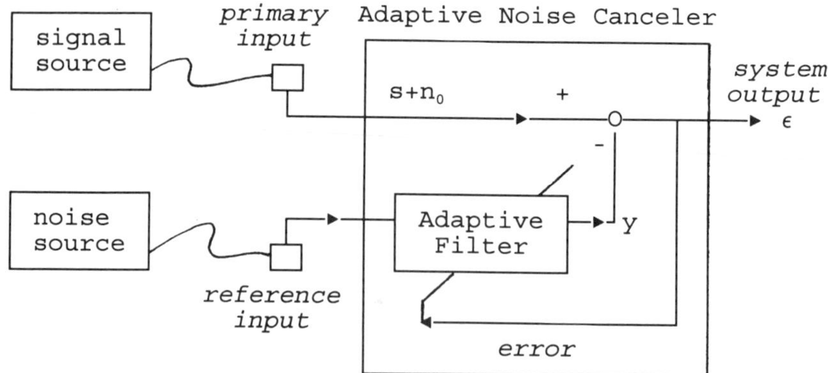

Figure 3.14 Separation of signal and noise by adaptive noise canceling approach (Reproduced from Widrow and Winter, 1988, © 1988 IEEE)

Finally it should be noted that the math-model segments describing the process fundamental principles, e.g. the energy and mass balances, can also be used to verify the reliability of the data.

3.4.5 Learning pace and generalization

Since 'learning' is a process of casting the training information into a knowledge base or so-called associative memory, the pace of the information casting significantly affects the resulting neural network model, particularly for 'short-term' or 'long-term' adaptive learning.

We have noted before that the convergent properties of the learning can be controlled by the adaptation of the learning rate. However, the problem of finding an appropriate learning rate can also be viewed as a specific issue of the 'stability–plasticity dilemma' that can be described as the following question (Grossberg, 1987): '*How can a learning system be designed to remain plastic in response to significant new events, yet also remain stable in response to irrelevant events?*'. From this dilemma we can realize that the learning pace should not be either too fast or too slow. Eventually, the pace should be determined in terms of the problem under study; for instance, if we hope that the resulting neural network model is very stable or slowly upgraded while the new patterns join the training set, then we should design a learning system with a slower learning pace. However, in contrast, if we hope that the neural network model can quickly adapt to the changes of the system environment, then a neural network with a more plastic property should be designed. Park *et al.* (1991) have developed an adaptive neural network in which the layered perception's weights adapt to a slowly varying non-stationarity. The general rule in this algorithm is to respond appropriately to previous training data if those data do not conflict with new training data, and to adapt to the new training data even when they are in conflict with portions of the old data.

It should be emphasized that if the information was cast into the neural network perception's weights through a very slow learning pace (long-term learning), correspondingly such stored information will also be upgraded at a very slow pace, even when the training information has been upgraded. In many industrial control problems, when the production environment is significantly changed, to quickly adapt to the new environment, besides upgrading the training patterns, we can also start retraining from the very beginning, i.e. furnish the weights with small random numbers; otherwise it will take a long time to perform the memory transition from old to new. This behavior of neural networks is similar to that of the human being; for instance it is almost impossible for an immigrant to forget his or her native language.

3.4.6 Concluding remarks

From an industrial application viewpoint, generalization is one of the most important factors in identifying if the neural network developed can be applied in practice. It should be noted that periodic comparison between the training and working data sets in their statistical properties may avoid poor generalization.

The evaluation and adjustment of generalization for dynamic systems are much more complicated and difficult, since the training data are represented by time sequential trajectories. The problem is the same as in the identification for nonlinear dynamic systems. To make a neural network really applicable in industrial control, the improvement of generalization has become an attractive research and development area.

Finally, it should be emphasized that the evaluation of generalization should be widely used in various forms of system modeling and associative memory.

3.5 REINFORCEMENT LEARNING

3.5.1 Introduction

The term 'reinforcement learning' was originally used in studies of AI and animal learning in psychology. Reinforcement learning can be defined as (Barto, 1992): 'If an action taken by a learning system is followed by a satisfactory state of affairs, then the tendency of the system to produce that particular action is strengthened or reinforced. Otherwise, the tendency of the system to produce that action is weakened.' Physically, the learning process in an associative reinforcement learning is to produce an associative I/O pattern mapping through 'trial and error' to maximize a desired scalar performance index, the so-called 'reinforcement signal'. In fact, in reinforcement learning the action taken can be viewed as a 'credit–assignment' or 'reward–penalty' problem.

From the basic concepts of reinforcement learning, we can see that many industrial control systems can be formulated as credit–assignment problems. However, due to fewer communications between control technology and neuroscience, studies on applications of reinforcement learning in industrial control are still at an earlier stage. The purpose of including reinforcement learning in this text is to provide the fundamental concepts and algorithms to be applied to industrial control.

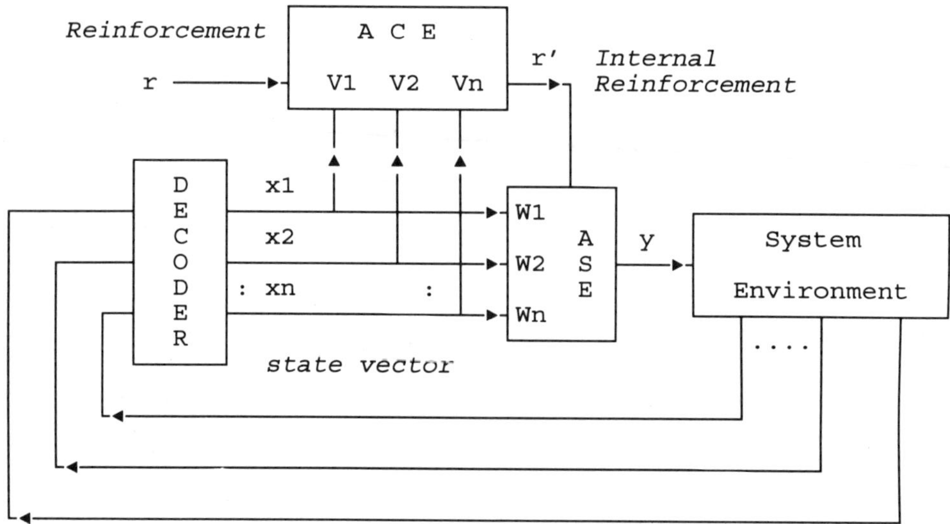

Figure 3.15 A reinforcement learning system with ASE and AEC (Reproduced from Barto *et al.*, 1992, © 1992 IEEE)

Functionally, a reinforcement learning system consists of two major components, i.e. 'associative search element (ASE)' and 'adaptive critic element (ACE)' schematically shown in Figure 3.15 (Barto, 1992). In this learning system, the primary reinforcement signal observed from a system environment is converted into a heuristic reinforcement scalar signal with a higher quality through a 'critic'. The knowledge base is upgraded by actions of a learning element. The function of the performance element is to select the actions randomly in terms of information received from both system environment and knowledge base. As a result, a mapping of 'input–output distribution-action' can be established through learning.

In general, reinforcement learning can be divided into 'associative' and 'non-associative' categories. The task of associative reinforcement learning is to build up an associative pattern mapping between the system environment inputs and outputs, such that the reinforcement signal is then driven to be maximally successful. Instead, in non-associative reinforcement learning, the learning system receives a reinforcement signal only from the system environment, and an optimal action is selected to maximize the desired reinforcement signal. Non-associative reinforcement learning has been intensively studied as a function optimization associated with a genetic algorithm (Holland, 1992b) and as stochastic learning automata (Narendra and Thathachar, 1989).

3.5.2 Reinforcement learning algorithms

To make our discussion more feasible and practical, here we introduce reinforcement learning algorithms with the pole balancing learning control problem (Barto *et al.*, 1983). Figure 3.16 shows a cart to which a rigid pole is hinged. The cart is free to move within the bounds of a one-dimensional track. The controller can apply an

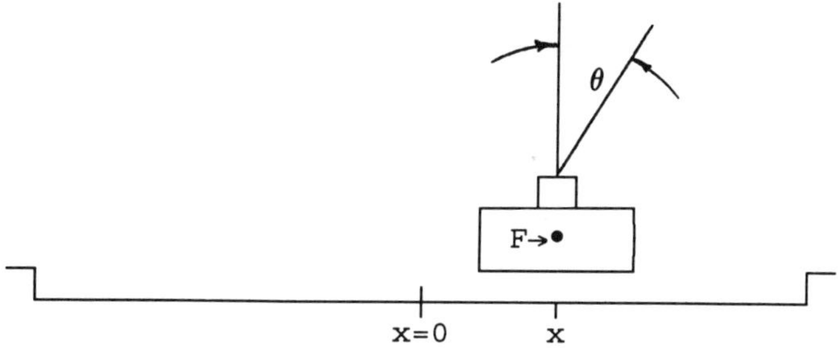

Figure 3.16 Illustration of cart-pole system

impulsive 'left' or 'right' force of fixed magnitude to the cart at discrete time intervals. A dedicated nonlinear dynamic math-model is given in (Cannon, 1967). There are the following four state variables in a cart-pole balancing system:

x, position of the cart on the track;
θ, angle of the pole with the vertical;
dx/dt, cart velocity;
$d\theta/dt$, rate of change of the angle.

The problem for investigating reinforcement learning control is to design an intelligent control system without having knowledge of the cart-pole system dynamics. In other words, the resulting control system must attempt to generate a controlling force to avoid the failure signal when the cart moves. The only evaluative feedback signals are the success and failure signals.

The algorithm of Associative Search Element (ASE)

The ASE element has a reinforcement input pathway, n pathways for the non-reinforcement input, and a single output pathway. Let $\{x_i(t), 1 \leq i \leq n\}$ and $y(t)$ denote the real-valued signal on the ith non-reinforcement input pathway and output at time t respectively. The element output can be determined as follows:

$$y(t) = f\left[\sum_i w_i(t)x_i(t) + \text{noise}(t)\right] \tag{3.52}$$

where noise(t) is defined as a real random variable with probability function, e.g. the zero mean Gaussian distribution with covariance, and $f(*)$ can be a threshold function:

$$f(x) = \begin{cases} +1 & \text{if } x \geq 0 \text{ (control action right)} \\ -1 & \text{if } x < 0 \text{ (control action left)} \end{cases} \tag{3.53}$$

The weights are upgraded in terms of both $r(t)$, real-valued reinforcement, and $e_i(t)$, eligibility of input pathway i, as follows:

$$w_i(t+1) = w_i(t) + \eta r(t) e_i(t) \tag{3.54}$$

where η denotes the learning rate, and positive and negative r represent the rewarding and punishing actions respectively. The exponentially decaying eligibility traces e_i can be formulated as follows:

$$e_i(t) = \delta e_i(t) + (1-\delta) y(t) x_i(t) \tag{3.55}$$

where δ $(0 \le \delta < 1)$ determines the trace decay rate.

The Adaptive Critic Element (ACE)

The task of an ACE in the learning system is to produce a future prediction of reinforcement, which is used to determine an internal reinforcement signal, r', being transferred to ASE. As a result of introducing ACE, the control actions will be taken before the occurrence of failure.

The prediction of eventual reinforcement can be described as a linear function of the input vector X, namely

$$p(t) = \sum_i v_i(t) x_i(t) \tag{3.56}$$

The prediction of $p(t)$ can converge to an accurate prediction by upgrading the weights v_i as follows:

$$v_i(t+1) = v_i(t) + \lambda [r(t) + \gamma p(t) - p(t-1)] x'_i(t) \tag{3.57}$$

where

λ = learning rate;
$r(t)$ = reinforcement signal provided by system environment;
$x'_i(t)$ = a trace of the input vector $x_i(t)$;
γ = a positive constant, $0 < \gamma < 1$, that makes predications decay in the absence of external reinforcement.

The improved or internal reinforcement signal, i.e. the output of ACE, can be calculated as follows:

$$r'(t) = r(t) + \gamma p(t) - p(t-1) \tag{3.58}$$

The decoder

A four-dimensional cart-pole state space can be divided into disjoint regions by quantizing the four state variables, for instance three grades for cart position, six for pole angle, three for cart velocity and three for pole angular velocity; as a result, 162 $(3 \times 3 \times 6 \times 3 = 162)$ regions in a four-dimensional space are constructed through the real physical signal coding.

Simulation results

The simulation runs have shown that the cart-pole system controlled by an ASE/ACE learning controller can achieve much longer runs than other control systems, e.g. box system and only ASE system. This is just because ACE provides reinforcement throughout trials.

Finally, based on the features of reinforcement, such as on-line learning, maximizing a scalar performance index, or the so-called reinforcement signal, and the ability of the adaptive heuristic critic, a more general architecture of a reinforcement learning system can be schematically shown in Figure 3.17 (Haykin, 1994). The *critic* converts the primary reinforcement signal observed from the system environment into a higher quality reinforcement signal, the so-called 'internal reinforcement signal', as provided by AEC, or *'heuristic reinforcement signal'*. In the learning system, there are three components: *learning element*, *knowledge base* and *performance element*. This kind of structure is similar to the common AI version, namely the function of learning is to upgrade the knowledge base, and the performance element is to select actions on the basis of a distribution determined by the knowledge base associated with the system environment. In addition, it should also be noted that the reinforcement learning addressed in this section belongs to the family of supervised learning; on the other hand, the reinforcement learning can also be implemented by unsupervised learning, though this topic is beyond the scope of this text. The reinforcement learning strategy can be applied in quality control, fault detection and diagnosis, optimization, production scheduling, and dynamic control as well.

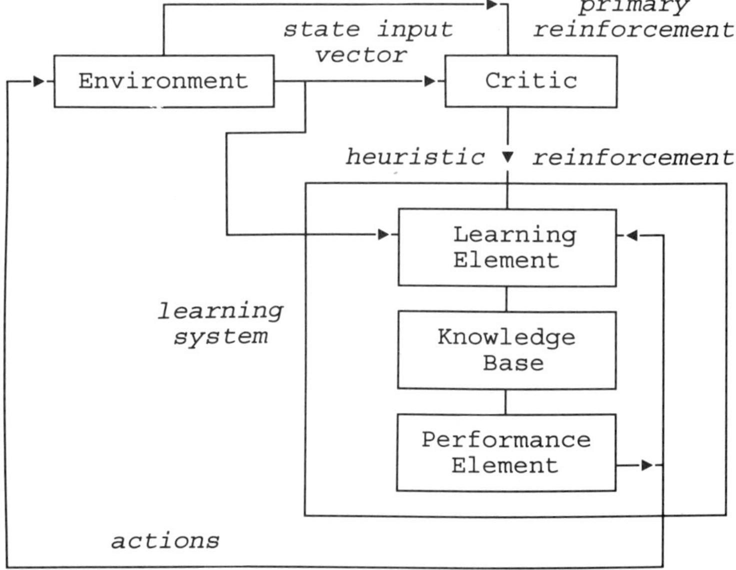

Figure 3.17 General configuration of reinforcement learning (Reproduced from Haykin, 1994, © 1994 IEEE)

3.6 SELF-ORGANIZED LEARNING

3.6.1 Introduction

'Self-organized learning' or so-called 'unsupervised learning' is another type of learning strategy in the machine learning paradigm. In comparison with the supervised learning (both BP and reinforcement learning) previously described, the key feature of self-organized learning is 'learning without a teacher', in other words, the purpose of unsupervised learning is not to establish a pattern mapping from the system input space to output space with the training samples. If we examine human intelligence, we will find that the abilities of comparison, clustering and regression make humans much smarter than any mathematical and/or AI tools. Humans can effectively classify birds from fishes through unsupervised learning.

From an industrial control application point of view, the objectives of self-organized learning are to explore correlations between input variables, and to construct the 'clusters' or the 'subspecies' within the input data space in terms of the data distribution in a multi-dimensional input space. To perform these objectives a specified optimization criterion is needed. In industrial processes, every day we work with a huge number of production and quality related data. How to extract a limited number of data that contain the maximum amount and quality of information is a critical and common problem in industrial control. For example, if we receive hundreds of customer orders with a huge number of steel grades, we might have to group them into a limited number of families with similar chemical compositions. As a result of clustering, the steelmaking production pace will be more consistently stable. Such data-driven clustering problems are quite common in job-machine assignment, production scheduling and optimization. In this text, we address only the fundamental concepts and algorithms of self-organized learning which have potential applications in industrial control. From the concepts behind self-organized learning, this technique is particularly applicable when the system has high input dimensions and the size of data set (number of input data patterns) is large.

In recent years, the implementation of self-organized learning has been moving from the classical approaches of statistical clustering (Ball, 1965) and k-means (MacQueen, 1967), etc., to neural network oriented unsupervised learning, such as feature mapping (Kohonen, 1988) and competitive learning (Rumelhart *et al.*, 1986), etc.

3.6.2 Kohonen's self-organizing feature maps

Kohonen self-organizing algorithm

The key function of Kohonen's algorithm (Kohonen, 1984) is to create a vector quantizer by adjusting the connection weights from N input nodes to M output nodes arranged in a two-dimensional grid as shown in Figure 3.18 (Lippmann, 1987). *After enough input vectors have been presented, the weights will specify cluster or vector centers that sample the input space such that the point density function of the vector*

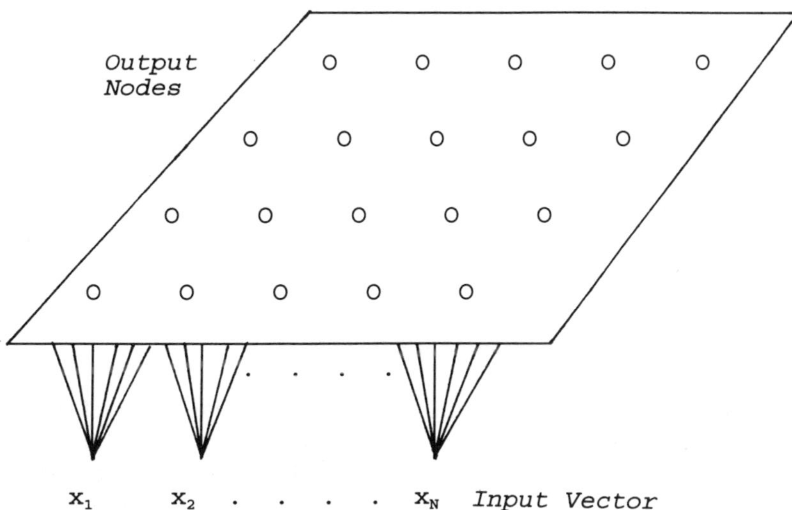

Figure 3.18 Architecture of neural network for Kohonen feature map (Reproduced from Lippmann, 1987, © 1987 IEEE)

centers trends to approximate to the probability density function of input vector probability (Kohonen, 1984). As a result, the input vectors will be clustered in terms of their physical similarity without any supervised information, in other words, in this kind of unsupervised learning, the continuous-valued input vectors are presented sequentially without specifying the corresponding desired output. Kohonen's algorithm can be described as follows (Kohonen, 1988; Lippmann, 1987):

(a) Let $t = 0$ and initialized weights $W_{ij}(t)$ from the N inputs to the M output nodes to be small real random values, and set the initial radius of the neighborhood.

(b) Present an input vector $X(t) = [x_1(t), \ldots, x_N(t)]$ to the corresponding input nodes.

(c) Compute the distance d_j between the input vector $X(t)$ and the weight vector $W_{ij}(t)$ for each output node j:

$$d_j = \sum_{i=1}^{N} (x_i(t) - W_{ij}(t))^2, \quad j = 1, M \tag{3.59}$$

(d) Select the node j^* as that output node with a minimum distance, namely

$$\min(d_1, d_2, \ldots, d_M) = d_{j^*} \tag{3.60}$$

(e) Upgrade the weights to node j^* and all nodes within the neighborhood as defined:

$$W_{ij}(t+1) = W_{ij}(t) + \eta(t)(x_i(t) - W_{ij}(t)), \quad \forall j \in NE_j(t), \quad i = 1, N \tag{3.61}$$

where $NE_j(t)$ denotes the defined neighborhood function centered around the winning neuron j^* and $\eta(t)$ $(0 \leq \eta(t) \leq 1)$ is the learning rate, both dynamically upgraded to get good learning results.

(f) Set $t = t + 1$ and return to step (b). The training will stop until the outputs reach a convergent status.

It can be seen that the goal of a self-organized network is to find the best-matching (winning) neuron by minimizing the Euclidean distance between the input vector and the synaptic weight vector.

Properties of self-organizing feature mapping

Let Ω_X be a continuous input space and Ω_j denote a spatially discrete output space, then the self-organizing feature map Φ can be viewed as a pattern mapping (or approximation) from the original large set of input vector $X \in \Omega_X$ to a smaller set of output vector $j \in \Omega_j$, namely, $\Phi : X \Rightarrow j(X)$, through adjusting the synaptic weights. The illustrative relationship between the feature map Φ and weight vector W_j of the winning neuron j^* is shown in Figure 3.19 (Haykin, 1994). In fact, the self-organizing feature map is based on *vector quantization theory,* the motivation of which is dimensionality reduction or data compression (Gray, 1984).

Another property is so-called 'density matching'. From a statistics point of view, the feature map also reflects variations in the statistics of the input distribution. In other words, the regions in the input space Ω_X from which sample vector X are drawn with high probability of occurrence are mapped onto the larger domains of

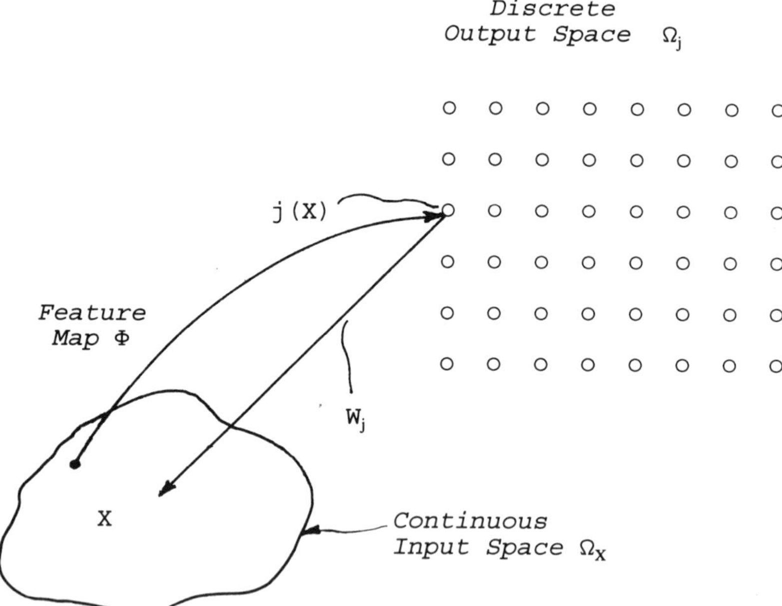

Figure 3.19 Illustration of a feature map from input space to output space (Reproduced from Haykin, 1994, © 1994 IEEE)

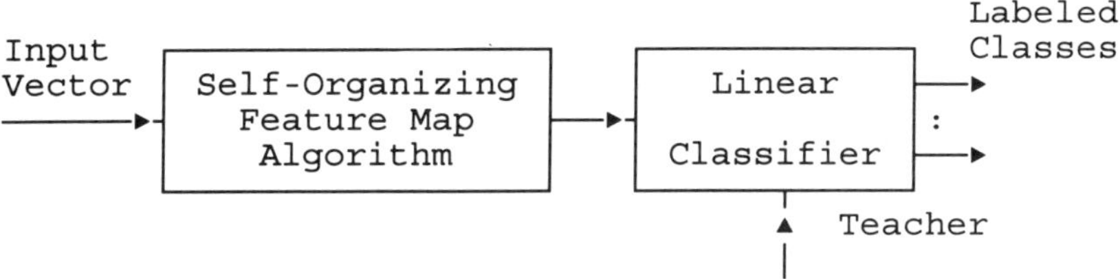

Figure 3.20 Scheme of a two-stage adaptive pattern classifier (Reproduced from Lippmann, 1987, © 1987 IEEE)

the output space Ω_j, and therefore with better resolution than regions in Ω_X from which the sample vector X is drawn with a low probability of occurrence (Haykin, 1994).

Adaptive pattern classification

The task of pattern classification is to classify the sets of signal patterns into a number of finite labeled classes, such that the patterns in the same class have a similar property. In addition to the non-parametric approach to self-organizing feature map, the Bayesian approach may also be used in the pattern classification, particularly for the patterns with a Gaussian distribution. Alternatively, a hybrid scheme (Lippmann, 1989) with a combination of feature mapping and supervised learning may provide the best results, if the teacher signals are available. The scheme of a two-stage adaptive pattern classifier is shown in Figure 3.20 (Haykin, 1994).

Potential applications in industrial control

We have learned that self-organized learning can automatically classify the system input vector with higher dimensionality into a number of labeled discrete clusters without having the supervised teaching signals. In industrial control, this kind of clustering is particularly applicable in discrete-event driven systems, for instance, grouping steel grades or an other quality of chemical products with chemical analysis, or operating conditions of a batch production system. The main potential applications of self-organized learning are pattern classification and clustering oriented quality prediction and control, fault detection and diagnosis, etc. Obviously, self-organized learning can also be effectively used in signal coding and processing for supervised learning systems.

3.7 ADALINE ADAPTIVE LEARNING

3.7.1 Introduction to Widrow–Hoff delta rule

Both adaptive signal processing and adaptive neural network have been developed with applications of the adaptive linear element (Adaline). A general scheme of an

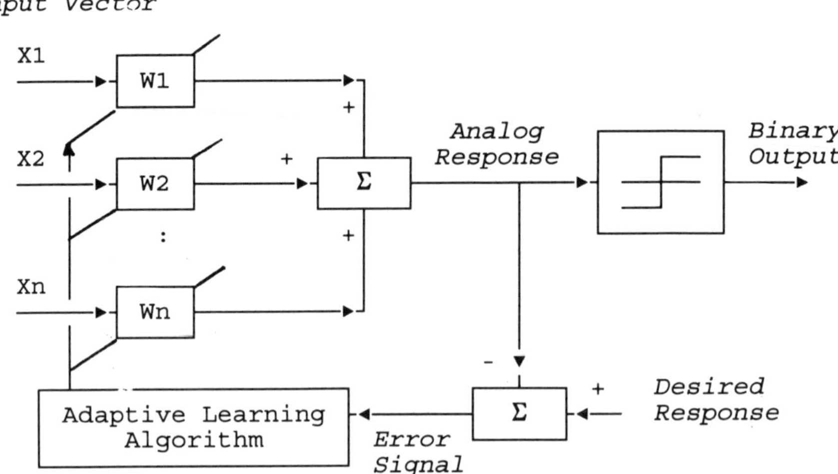

Figure 3.21 Scheme of adaptive neural network (Reproduced from Widrow and Lehr, 1990, © 1990 IEEE)

Adaline neural network is illustrated in Figure 3.21 (Widrow and Lehr, 1990). The error signal, e, being the difference between the desired response, r, and Adaline output, y, serves as an adaptive learning signal. To minimize the mean square of the error signal, the connecting weights between Adaline inputs and output are upgraded with a least mean square (LMS) learning algorithm, e.g. the Widrow–Hoff delta rule (Widrow and Lehr, 1990).

Let X and W be an input vector and a weight vector from X to y respectively; then the LSM learning algorithm can be represented as

$$W_{k+1} = \begin{cases} W_k + \eta e_k X_k / |X_k|^2 & \text{if } |X_k|^2 \neq 0 \\ W_k & \text{if } |X_k|^2 = 0 \end{cases} \tag{3.62}$$

where η denotes the learning rate, and the error signal is

$$e_k = r_k - X_k^T W_k \tag{3.63}$$

The change of the error can be represented as

$$\Delta e_k = \Delta(r_k - X_k^T W_k) = -X_k^T \Delta W_k \tag{3.64}$$

From equation 3.62, we can see that

$$\Delta W_k = \eta e_k X_k / |X_k|^2 \tag{3.65}$$

Substituting equation 3.65 into 3.64 yields

$$\begin{aligned} \Delta \varepsilon_k &= -X_k^T \eta e_k X_k / |X_k|^2 \\ &= -\eta e_k X_k^T X_k / |X_k|^2 = -\eta e_k \end{aligned} \tag{3.66}$$

From equation 3.66 we can see that error e_k will be convergent if we select a small learning rate, i.e. $0 < \eta < 2$. In fact, η controls the convergence rate of the error response.

3.7.2 Two-layer adaptation algorithm

In practical applications, a multi-layer Adaline neural network with more generalized signal function can also be developed, and the algorithms can be described as follows.

Generalized algorithm

Let $S(X)$ be a signal function vector; then the Adaline weight vector is adapted according to the rule:

$$W_{k+1} = \begin{cases} W_k + \eta e_k S(X) & \text{if } X^T S(X) \neq 0 \\ W_k & \text{if } X^T S(X) = 0 \end{cases} \tag{3.67}$$

With $0 < \eta < 2$, the error e_k converges asymptotically to 0 at the rate of $1 - \eta$. It can be seen that if $S(X) = X$, then the generalized algorithm 3.67 becomes the delta rule 3.62. On the other hand, the signal function vector can also be defined as

$$S(X) = \begin{bmatrix} sgn(x_1) \\ \vdots \\ sgn(x_n) \end{bmatrix} \tag{3.68}$$

where

$$sgn(X) = \begin{cases} +1 & \text{if } x > 0 \\ -1 & \text{if } x < 0 \end{cases} \tag{3.69}$$

Two-layer weight adaptation algorithm

A two-layer network is shown in Figure 3.22 (Kushewski, *et al.*, 1993); the weight matrix update algorithm is given as follows (Kuschewski *et al.*, 1993):

$$W_{i,k+1} = W_{i,k} + \Delta W_{i,k}, \quad i = 1, 2 \tag{3.70}$$

where

$$\Delta W_{1,k} = \begin{cases} \dfrac{-2Z_{1,k} S_1^T(X)}{S_1^T(X)X} & \text{if } S_1^T(X)X \neq 0 \\ 0 & \text{if } S_1^T(X)X = 0 \end{cases} \tag{3.71}$$

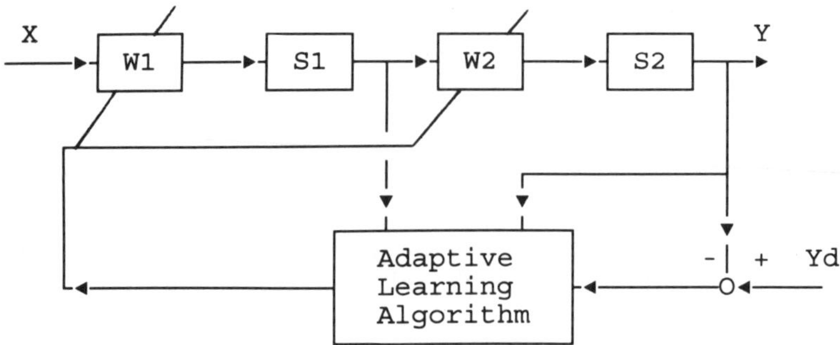

Figure 3.22 Block diagram of a two-layer adaptive neural network

$$\Delta W_{2,k} = \begin{cases} -2W_{2,k} - \dfrac{Ae_k S_2^T(Y_{1,k})}{S_2^T(Y_{1,k})Y_{1,k}} & \text{if } S_2^T(Y_{1,k})Y_{1,k} \neq 0 \\ 0 & \text{if } S_2^T(Y_{1,k})Y_{1,k} = 0 \end{cases} \tag{3.72}$$

where A is defined as an error reduction matrix with a diagonal form to be selected.

Because of the good dynamic performance of the adaptive neural network with LMS learning, Adaline networks have been widely used in adaptive signal processing, system modeling and statistical prediction, adaptive pattern recognition and adaptive control; for details refer to Widrow and Winter (1988) and Kuschewski *et al.* (1993).

3.8 SUPERVISED LEARNING II: LEARNING IN FUZZY SYSTEMS

3.8.1 Introduction

'Learning' has also been playing an important role in making a fuzzy model or a fuzzy control law robust and adaptive, while the system works under an uncertain fuzzy environment. In general, learning in a fuzzy system can be divided into two categories: 'parameter learning' and 'structure learning', corresponding to the update of the fuzzy model parameters, e.g. membership, and the fuzzy model structure, e.g. fuzzy rules, respectively. The functions of learning in a fuzzy system involve knowledge update or knowledge acquisition. A general architecture of the fuzzy system's supervised learning can be illustrated in Figure 3.23. The applications of neural networks in fuzzy system learning and knowledge acquisition have been showing great opportunities to make fuzzy system learning more efficient and simple. In this section, we will present some major learning strategies for fuzzy systems.

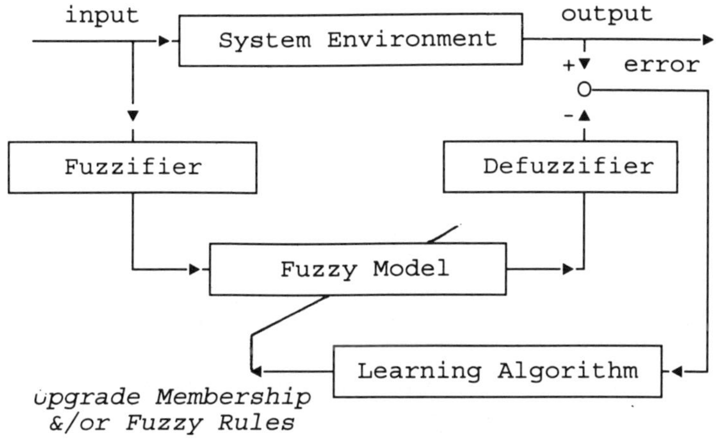

Figure 3.23 Architecture of fuzzy system's supervised learning

3.8.2 Membership function self-tuning

The self-tuning of the membership function is a major learning strategy in fuzzy system learning. Here we introduce an optimization oriented membership function learning algorithm (Chen and Spanos, 1992). Without loss of generality, suppose a fuzzy model with two inputs $\{x_1, x_2\}$ and one output y and the fuzzy rules can be described as:

If x_1 is A_{1i} and x_2 is A_{2j}, Then y is W_{ij}, $i = 0, m; j = 0, n$

where i, j denote the rule numbers, A_{1i} and A_{2j} are the linguistic values of x_1 and x_2 represented by the corresponding membership values m_{1i} and m_{2j}, and W_{ij} is an initial fuzzy value of y with respect to the input combination.

The membership values are governed by the following equation:

$$m_{1,i} = \begin{cases} \dfrac{x_1 - a_{1(i-1)}}{a_{1i} - a_{1(i-1)}} & \text{if } a_{1(i-1)} < x_1 \le a_{1i} \\ \dfrac{a_{1(i+1)} - x_1}{a_{1(i+1)} - a_{1i}} & \text{if } a_{1i} < x_1 \le a_{1(i+1)} \\ 0 & \text{otherwise} \end{cases} \quad i = 1, 2 \tag{3.73}$$

While the products of m_{1i} and m_{2j} are used as the corresponding output membership m_{yij}, the output y can be represented by a weighted average:

$$y = \sum_{i,j} m_{yij} W_{ij} = \sum_{i,j} m_{1i} m_{2j} W_{ij} \tag{3.74}$$

Note that $\sum m_{yij} = \sum m_{1i} m_{2j} = 1$.

The goal of self-tuning is to determine the values $\{a_{1i}, a_{2j}, W_{ij}\}$ which minimize a cost function:

$$\min E = \frac{1}{2}\sum_{k}(y_k^p - y_k)^2$$
$$= \frac{1}{2}\sum_{k}\left(\sum_{i,j} m_{1i}m_{2j}W_{ij} - y_k\right)^2 \qquad (3.75)$$

with a set of learning data $\{x_{1k}, x_{2k}, y_k\}$, where y_k^p denotes the predicted output given by the fuzzy model.

To solve this problem many optimization techniques can be used; for details refer to Hiroyoshi *et al.* (1992) and Chen and Spanos (1992).

In addition, self-learning in fuzzy systems can also be performed by modifying the fuzzy rules (fuzzy relations and rule structure) with fuzzy system identification (Xu and Lu, 1987).

An application of fuzzy membership self-tuning in a low pressure chemical vapor deposition (LPCVD) process for quality modeling will be given in section 8.5.2.

3.8.3 Neuro-fuzzy learning systems

The combination of a fuzzy system with a neural network has been considered as a great opportunity to enhance system learning. During the last few years, some issues addressed in this direction involve automatic design and tuning of the membership functions, knowledge acquisition and representation, fuzzy cognitive maps, clustering and pattern recognition, etc. (Tagaki, 1990). The joint applications of the neural network and the fuzzy logic technologies in consumer products are summarized in the following ways (Tagaki *et al.*, 1992).

Neural networks as development tools for fuzzy systems

We know that fuzzy logic can encode expert (qualitative or linguistic) knowledge directly using fuzzy rules; however, the most difficult and time consuming task in developing a fuzzy system is to determine its membership functions which minimize the error between the fuzzy model and the actual system outputs. The applications of neural networks can first extract the fuzzy rules through data clustering with unsupervised (or self-organized) learning as we discussed before. Then another neural network will be further used to construct a fuzzy associative memory (or so-called nonlinear pattern mapping) with an optimized multi-dimensional membership function. As a result, the neural networks in a fuzzy system can serve as either a development tool or an adaptive component. The application of the neural network in determining the membership function for a fuzzy system is shown in Figure 3.24.

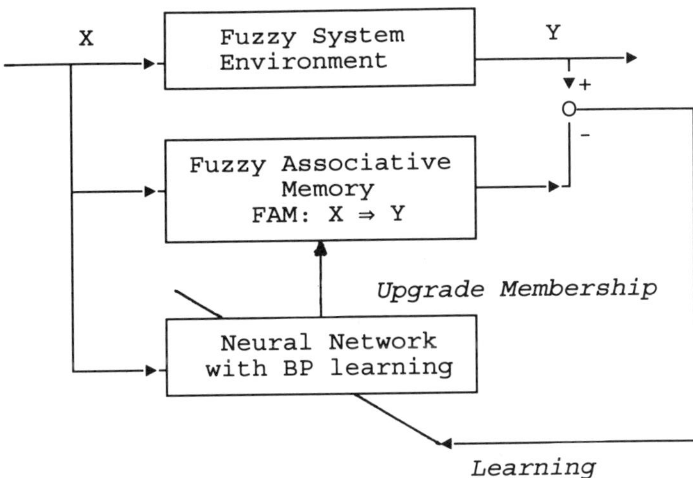

Figure 3.24 Application of neural network in determining membership function for fuzzy systems

Independent use of neural networks and fuzzy logic

A nonlinear system modeling or control problem can be solved by either fuzzy or neural network techniques independently. In general, a neural network can handle such kinds of task better, while the system under study is able to provide the I/O training patterns with rich information. However, on the other hand, if the system has rich linguistic knowledge, e.g. fuzzy rules, then a fuzzy system can be applied.

Neural networks as correcting mechanisms for fuzzy system

Based on the more informative and learning capabilities of neural networks, neural networks can serve as a correcting component for a fuzzy system as schematically shown in Figure 3.25. It should be pointed out that to train a neural network, the desired correction should be used; however, it is not necessary to have exactly the same input variables for both fuzzy system and neural network models; for instance, the fuzzy system may include more qualitative inputs.

Cascade combination of neural networks and fuzzy systems

A neuro-fuzzy system can also be constructed with a serial connection of a neural network and a fuzzy system. In general, they are in charge of different tasks in an integrated system. For instance, one can be used as a controller at a lower level, and the other may be used as a self-tuning or a supervisory mechanism.

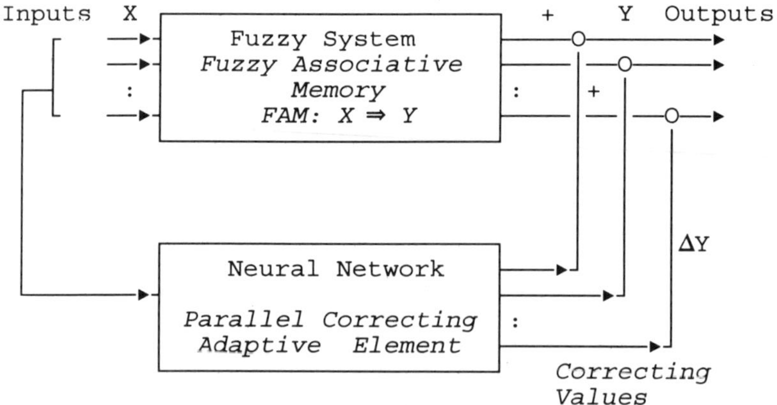

Figure 3.25 Application of neural network as parallel correction for fuzzy system

Finally, it should be noted that the fundamentals and strategies behind the applications in consumer products provide us with a general idea why and how to apply these two technologies in solving a practical control related problem.

3.9 CONCEPTUAL LEARNING

3.9.1 Introduction

An expert system can be viewed as a problem solving program to solve a sophisticated and difficult problem with some human-like behavior in a specialized problem domain. From an application point of view, both knowledge acquisition and learning are the bottleneck in developing an expert system under an environment with incomplete and/or imprecise information and knowledge. In fact, learning plays an important role in either knowledge acquisition or knowledge base update. Similar to learning in a neural network or a fuzzy system, the concepts of learning from examples offer potential for domain experts to interact directly with a machine to transfer knowledge. The concept of learning methods can be divided into *similarity-based, hierarchical, function induction,* and *explanation-based knowledge intensive techniques* (Witten and MacDonald, 1990). On the other hand, knowledge representation is another important dimension in an intelligent system. The traditional forms of knowledge representation can be *production system, semantic networks, frames and scripts, logic procedural representation,* etc. During the last decade, both neural network and fuzzy rules have been becoming the alternative schemes of knowledge representation. Obviously, the learning policy designed should depend on the form of the knowledge representation.

Table 3.1 Sample I/O quality related production data

	Operating conditions					Product quality
Case	X_1	X_2	X_3	X_4	X_5	Y
1	1	2	2	1	1	3
2	3	2	1	2	2	1
⋮			⋮			⋮
N			⋮			⋮

3.9.2 Knowledge acquisition through rule learning

To apply the learning concept in solving an industrial control problem, the construction of a decision tree or the collection of rules which discriminates positive from negative examples through learning has great opportunities in practical applications. Let us first examine a quality prediction, diagnosis and control problem in discrete manufacturing, batch or continuous production systems. Suppose we have a set of production data describing the operating conditions and the relevant quality of the finished product as shown in Table 3.1. From this table we can see that instead of crisp values, both operating and quality parameters are ranked with a few discrete levels. Now the problem is *how can we extract a decision tree and the rules from the data given in Table 3.1 through a learning process*. In this section, we only focus on the general concepts of rule creation, inductive learning and the combination of neural network and rule based system. The detailed algorithms are beyond the scope of this text.

Here we take the professional judgment of the optician as an example (Witten and MacDonald, 1990) to describe how to construct a decision tree of production rules from the learning cases.

Example 3.3

The decision making for fitting contact lenses is ranked into three groups, namely 1, 2, and 3, which correspond to the decisions of fitting hard contact lenses, soft contact lenses and lenses not recommended, respectively. The learning example cases are listed in Table 3.2 to describe the status of the patient with four factors and the decision making for purchasing contact lenses. The ID3 algorithm (Quinlan, 1986) can be used to construct a decision tree with a minimal number of nodes in the tree on the basis of the given learning examples. In fact, *it uses an information-theoretic heuristic to determine which attribute should be tested at each node, looking at all members of the training set which reach that node and selecting the attribute that most reduces the entropy of the positive/negative decision* (Witten and MacDonald, 1990). Similar to a neural network or a fuzzy logic system, the 'generalization' is highly dependent on the selected training samples. The training samples shown in Table 3.2 can be converted into a decision tree (Figure 3.26) and the relevant production rules (Table 3.3) with the ID3 algorithm. Moreover, the production rules (Table 3.3) can be

Table 3.2 I/O data for producing rules and decision tree (from Cendrowski, 1988)

Case	a	b	c	d	Decision
1	1	1	1	1	3
2	1	1	1	2	2
3	1	1	2	1	3
4	1	1	2	2	1
5	1	2	1	1	3
6	1	2	1	2	2
7	1	2	2	1	3
8	1	2	2	2	1
9	2	1	1	1	3
10	2	1	1	2	2
11	2	1	2	1	3
12	2	1	2	2	1
13	2	2	1	1	3
14	2	2	1	2	2
15	2	2	2	1	3
16	2	2	2	2	3
17	3	1	1	1	3
18	3	1	1	2	3
19	3	1	2	1	3
20	3	1	2	2	1
21	3	2	1	1	3
22	3	2	1	2	2
23	3	2	2	1	3
24	3	2	2	2	3

a: age of the patient
1 – young
2 – pre-presbyopic
3 – presbyopic

b: her spectacle prescription
1 – myope
2 – hypermetrope

c: whether astigmatic
1 – no
2 – yes

d: tear production rate
1 – reduced
2 – normal

Decision:
1 – fit hard contact lenses
2 – fit soft lenses
3 – lenses not recommended

Table 3.3 Production rules from ID3 (from Cendrowski, 1988)

Rule						
Rule 1:	d=1				⇒	decision 3
2:	a=1	b=1	c=1	d=2	⇒	decision 2
3:	a=2	b=1	c=1	d=2	⇒	decision 2
4:	a=3	b=1	c=1	d=2	⇒	decision 3
5:	b=2	b=1	d=2		⇒	decision 2
6:	b=1	c=2	d=2		⇒	decision 1
7:	a=1	b=2	c=2	d=2	⇒	decision 1
8:	a=2	b=2	c=2	d=2	⇒	decision 3
9:	a=3	b=2	c=2	d=2	⇒	decision 3

further reduced to a simpler and more general one (Table 3.4) with the PRISM procedure (Cendrowski, 1988).

On the other hand, rule based inference can also be performed by a neural network associative memory. Based on the cases listed in Table 3.2, a multilayer feedforward neural network (with four input neurons, three output neurons and a single hidden layer) is designed (Lu, 1994). Each input neuron corresponds to the relevant condition. The digital number {1, 2, 3} of the condition (a, b, c, d) can be coded into the neurons' inputs with $\{X(i) = 0.2, 0.5, 0.8; i = 1, 4\}$, where $i = 1, 4$

Table 3.4 Production rules from PRISM (from Cendrowski, 1988)

Rule 1:	d=1			⇒	decision 3
2:	a=1	c=1	d=2	⇒	decision 2
3:	a=2	c=1	d=2	⇒	decision 2
4:	a=3	b=1	c=1	⇒	decision 3
5:	b=2	c=1	d=2	⇒	decision 2
6:	b=1	c=2	d=2	⇒	decision 1
7:	a=1	c=2	d=2	⇒	decision 1
8:	a=2	b=2	c=2	⇒	decision 3
9:	a=3	b=2	c=2	⇒	decision 3

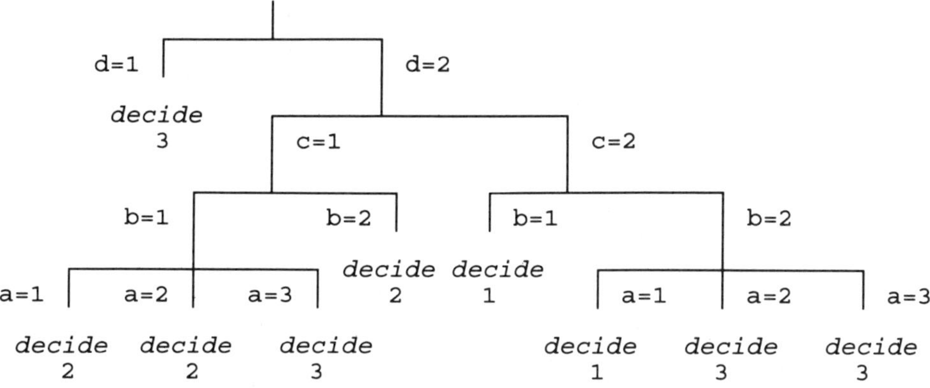

Figure 3.26 Decision tree produced from Table 3.2

corresponds to $\{a, b, c, d\}$, and the digital number of the decision can be coded as follows:

$$\text{If } \{Y(k), k = 1, 3\} = \begin{cases} 0.9 & \text{if } k = k^*, k^* = 1, 3 \\ 0.1 & \text{otherwise} \end{cases}$$

where k and k^* indicate the position of the output neuron and the status of the decision. Then the I/O data shown in Table 3.2 can be represented by a set of training data for neural network training as shown in Table 3.5.

After the training we establish a neural network associative memory to describe the mapping between the conditions and decisions. The neural inference can be performed through the 'recall' of the associative memory, and the decision can be inferred with the status of the conditions through the network output decoding as follows:

$$Y_{\text{out}}(k^*) = \max\{Y_{\text{out}}(k), k = 1, 3\} \tag{3.76}$$

where $Y_{\text{out}}(k)$ denote the kth neuron's output, and the k^* indicates the decision number. In fact, the decision is dependent on the position of the neuron which

Table 3.5 I/O data with coding for neural network training

Case	$X(1)$	$X(2)$	$X(3)$	$X(4)$	$Y(1)$	$Y(2)$	$Y(3)$
1	0.2	0.2	0.2	0.2	0.1	0.1	0.9
2	0.2	0.2	0.2	0.5	0.1	0.9	0.1
3	0.2	0.2	0.5	0.2	0.1	0.1	0.9
4	0.2	0.2	0.5	0.5	0.9	0.1	0.1
5	0.2	0.5	0.2	0.2	0.1	0.1	0.3
⋮							⋮
24							

has the maximum output. When applying the neural associative memory in rule reasoning, the neural network can perfectly matching the results of Tables 3.3 and 3.4. For instance, if we set $d = 1$ and set a, b and c with any random integer of the set $\{1, 2, 3\}$, the neural network will always provide $k^* = 3$, i.e. the decision is 3. From the application of neural memory/inference in this simple example, we can see that we have a great opportunity to apply neural networks in rule based system learning and reasoning.

3.9.3 Combination of neural network and rule learning

A general framework with combination of a neural network and a rule based system has been considered as an attractive form in developing a pattern recognition or decision making system in industrial and quality control. The main advantages of

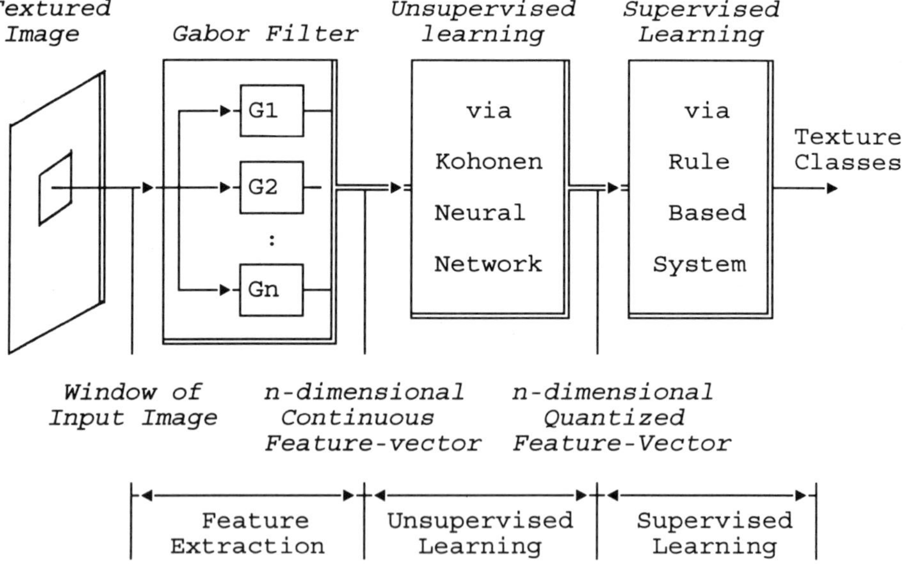

Figure 3.27 Architecture of neuro-rule based system for texture-analysis task

this framework can be summarized as follows:

- jointly using unsupervised and supervised learning;
- providing the user with system structured knowledge, e.g. the production rules, such that more physical explanation of the system may be available;
- using the neural network to extract (continuous and quantized) feature vectors with system input data or image.

An example structure of a neuro-rule-based system for the texture-analysis task is presented in Figure 3.27 (Greenspan *et al.*, 1992). In the unsupervised learning phase, a neural network (Kohonen network) is used for input vector quantization, and then in the supervised learning (rule system) phase, the labeling of the quantized attributes is achieved using a rule based system. *This information-theoretic*

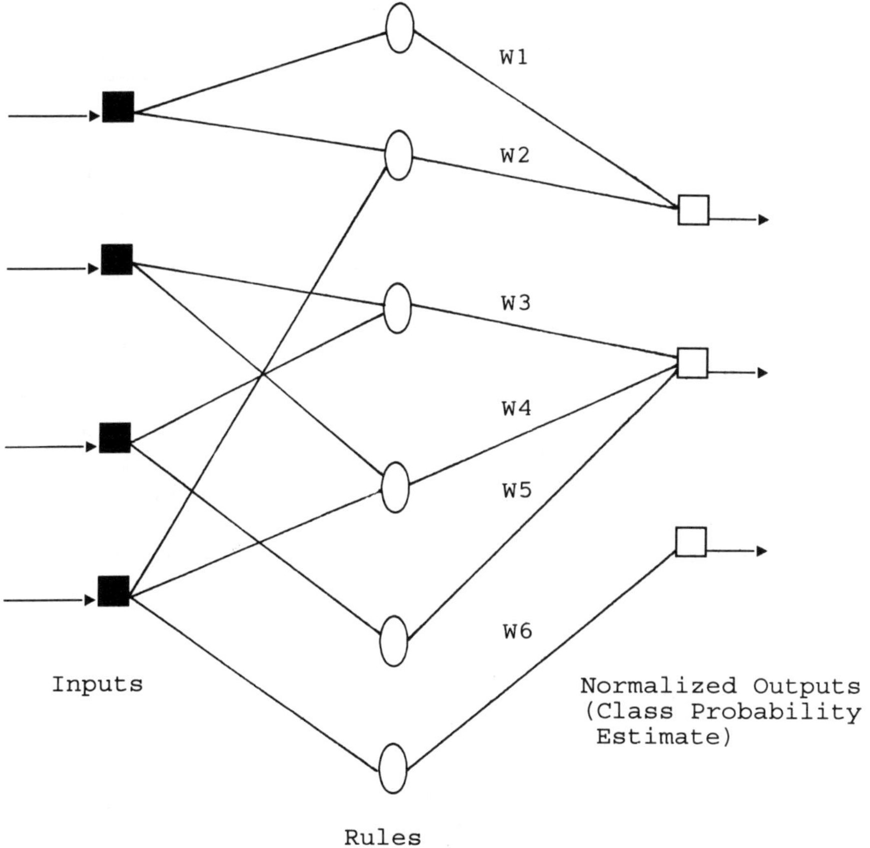

Figure 3.28 Structure of rule-based network

technique is utilized to find the most information correlations between the attributes and the pattern class specification, while providing probability estimates for the output classes (Greenspan *et al.*, 1992).

In supervised learning via a rule based system, the goal is to establish the feature maps between the input labeling attribute and the output classification. A neural network for the rule-based classification can be a three-layered feedforward architecture as shown in Figure 3.28 (Greenspan *et al.*, 1992). The rule-based network can be applied in solving many industrial control problems, such as fault detection and diagnosis, quality analysis and control, decision making, etc.

The more detailed introduction to the marriage of a neural network and rule base system can refer to Fu (1990) and Fujioka and Tanaka, 1993.

3.9.4 Rule learning through deep reasoning

General concepts of rule learning and deep reasoning

The solutions of an expert system are usually provided through the heuristic research and the rule match based inference. However, sometimes the solution might be 'empty' due to incomplete knowledge and/or strict constraints. For instance, while solving a production scheduling or job assignment problem with an expert system, we might have a result of 'empty of solutions' or 'ask for help' with some explanation. This may happen particularly when the production working conditions significantly change from what we considered in the system design phase. It has become an attractive R&D area to design an expert system with rule learning and deep reasoning.

While applying rule learning and deep reasoning in problem solving, if an expert system can not find a feasible solution or rule match with the regular heuristic search and reasoning, then the deep reasoning and the rule learning will automatically work to find the solution by changing some problem conditions, e.g. relaxing or giving up some soft constraints. Moreover, when a solution is finally obtained through rule learning and deep reasoning, the new rules (i.e. previously unknown rules) will be automatically embedded and reorganized into the knowledge base. As a result, if a similar status happens again, these new rules in the upgraded rule base will be effectively used in problem solving. In fact, rule learning and deep reasoning perform a function of knowledge self-acquisition; expert systems work under the environment with incomplete knowledge and uncertainty.

'Layered-generate-test' method for problem solving

Problem solving generally can be divided into two phases: 'heuristic rule solving' and 'deep reasoning', as schematically shown in Figure 3.29. Problem solving starts the regular heuristic search to find a solution in terms of the state match and heuristic rule search. However, if the 'heuristic rule search' fails, this means the system can not find a solution with the existing knowledge, i.e. the match between

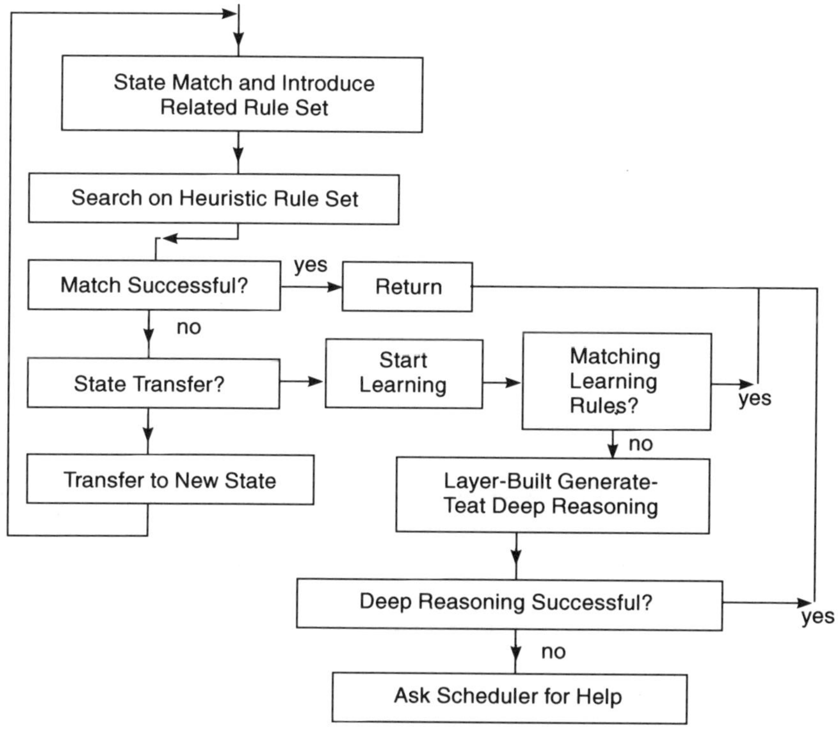

Figure 3.29 Problem solving by deep reasoning and rule learning

the process state and rules; then the system starts to run 'deep reasoning' to explore the new knowledge (new rules) which may be used to find the solution based on a given system environment, namely criteria and corresponding constraints.

The heuristic rules in a hybrid knowledge base are based mainly on human experience and are divided into a number of sub-rule sets. An approach to 'layered-generate-test' (Lu *et al.*, 1995) has been developed in dealing with deep reasoning based problem solving with the 'generate-and-test' approach (Steel, 1985). The 'generate-and-test' method can be effectively used in solving problems with a large and decomposable search solution space. In general, the 'generate-and-test' is divided into two components: 'generator' and 'tester'. The 'generator' produces all possible solutions and the 'tester' eliminates those solutions which do not satisfy the given constraints. When the 'generator' and 'tester' iteratively work together, the search becomes more effective in cooperatively working with a knowledge base. The main improvements of the 'layered-generate-test' method are as follows:

- The solution space can be abstractly divided into n-sub-spaces and the 'layered-generate-test' method is used to make the search in the individual sub-space by using the uncertain factor reasoning. As a result of applying the 'layered-generate-test' method, cutting too many rules in the earlier stages can be avoided, and then the search will be more efficient.

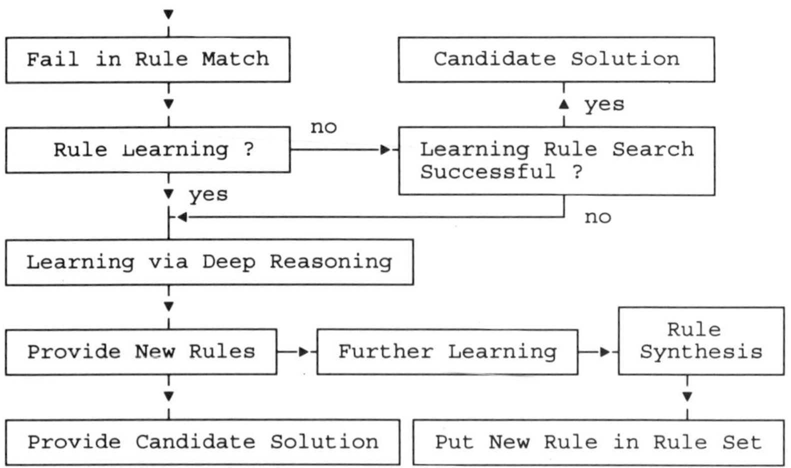

Figure 3.30 Reasoning process in rule learning

- The other problem in the 'generate-and-test' method is the knowledge distribution between 'generator' and 'tester'. The 'layered-generate-test' method can effectively apply these eliminated rules to the earlier generating process. Dutta (1985) gave the approaches to reasoning with imprecise knowledge in expert systems.

It should be emphasized that using only these two measures together, the search efficiency can then be significantly improved with a much lower possibility of losing the feasible solutions.

Rule learning through deep reasoning

We noted before that the solution of the reasoning might be 'empty', since the design rules are not complete or the given constraints are too restrictive to provide a feasible solution. To solve these problems, the rule learning approach to 'gradual precision–circumscriptive reasoning' (Steel, 1985) can be introduced in the expert systems. Generally, the learning rules consist of the following categories:

- *Creating new rules through rule learning based deep reasoning.* In deep reasoning, the system tries to track the symptom and then further find the error sources, and finally produces the new production rules through the corresponding conditions. However, the created new rules should work with the existing rules, hence the rule synthesis is required.
- *Embedding new rules into the related rule sets, or so-called 'rule synthesis'.* Based on the function and characteristics of the new rules, the new rules can be organized into a particular rule set.

The detailed reasoning process in the rule learning is illustrated in Figure 3.30. In practice, if the solution is empty, some soft constraints and/or human willingness may be released through the deep reasoning and the rule learning to provide a 'satisfactory', or at least a 'feasible' solution. The application of rule learning and deep reasoning in a production scheduling system will be addressed in section 6.5.

3.10 REMARKS

Learning systems have been popularly used in solving signal processing, pattern recognition and other information processing problems. In contrast, applications of learning systems in industrial control are still at an earlier stage. In this chapter we have introduced the fundamental concepts, strategies and algorithms of learning processes. However, since learning or so-called 'machine learning' is such a wide topic, it will be difficult to cover all learning approaches in a single chapter. The goal of this chapter is to provide the reader with the basic background and the major methodologies of the learning processes in the design of industrial intelligent systems. It should also be emphasized that hybrid learning strategies are particularly applicable in solving a complex industrial control problem. The detailed applications of learning systems in industrial control will be investigated in conjunction with various topics, e.g. process modeling, estimation, dynamic control, optimization, etc., in the second part of this book (Chapters 4–8).

4 System modeling and estimation

4.1 INTRODUCTION

The applications of system modeling and estimation techniques have been one the most attractive areas in industrial control. During the last few decades, many efforts have been made in control societies to move math-model based system identification and estimation from academic research to industrial applications. However, the existing methods either require a known system model structure or are too complicated to be applied in a complex industrial system with nonlinear and uncertain characteristics. As a result, it is highly demanding to develop math-model free modeling and estimation technology with intelligent system methodology. The aim of this chapter is to investigate applications of intelligent systems in industrial system modeling and estimation.

This chapter starts with a review of the math-model based optimal estimation technique and its industrial applications, and then introduces the basic concepts, architectures and algorithms of math-model free estimation techniques with fuzzy logics, pattern recognition and neural networks. A number of industrial worked examples are included to illustrate the detailed application procedures and results.

We noted in section 1.2.3 that the tasks of system estimation are to provide more precise and/or complete information in terms of the system observations (measurements). Traditionally, the techniques of system estimation are based on a mathematical model, or so-called *'math-model based estimation'*. The estimation method can be divided into two categories: *'state observer'* (Luenburger, 1964) and *'optimal estimator'*, e.g. the Kalman filter (Kalman and Bucy, 1960) for deterministic and stochastic systems respectively. The design of a state observer is based on the required dynamics of the estimation error responses. The pole and zero assignment approaches can be used to develop a transfer function of the observer. In contrast, the design of an optimal estimator is to develop an optimal estimation law which minimizes the estimation errors under a stochastic environment. Based on the functions, the estimation problem can be grouped into the following three categories illustrated in Figure 4.1 (Gelb, 1974):

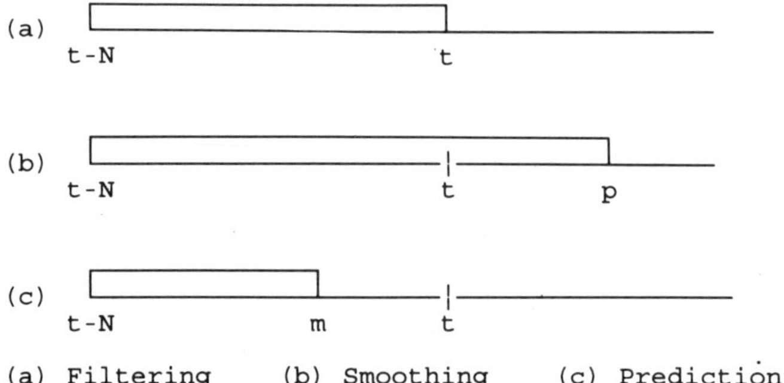

Figure 4.1 Illustration of system estimation problems, © MIT Press

- *Filtering*: The estimate of $X(t)$ is provided based on the observation $\{X(t), t = t - N, t\}$. The filtering refers to estimating the state vector at the current time and can be used as a real-time signal filter for noisy signal processing or state estimation for providing the system unmeasurable and internal information. The software oriented sensors may be developed using the filtering techniques.
- *Smoothing*: The estimate of $X(t)$ is provided based on the observation $\{X(t + p), t = t - N, t\}$. The task of the smoothing is to estimate the value of the state at some prior time based on all measurements taken up to the current time. The smoothing can be used as a post-experiment algorithm in signal processing;
- *Prediction*: The estimate of $X(t)$ is provided based on the observation $\{X(t - m), t = t - N, t\}$. The prediction refers to estimating the state at a future time. The prediction can be used as a predictor for both system output and state variables. A predictive control can be performed with an early predicted information.

Applications of math-model based estimation techniques have been studied in nuclear medicine, statistical image enhancement, estimation of traffic densities, chemical process control, estimation of river flow, power system load forecasting, nuclear reactor parameter identification, etc. As a function estimator, math-model based estimation provides a numerical solution with minimum error between actual and estimated values, only if both math-model and statistical properties of system observations and dynamics are available. However, in fact, an optimal estimation system is sensitive to the imprecision and uncertainty of the knowledge of the system mathematical dynamics and statistics. This becomes one of the key factors to make applications of optimal estimation techniques in complex industrial systems more difficult.

The introduction of machine intelligence in estimation systems has opened a new path to develop math-model free estimation techniques. Kosko (1992) defines adaptive math-model free estimation as *'an intelligent system which adaptively estimates a continuous function from data without specifying mathematically how outputs depend on inputs.'*

4.2 MATH-MODEL BASED OPTIMAL ESTIMATION

Since math-model free estimation has a similar architecture to the math-model based estimation, knowledge of math-model based estimation will help us to learn math-model free estimation techniques, particularly for the reader who has less knowledge of the math-model estimation. Here we take the application of Kalman filtering in linear system estimation as an example to describe the basic structure and algorithm of optimal estimation. Kalman filtering has received extensive studies since the publication of the classic paper by Kalman and Bucy (1960). In an optimal estimation problem, the objective is to calculate full state estimates X_k from noisy measurements of the system I/O, such that the estimator minimizes a specific performance index. The algorithm of Kalman filtering can be formulated as follows.

Consider a linear, time-invariant system governed by the following state space model with the corruption of both system and measurement random noises, W_k and V_k respectively, such that

$$X_{k+1} = AX_k + BU_k + DW_k \tag{4.1}$$

$$Y_k = CX_k + V_k \tag{4.2}$$

where A, B and C are the constant matrices describing the system dynamics, W_k and V_k are zero-mean, white, Gaussian noise sequences which represent the model uncertainties under a random environment, and the measurement noises being independent of W_k respectively. Then the corresponding covariance matrices of W_k and V_k can be written as follows:

$$Cov[W_k W_k^T] = E[W_k W_k^T] = Q_k \delta_{kj} \tag{4.3}$$

$$Cov[V_k V_k^T] = E[V_k V_k^T] = R_k \delta_{kj} \tag{4.4}$$

$$Cov[W_k V_k^T] = E[W_k V_k^T] = 0 \tag{4.5}$$

where Q_k and R_k are the covariance matrices for process and measurement noises respectively. If the system is assumed to be a statistical stationary case, then Q_k and R_k can be replaced by the constant matrices Q and R. δ_{kj} is the Kronecker delta. A linear Kalman filter reconstructs the estimate of the state X_k, which minimizes the following performance index:

$$J = E\{[X_k^p - X_k]^T [X_k^p - X_k]\}, \quad k = k - N, k \tag{4.6}$$

The optimal estimate of the state, X_k^p can be solved from a discrete-time Kalman filtering algorithm (Hamilton, 1973):

$$X_k^p = \overline{X}_k + K_k(Y_k - C\overline{X}_k) \tag{4.7}$$

where $\overline{X}_k$ can be calculated from a deterministic model as follows:

$$\overline{X}_k = AX^p_{k-1} + BU_{k-1} \tag{4.8}$$

The filter gain matrix K_k can be solved from the following recursive Riccati matrix algorithm:

$$K_k = M_k C^T (CM_k C^T + R)^{-1} \tag{4.9}$$

$$P_k = (I - K_k C)M_k \tag{4.10}$$

$$M_{k+1} = AP_k A^T + DQD^T \tag{4.11}$$

The initial condition of the recursive calculation can be given based on the guessed initial estimated error, namely:

$$M_0 = E\{[X^p_0 - X_0][X^p_0 - X_0]^T\} \tag{4.12}$$

If the convergent time of the matrix K_k is much less than the total estimation time, then the filter gain K_k can be replaced by a constant matrix K which is equal to the convergent value of K_k from the off-line calculation of 4.9–4.11, such that 4.7

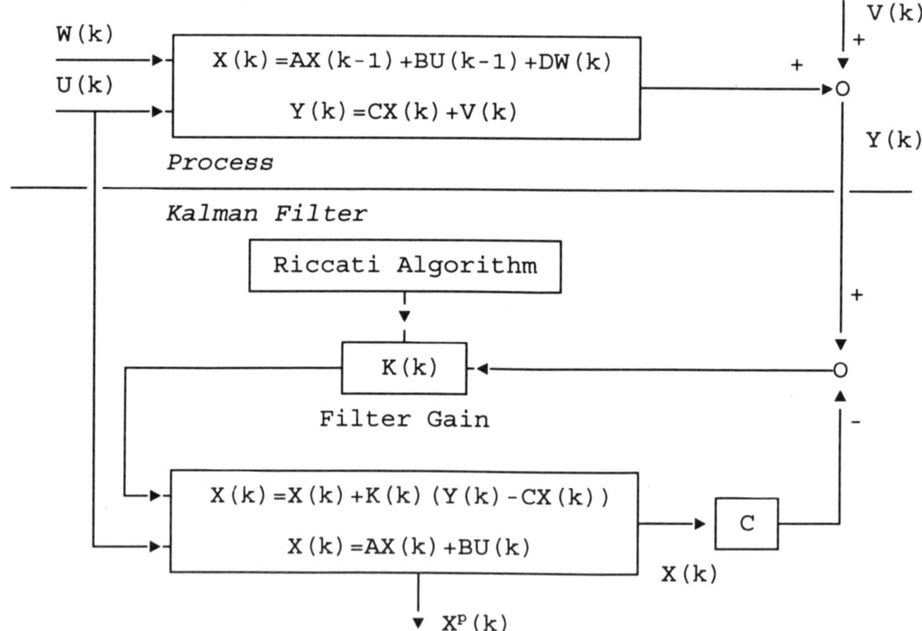

Figure 4.2 Block diagram of a Kalman filter

becomes a stationary form of the Kalman filtering as follows:

$$X_k^p = \overline{X}_k + K(Y_k - C\overline{X}_k) \tag{4.13}$$

As a result, the computer time and memory can be significantly reduced by using a stationary filter gain. The block diagram of a linear discrete-time Kalman filter is shown in Figure 4.2.

In addition, the Kalman filtering can also be applied to nonlinear systems with the linearization of the system model about an operating point; then an extended Kalman filter can be developed (Gelb, 1974). Now an industrial simulation example is given to describe the application of the Kalman filter in practice.

Example 4.1

An application is given of a Kalman filter in the estimation of the internal temperature profile of steel ingots in a soaking pit (Lu and Williams, 1983).

The temperature profile of the steel ingots in the soaking pit is not available for on-line sensory measurement. The task is to estimate the real-time internal temperature profile of the ingots based on a simplified dynamic model and the observations. A reduced order linear state space model was developed from a partial differential equation and the related boundary conditions which describe the unsteady state heat transfer in a steel ingot as follows:

$$X_{k+1} = AX_k + BU_k + DW_k \tag{4.14}$$

$$Y_k = CX_k + V_k \tag{4.15}$$

where

$X = [T_1 \ldots T_5 | T_W]^T$;
$U = T_g$

$$A = \begin{bmatrix} 0.955\,22 & 0.004\,48 & & & & \\ 0.011\,19 & 0.966\,42 & 0.022\,39 & & 0 & \\ & 0.014\,93 & 0.962\,69 & 0.022\,39 & & \\ & & 0.016\,79 & 0.960\,82 & 0.022\,39 & \\ & 0 & & 0.024\,88 & 0.923\,84 & 0.014\,67 \\ & & & & 0.003\,52 & 0.967\,28 \end{bmatrix}$$

$B = [0 \quad 0 \quad 0 \quad 0 \quad 0.036\,61 \quad 0.029\,20]^T$
$C = [0 \quad 0 \quad 0 \quad 0 \quad 0 \quad 1]$

and $\{T_i, i = 1, 5\}$ denotes the ingot temperature profile from the center to the surface; T_W, T_g denote the furnace wall and the gas temperatures, of which the latter can be represented as a function of fuel flow rate.

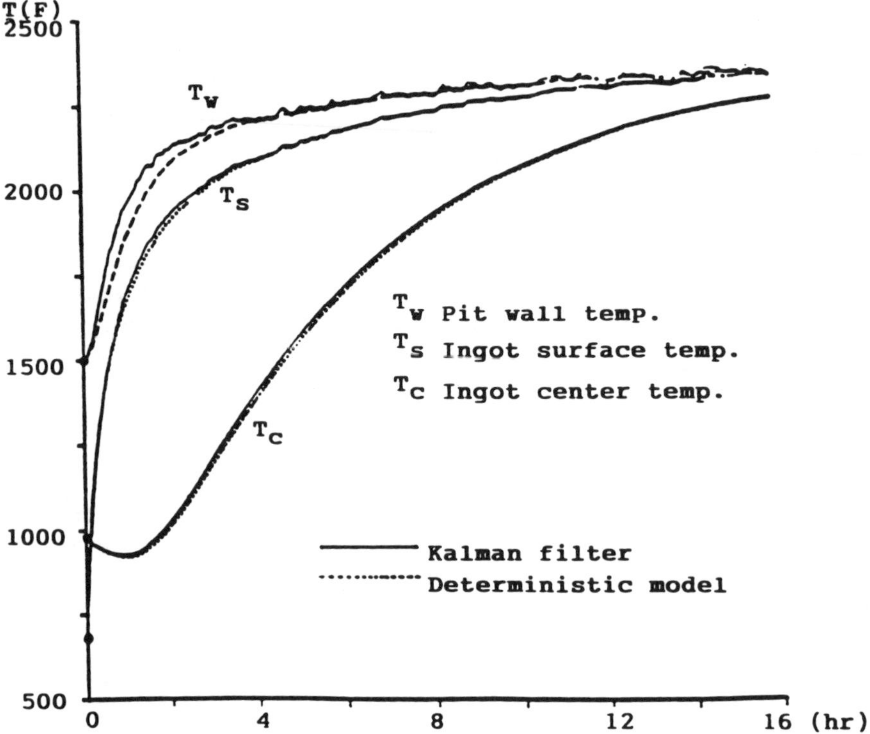

Figure 4.3 Estimated results given by Kalman filter for ingot temperature profile prediction

Select an initial condition as follows:

$$P(0) = M(0) = diag[900 \quad 625 \quad 225 \quad 200 \quad 25]$$

The solution of the Riccati algorithm (equations 4.9 to 4.11) shows the convergent time of the filter gain matrix is much less than the required total heating time. A steady state Kalman filter gain matrix can be used. The estimated results given by stationary Kalman filtering for both center and surface ingot temperatures are given in Figure 4.3. It can be seen that the estimated temperatures are very close to the temperatures produced by a simulated furnace-ingot model, with high dimensions.

If a Kalman filter is designed for a system with significant uncertainties, an adaptive Kalman filter is then required. In an adaptive Kalman filtering system, the filtering gain matrix should be recursively upgraded with equations 4.9–4.11 and 4.13 based on the latest knowledge of the system dynamics and the statistical properties. In fact, this is a math-model based learning process, which however, requires much more computer CPU time and memory, particularly for a high dimension system.

In the math-model free estimation, the system keeps the same structure, but the mathematical model and the estimation law are replaced by a non-math-know ledge base or an associative memory with fuzzy logic, pattern recognition or neural network model.

4.3 MATH-MODEL FREE ESTIMATION

Let us also take the filtering as an example to describe the math-model free estimation problem and its architecture. The results can be extended to both prediction and smooth problems. Suppose we have a nonlinear dynamic system with the following generic functional description:

$$X_{k+1} = \Phi(X_k, U_k, k) + W_k \tag{4.16}$$

$$Y_k = \Psi(X_k, U_k, k) + V_k \tag{4.17}$$

where X, Y and U are defined as system state, output and input vectors. They could be numerical, multivalued or symbolic forms. W_k and V_k represent the uncertainties of system model and observations in the randomness or fuzziness sense respectively. $\Phi(*)$ and $\Psi(*)$ denote the functional relations or other knowledge representations, such as rule base, pattern mapping or neural network describing the system behavior:

$$\Phi(*) : X_k, U_k \rightarrow X_{k+1} \tag{4.18}$$

$$\Psi(*) : X_k, U_k \rightarrow Y_k \tag{4.19}$$

The math-model free estimation problem can be defined as: develop an associative memory which can be used to discover the real-time information (state), $\{X_k, k = k - N, k\}$ with minimum error between the estimated states, X_k^p and the actual states, X_k based on the system observed I/O data, initial state X_0, and the infrequent observation of the state variables. The estimation problem can be written in the following mathematical form:

$$\min_{\mathcal{E}} \sum_{k-N}^{k} \| X_k - X_k^p \|^2 \tag{4.20}$$

subject to 4.18 and 4.19 and other system related constraints, where $\mathcal{E}$ denotes the associative memory for estimation:

$$\mathcal{E} : \{U_k, Y_k, X_0, X_k\} \rightarrow X_{k+1} \tag{4.21}$$

In industrial systems, the physical illustration of the relations between the system output vector Y and the state vector X can be described as follows:

- the system outputs are equal to the time delayed states, namely $Y(k) = X(k - L)$, i.e. a time-delayed state observation;

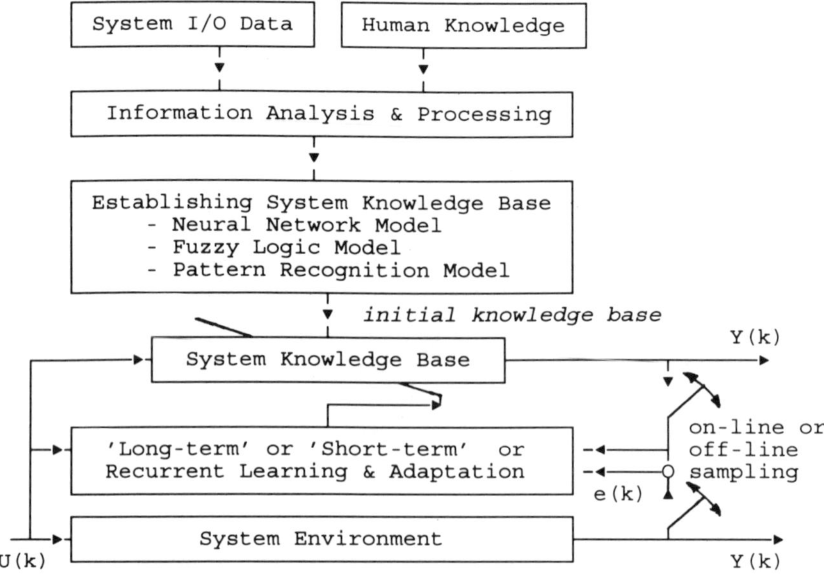

Figure 4.4 Adaptive math-model free estimation

- the system output vector is a subset of the system state vector, namely $Y(k) \subset X(k)$, in other words, only part of the state variables can be observed on-line;
- the system state variables can only be infrequently observed off-line, for instance analysis data from a quality control laboratory.

How to produce quality products through quality control and process control as well as production management has been one of the most critical issues in any enterprise. However, it should be kept in mind that many quality variables in industrial processes, for instance polybutadiene, cement, rubber, food processing, biochemical production, etc., cannot be measured on-line, but can only be observed infrequently through laboratory analysis. Obviously, such kind of long-term, infrequent system information must be too late to provide a timing control action in producing a quality product. The application of state estimation techniques may construct the real-time estimate of the quality information, which can be used as a feedback or a feedforward signal for real-time control purposes. In general, infrequent, off-line observed data may be used as a learning feedback signal to upgrade an associative memory of the state estimation when the system works under a variable environment. The architecture of a math-model free estimator with self-learning is shown in Figure 4.4. The math-model free estimation techniques to be investigated in this chapter basically follow this generic structure. Math-model free estimation techniques can be widely applied in state estimation, parameter estimation, quality prediction and control, fault detection and diagnosis, etc. Now we will introduce a number of math-model free estimation techniques in the following sections.

4.4 FUZZY IDENTIFICATION AND ESTIMATION

4.4.1 Introduction

In general, the behavior of a fuzzy system, both static and dynamic, can be represented by a set of fuzzy linguistic rules or a fuzzy relation model. In practice, a set of fuzzy rules with structured knowledge under a symbolic framework can be encoded into a fuzzy relation model under a numerical framework through fuzzy identification and system nonfuzzy I/O data. As a result, the fuzzy model can be used in continuous function estimation, eventhough initially we had only structured knowledge, e.g. the fuzzy rules. This is one of the major advantages of the fuzzy system over the regular rule base in expert systems.

The fuzzy identification may be carried out in two ways: the linguistic approach (Tong, 1978; Pedrycz, 1984) and the approach to resolving the fuzzy relational equations (Xu and Lu, 1987; Pedrycz, 1983). In this section, we will introduce the fuzzy identification techniques to establish a fuzzy relational model. Tong (1978) proposed a *'logical examination'* method to solve a linguistic identification problem, and Li and Liu (1980) further proposed an approach to an adaptive algorithm for fuzzy model identification based on a 'decision table'. However, the proposed algorithms can not be used for multivariable systems with high dimensions due to the large amount of computer memory and time required.

This section introduces a generic fuzzy model identification algorithm for MISO (multi-input/single-output) systems and a related self-learning algorithm. Two numerical examples show that the proposed identification algorithm can build a fuzzy model with a fairly high accuracy. On the other hand, a proposed self-learning algorithm makes a fuzzy model more robust and accurate. Note that the self-learning algorithm can also readily be converted into a real-time algorithm for on-line applications.

4.4.2 Problem formulation

A discrete-time fuzzy relational model for a MISO system with p inputs can be written in the following general form:

$$\begin{aligned} y(k) = y(k-\tau) \circ y(k-\tau-1) \circ \cdots \circ y(k-\tau-n_y) \circ \cdots \circ \\ u_1(k-\tau_1) \circ \cdots \circ u_1(k-\tau_1-n_1) \circ \cdots \circ \\ u_p(k-\tau_p) \circ \cdots \circ u_p(k-\tau_p-n_p) \circ R \end{aligned} \tag{4.22}$$

where output $y(*)$ and inputs $\{u_i(*),\ i=1,p\}$ are all fuzzy variables, and R is the fuzzy relation between the inputs and output. The symbol '$\circ$' denotes the fuzzy composition operator. $\{\tau_i, i=1,p\}$ and $\{n_y \;\&\; n_i, i=1,p\}$ represent the time delays and system orders respectively. Similar to the conventional math-model based identification, the fuzzy model identification usually also involves both *'structure identification'* and *'parameter estimation'*. In fact, the structure identification for the problem under study is to determine the time delays, $\{\tau_i, i = 1, p\}$, and system orders, $\{n_y \;\&\; n_i, i = 1, p\}$. The task of the parameter estimation is to determine the fuzzy relation R. For convenience, let

$$\begin{cases} x_1(k) = y(k-\tau) \\ x_2(k) = y(k-\tau-1) \\ \vdots \\ x_{n_y+1}(k) = y(k-\tau-n_y) \\ x_{n_y+2}(k) = u_1(k-\tau_1) \\ \vdots \\ x_n(k) = u_p(k-\tau_p-n_p) \end{cases} \tag{4.23}$$

where

$$n = n_y + 1 + \sum_{i}^{p}(\tau_i + 1) \tag{4.24}$$

Substituting 4.24 into 4.23, yields

$$y(k) = x_1(k) \circ x_2(k) \circ \cdots \circ x_n(k) \circ R \tag{4.25}$$

It should be emphasized that here the fuzzy relational model 4.25 is based on the concept of the 'referential fuzzy sets' popularly used. A composition rule has been adapted such that the fuzzy relation R is related only to the referential fuzzy sets in each universe. The fuzzy model identification problem can then be described as determining the fuzzy relation R, model structure and fuzzy reference set with

$$\min J = \frac{1}{N - \tau_{\max} + 1} \sum_{k=\tau_{\max}}^{N} (y(k) - y^p(k))^2 \tag{4.26}$$

where

$$\tau_{\max} = \max(\tau, \tau_1, \ldots, \tau_p) + 1 \tag{4.27}$$

subject to the I/O data sequence $\{x(k), y(k), k = 1, N\}$.

4.4.3 Determination of referential fuzzy sets

Let $\{Y, X_i, i = 1, n\}$ be $(n+1)$ universes of discourse and $\{y, x_i, i = 1, n\}$ be their generic elements respectively. Assume each universe contains the same number $(= r)$ of the referential sets which are

$$\begin{cases} A_{11}, \ldots, A_{1r} \in F(X_1) \\ A_{n1}, \ldots, A_{nr} \in F(X_n) \\ B_1, \ldots, B_r \in F(Y) \end{cases} \tag{4.28}$$

where $F(*)$ stands for the set of all fuzzy sets on '$*$'. The referential fuzzy sets have their linguistic meanings, for instance, if X_1 represents the temperature, then one may define A_{11} as *'low temperature'*, A_{1i} $(1 < i < r)$ as *'medium temperatures'*, and A_{1r} as *'high temperature'*, etc. These referential fuzzy sets are characterized by their membership functions:

$$\begin{cases} A_{ij}(x_i) : X_i \Rightarrow [0,1], & i = 1, n, \ j = 1, r \\ B_j(y) : Y \Rightarrow [0,1], & j = 1, r \end{cases} \tag{4.29}$$

To determine the memberships of the referential fuzzy sets, statistics (Li and Liu, 1980), clustering (Yager, 1983), or subjective (Tong, 1979) methods can be used. To ensure the performance of a fuzzy model and to provide a uniform basis for further study, it is required that all referential sets should be normal, convex and satisfying the following completeness conditions:

$$\text{for all} \quad x_i \in X_i, \exists j \in r', \ A_{ij}(x_i) > 0, i = 1, \quad n \tag{4.30}$$

$$\text{for all} \quad y \in Y, \exists j \in r', \ B_j(y) > 0 \tag{4.31}$$

where $r' = [1, 2, \ldots, r]$.

The number of referential fuzzy sets in each universe, r, should be selected according to experience and/or tests. In general, the model accuracy may be improved by increasing r; however, on the other hand, a larger r will certainly require more computer memory and CPU time.

4.4.4 Identification algorithm

After determining the referential fuzzy sets, the second task in fuzzy identification is to determine fuzzy model structure parameters, such as orders and time delays, which are similar to the methods in regular system identification. We will discuss this in association with the numerical examples to be given later.

The final and most important task in fuzzy identification is to determine the fuzzy relation R, from the system nonfuzzy I/O data sequences. A fuzzy relation R based on the fuzzy referential sets is different from that based on the elements of universes, both in form and in meaning. Let the memberships of R be $R(s_1, s_2, \ldots, s_n, s)$, which represents the connection among the referential sets, $A_{1s_1}, \ldots, A_{ns_n}, B$ $(s_1, s_2, \ldots, s_n, s = 1, r)$. Based on the physical meaning of the fuzzy relation as defined, the membership of R represents the confidence of the rules

$$\text{If} \quad (A_{1s_1}, \ldots, A_{ns_n}), \quad \text{then } B_s \tag{4.32}$$

namely,

$$\begin{aligned} &\text{If } x_1 \text{ is } A_{1s}, \text{ and } , \ldots, \text{ and } x_n \text{ is } A_{ns_n}, \\ &\text{then } poss\ (y|B_s) = A_R(s_1, s_2, \ldots, s_n, s) \end{aligned} \tag{4.33}$$

where s_i and $s = r, i = 1, n$.

Since fuzzy identification is based on system fuzzy I/O data, the nonfuzzy input data should be first encoded to the fuzzy variables and the resulting fuzzy output should be then decoded back to the nonfuzzy data. The methods of *'fuzzification'* and *'defuzzification'* been introduced in section 2.1.3. The fuzzy identification algorithm can be summarized as follows (Xu and Lu, 1987):

(1) Calculate the possibility distribution of the kth $(k = 1, N)$ data pair $[x_1(k), \ldots, x_n(k), y(k)]$ on the corresponding referential fuzzy sets:

$$\begin{cases} p_{1j}(k) \triangleq poss(A_{1j}|x_1(k)) = \sup \min_{x_1}[A_{1j}(x_1), X_1(k, x_1)] \\ \vdots \\ p_{nj}(k) \triangleq poss(A_{nj}|x_n(k)) = \sup \min_{x_n}[A_{nj}(x_n), X_n(k, x_n)] \\ p_j(k) \triangleq poss(B_j|y(k)) = \sup \min_y[B_j(y), Y(k, y)] \end{cases} \quad j = 1, r \tag{4.34}$$

where $X_i(k, x_i)$ is the memberships of $x_i(k)$, $i = 1, n$.

(2) Construct the vectors

$$\begin{cases} p_{x_1}(k) \triangleq [p_{11}(k), \ldots, p_{1r}(k)] \\ \vdots \\ p_{x_n}(k) \triangleq [p_{n1}(k), \ldots, p_{nr}(k)] \\ p_y(k) \triangleq [p_1(k), \ldots, p_r(k)] \end{cases} \quad k = 1, n \tag{4.35}$$

The sub-relation R_k can then be constructed based on $(P_{xi}(k), i = 1, n), p_y(k)$:

$$R_k = p_{x1}(k) \times \cdots \times p_{xn}(k) \times p_y(k) \tag{4.36}$$

where $\times$ denotes the Cartesian product, i.e.

$$R_k(s_1, \ldots, s_n, s) = \min[p_{1s_1}(k), \ldots, p_{ns_n}(k), p_s(k)], \quad \forall s_1, \ldots, s_n, \ s \in r \tag{4.37}$$

(3) Calculate fuzzy relation R:

$$R = \cup_k R_k = R_1 \cup \cdots \cup R_N \tag{4.38}$$

where '$\cup$' denotes the union operation, i.e.

$$R(s_1, \ldots, s_n, s) = \bigvee_k R_k(s_1, \ldots, s_n, s), s_1, \ldots, s_n, \quad s \in r \tag{4.39}$$

where '$\bigvee$' = 'max'.

4.4.5 Use of fuzzy model

If the fuzzy relation R and the proposition $\{x_i(k), i = 1, n\}$ in 4.25 are given, $y(k)$ can then be solved. First, find referential fuzzy sets, $\{A_{i\lambda_i}, i = 1, n\}$, which are 'closest' to

$\{x_i(k), i = 1, n\}$, where

$$\begin{cases} \lambda_1 = \{ j | p_{1j}(k) > q, j \in r' \} \\ \quad \vdots & \quad 0 < q < 1 \\ \lambda_n = \{ j | p_{nj}(k) > q, j \in r' \} \end{cases} \tag{4.40}$$

and q is a preselected threshold. If $\{\lambda_i, i = 1, n\}$ are unique, $y(k)$ can then be calculated in terms of its membership $Y(k, y)$ as follows:

$$Y(k, y) = \max_s \min[R(\lambda_1, \ldots, \lambda_n, s), B_s(y)] \tag{4.41}$$

If $\{\lambda_i, i = 1, n\}$ are not unique and are denoted by $\lambda_1^{(1)}, \ldots, \lambda_1^{(k1)}, \ldots, \lambda_n^{(1)}, \ldots, \lambda_n^{(kn)}$, respectively, then $Y(k, y)$ can be solved from

$$Y(k, y) = \max_{i_1 \in K_1} \max_{i_2 \in K_2} \ldots \max_{i_n \in k_n} \max_s \min[R(z_1^{(i_1)}, \ldots, z_n^{(i_n)}, s), B_s(y)] \tag{4.42}$$

Note that 4.41 and 4.42 correspond to a 'max–min' composition. If a 'max–product' composition is used, the 'min–operation' in the equations should be replaced by product operation. Finally, it should be noted that both fuzzification and defuzzification processes should be taken into account either in constructing or in using a fuzzy model.

It can be seen that the fuzzy model is governed by a fuzzy relation R, associated with the referential fuzzy sets. The I/O data used in fuzzy identification are viewed as linguistic data and thus make their contribution to the fuzzy relation R according to their conditional possibility distribution on the corresponding referential fuzzy sets. The model is thus equivalent to a rule set and each entry in R corresponds to a specific rule:

$$\begin{aligned} &\text{if} \quad (x_i(k) = A_1 s_1 \text{ and } \cdots \text{ and } x_n(k) A_n s_n) \\ &\text{then} \quad poss(B_s | y(k)) = R(s_1, \ldots, s_n, s), \\ &\qquad \{s_i, i = 1, n\} \quad \text{and} \quad s \in r' \end{aligned} \tag{4.43}$$

4.4.6 Self-learning in fuzzy modeling

A fixed fuzzy model cannot provide a satisfactory result when the system environment significantly changes, such that an adaptive algorithm of the fuzzy model through self-learning should be further developed to give the resulting fuzzy model with robust and adaptive behavior. The schematic representation of a fuzzy model with real-time self-learning is shown in Figure 4.5. It can be seen that the performance criterion serving as a feedback driving force in the learning is the error between the actual output $y(k)$ and the predicted output by the fuzzy model, $y^p(k)$.

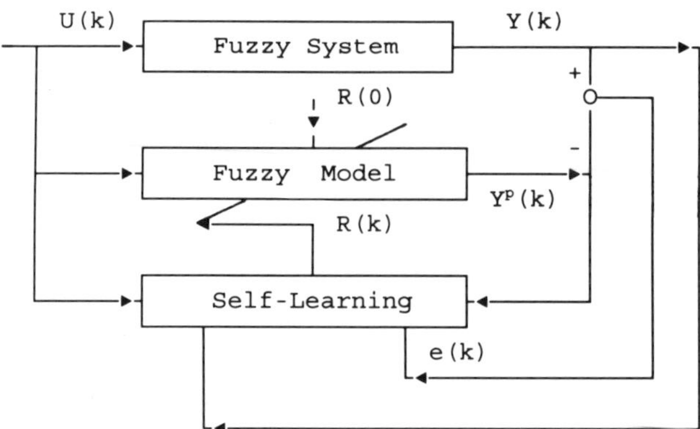

Figure 4.5 Fuzzy model with real-time learning

A fuzzy model usually consists of many rules. The referential fuzzy sets on different universes form the linguistic rules through a connection of R. Each element in R corresponds to a specific rule. The model provides the prediction $y^p(k)$ which is based on the system input $\{x_i(k), i = 1, n\}$ related to the propositions in a rule set. It should be pointed out that only some of the rules in the rule set are used to produce $y^p(k)$; for instance, when $\{\lambda_i, i = 1, n\}$ are unique, only r rules are used, thus only these rules should be modified through self-learning, if the predicted results are not satisfied.

In fact, the self-learning algorithm is based on the modification. To modify the rules, one can either modify the corresponding elements in R or modify the membership of referential fuzzy sets. However, if modifying the latter the whole rule set will be disturbed, since the referential fuzzy sets are used not only by the rules which are needed to be modified, but also by all of the others. For this reason, only the fuzzy relation R will be modified in the self-learning algorithm. What follows is the self-learning algorithm in its off-line form, but which can be readily upgraded to the on-line version for real-time application purposes.

Suppose the initial fuzzy relation $R^{(0)}$ is given, now we will introduce modification of the fuzzy rules based on the prediction error, and the I/O data through self-learning. The procedures are as follows:

(a) Set $k = 0$ and an initial fuzzy relation $R^{(0)}$.

(b) $k + 1 \Rightarrow k$.

(c) Use the fuzzy model $R^{(k-1)}$ and the fuzzy input data $\{x_i(k), i = 1, n\}$ to produce a predicted output $y^p(k)$ and let $e(k) = y(k) - y^p(k)$. If $|e(k)| \leq \epsilon$ (a pre-specified small number), then $R^{(k-1)} \Rightarrow R^{(k)}$ and return to (b), otherwise continue.

(d) Calculate $\{p_{ij}(k), p_j(k), i = 1, n, j = 1, r\}$ according to 4.34. Then calculate $\{\lambda_i, i = 1, n\}$. It is obvious that the following r rules concerned with

$\{A_{i\lambda_i i}, i = 1, n\}$ are responsible for the prediction $y^p(k)$:

$$\begin{cases} \text{If} \quad (A_{1\lambda_1}, \ldots, A_{n\lambda_n}), \quad \text{then} \quad poss(B_1|\, y(k)) = R^{(k-1)}(\lambda_1, \ldots, \lambda_n, 1) \\ \qquad \vdots \\ \text{If} \quad (A_{1\lambda_1}, \ldots, A_{n\lambda_n}), \quad \text{then} \quad poss(B_r|\, y(k)) = R^{(k-1)}(\lambda_1, \ldots, \lambda_n, r) \end{cases} \tag{4.44}$$

Thus only r rules are required to be modified. In other words, only r elements in the fuzzy relation R concerned with the memberships $\{A_{iz_i}, i = 1, n\}$ should be modified.

(e) Define the modification quantity $d_s^{(k)}$:

$$d_s^{(k)} \triangleq p_{1\lambda_1} * \cdots * p_{n\lambda_n}(k) * p_s(k), \quad s = 1, r \tag{4.45}$$

where '$*$' = min for 'max–min' composition and '$*$' = product for the 'max–product' case. Note that in 4.44,

$$p_s(k) = poss(B_s|y(k)), \quad s = 1, r \tag{4.46}$$

(f) Modifying $R^{(k-1)}$ results in $R^{(k)}$:

$$R^{(k)}(S_1, \ldots, s_n, s)$$

$$= \begin{cases} a_s d_s^{(k)} + (1 - a_s)R^{(k-1)}(s_1, \ldots, s_n, s), & \text{if } \{s_i = \{\lambda_i, i = 1, n\} \\ R^{(k-1)}(s_1, \ldots, s_n, s) & \text{otherwise,} \end{cases}$$

$$a_s \in [0, 1], \quad s = 1, r \tag{4.47}$$

(g) If $k = N$, end, otherwise return to (b).

It can be seen that the rules in 4.44 are replaced by

$$\begin{aligned} &\text{If} \quad (A_{1\lambda_1}, \ldots, A_{n\lambda_n}), \quad \text{then} \quad poss(B_1|\, y(k)) = d_1^{(k)} \\ &\text{If} \quad (A_{1\lambda_1}, \ldots, A_{n\lambda_n}), \quad \text{then} \quad poss(B_r|\, y(k)) = d_r^{(k)} \end{aligned} \tag{4.48}$$

Since $\{d_1^{(k)}, \ldots, d_r^{(k)}\}$ are calculated from the observed $y(k)$, 4.48 will attempt to make the following inference:

$$\text{If} \quad (x_1(k), \ldots, x_n(k)), \quad \text{then } y(k) \tag{4.49}$$

The modification just means that in a learning algorithm, if for all $s, a_s = 0$, the modification no longer exists, and if for all $s, a_s = 1$, the rule set 4.48 will completely replace rule set 4.43 in R. Since $y(k)$ is usually corrupted by noise, it is reasonable to keep $a_s < 1$ to reduce the effect of the measurement noises on the model. As a result, a compromise is made between the old rule set which results in $y^p(k)$ and the new rule which attempts to infer $y(k)$ from $\{x_i(k), i = 1, n\}$. It can be seen this is somewhat like the 'step length' in rule updating. A larger a_s will result in a higher self-learning rate; however, the model will be more sensitive to noise.

In practical application, $\{a_s, s = 1, r\}$ are determined by the following two factors:

- the amplitude of $|e(k)|$: in general a larger $|e(k)|$ should correspond to a larger a_s, and if $e(k) = 0$, then a_s should be equal to 0;
- the relative contributions of each of the r rules to $y^p(k)$, namely a rule which provides more contributions to $y^p(k)$ should undergo more modification (a larger a_s).

4.4.7 Recursive fuzzy estimator

The fuzzy relational model developed can also be applied to the real-time state estimation. The applications can be grouped into two categories: 'system output prediction' and 'state estimation'. It is well known that time delays are a major concern when we design a control system, since delayed feedback signals will cause poor performance or even an unstable response. The function of a fuzzy estimator can provide the prediction of the system output signals without time delay. A recursive fuzzy estimator has been proposed by Shaw and Kruger (1992) based on the conditional probabilities of an output, the combined probabilities of all inputs and internal states and the moving window type fuzzy identifier. As a result, the I/O formulation of the compositional rule of inference can be governed by a set of Nth order state transition equations:

$$\begin{aligned} x_{1k} &= (u_{1k} \circ \cdots \circ u_{Mk}) \circ (x_{1,k-1} \circ \cdots \circ x_{N,k-1}) \circ R \\ x_{2k} &= x_{1,k-1} \\ &\vdots \\ x_{Nk} &= x_{N-1,k-1} \end{aligned} \tag{4.50}$$

The output equation is:

$$y_k = x_{1k} \tag{4.51}$$

The input terms are:

$$u_{mk} = u_{mk-\tau_{um}}, \quad m = 1, M \tag{4.52}$$

The estimated fuzzy state vector may be written in a conventional form: $[x_{1k}, \ldots, x_{Nk}]^T$. Suppose $k = L, \ldots, Q$; since the fuzzy relation R represents the model's past experience during the first $L-1$ samples which is used in estimating every new value starting with sample L, the fuzzy identifier is actually a batch one whereas the estimator is a recursive one. The recursive fuzzy state estimator

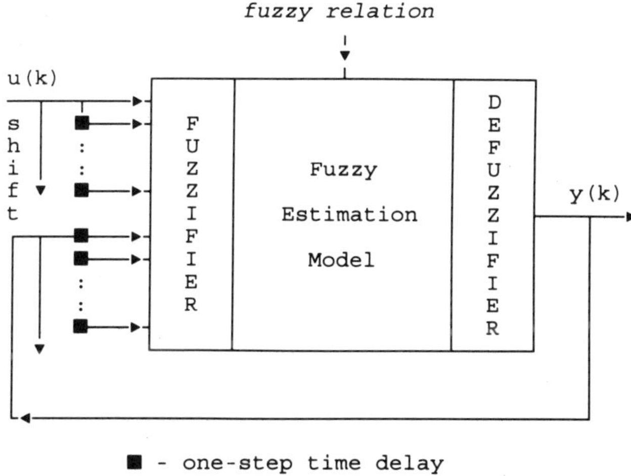

Figure 4.6 Recursive fuzzy state estimation algorithm

algorithm corresponding to the state space equations 4.50–4.52 is illustrated in Figure 4.6. It can be seen that the predicted results of the output are recursively provided in terms of the previous observations. In addition, the fuzzy relation R can also be recursively upgraded, thus an adaptive recursive estimator can then be performed.

Similar to an optimal state estimator as we described before, the fuzzy estimator can also be designed to estimate the internal states which are not available for sensory measurements. In order to do so, a fuzzy model describing the fuzzy relation between the system input, output and state needs to be established based on the off-line batch data; then a similar recursive fuzzy state estimator can be used to reconstruct the system state. However, in this case, the self-learning signals will be related to the error between the predicted states and the actual states which may be obtained off-line on a long-term basis. As a result, the update of the fuzzy model R then takes place less as frequently then as the output prediction we just introduced.

4.4.8 Numerical examples

The following numerical examples (Xu and Lu, 1987) illustrate how to apply the proposed fuzzy identification algorithms and the relevant results.

Example 4.2

This example is of fuzzy identification for a bilinear dynamic system. Let a bilinear system be governed by the following difference equation:

$$y(k) = 0.8y(k-1) + 0.5u_1(k-1)y(k-2) + u_2(k-4) + ae(k) \tag{4.53}$$

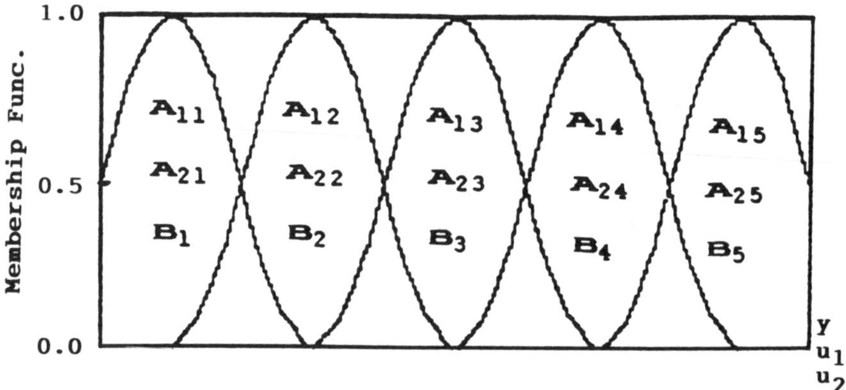

Figure 4.7 Fuzzy membership of fuzzy referential sets (Reproduced from Xu and Lu, 1987, © 1987 IEEE)

which is used as a simulated plant to provide the I/O data sequences as follows:

data 1 $(a = 0) : \{y(k), u_1(k), u_2(k), k = 1, 400\}$
data 2 $(a = 1) : \{y(k), u_1(k), u_2(k), k = 1, 400\}$

In the simulated model 4.53, $e(t)$ is an uncorrelated random noise with a uniform distribution on $[-0.08, 0.08]$. Inputs $u_1(k)$ and $u_2(k)$ are both uncorrelated random sequences uniformly distributed on [0.1, 0.9]. The fuzzy model identification procedures can be expressed as follows:

(a) Determine the universes Y, U_1 and U_2.

(b) Determine the referential fuzzy sets. Here we select $r = 5$. The memberships of the referential fuzzy sets, $\{A_{11}, \ldots, A_{15}, A_{21}, \ldots, A_{25}, B_1, \ldots, B_5\}$, are shown in Figure 4.7, and these memberships are all convex, normal and complete.

(c) Suppose that the rule has the following structure:

$$(y(k-\tau), u_1(k-\tau_1), u_2(k-\tau_2)) \Rightarrow y(k) \tag{4.54}$$

where the time delays, τ, τ_1, τ_2, will be determined later.

(d) Determine a performance index

$$J_{1,2} = \frac{1}{390} \sum_{k=11}^{400} [y(k) - y^p(k)]^2 \tag{4.55}$$

where J_1 and J_2 correspond to data 1 and data 2 respectively.

The time delays τ, τ_1, τ_2, composition and defuzzification methods will be determined by minimizing J_1 or J_2. It may be intuitively seen that the effect of

Table 4.1 Max–product composition and deF_2 (Reproduced from Xu and Lu, 1987, © 1987 IEEE)

τ	1	1	1	1	1	1	1	1	1	1	1	1
τ_1	0	0	0	0	0	0	0	0	0	0	0	0
τ_2	2	3	4	5	6	4	4	4	4	4	4	4
$J_1 \times 10^{-2}$	8.53	7.98	3.31	8.43	8.27	9.27	11.29	12.86	8.55	9.32	8.58	8.75
$J_2 \times 10^{-2}$	8.92	8.33	3.69	8.58	8.18	9.89	12.49	13.13	8.93	9.70	9.22	9.26

Table 4.2 Max–min composition and deF_2 (Reproduced from Xu and Lu, 1987, © 1987 IEEE)

τ	1	1	1	1	1	1	1	1	1	2	3	4
τ_1	0	0	0	0	0	1	2	3	4	0	0	0
τ_2	2	3	4	5	6	4	4	4	4	4	4	4
$J_1 \times 10^{-2}$	9.32	9.25	3.93	9.18	8.66	8.63	9.88	9.86	10.2	9.47	12.1	12.7
$J_2 \times 10^{-2}$	9.34	9.38	4.29	9.33	9.00	9.02	10.7	10.5	10.2	10.4	12.4	13.1

Table 4.3 $\tau = 1,\ \tau_1 = 0,\ \tau_2 = 4$

Composition		Defuzzification deF$_1$	deF$_2$	deF$_3$
Max–min	$J_1 \times 10^{-2}$	6.63	3.93	4.42
	$J_2 \times 10^{-2}$	6.52	4.29	4.68
Max–product	$J_1 \times 10^{-2}$	7.80	3.31	3.28
	$J_2 \times 10^{-2}$	8.21	3.69	3.64

delays τ, τ_1, τ_2, on J_1 or J_2 is not related to the composition and selected defuzzification methods. To verify this, let us look for τ, τ_1, τ_2, that minimize J_1 (J_2) under two different combinations of composition and defuzzification methods. The results are shown in Tables 4.1 and 4.2, from which it can be seen that when $\tau = 1, \tau_1 = 0$ and $\tau_2 = 4$, J_1 (J_2) are minimum. This result shows that in defuzzification we can first choose appropriate time delays by minimizing J_1 (J_2) using any composition and defuzzification methods, then fix the delays to their best value and minimize J_1 (J_2) in terms of the composition and the defuzzification methods. The effect of the composition and the defuzzification methods on J_1 (J_2) is shown in Table 4.3; here the best choices are 'max–product' composition and deF$_1$.

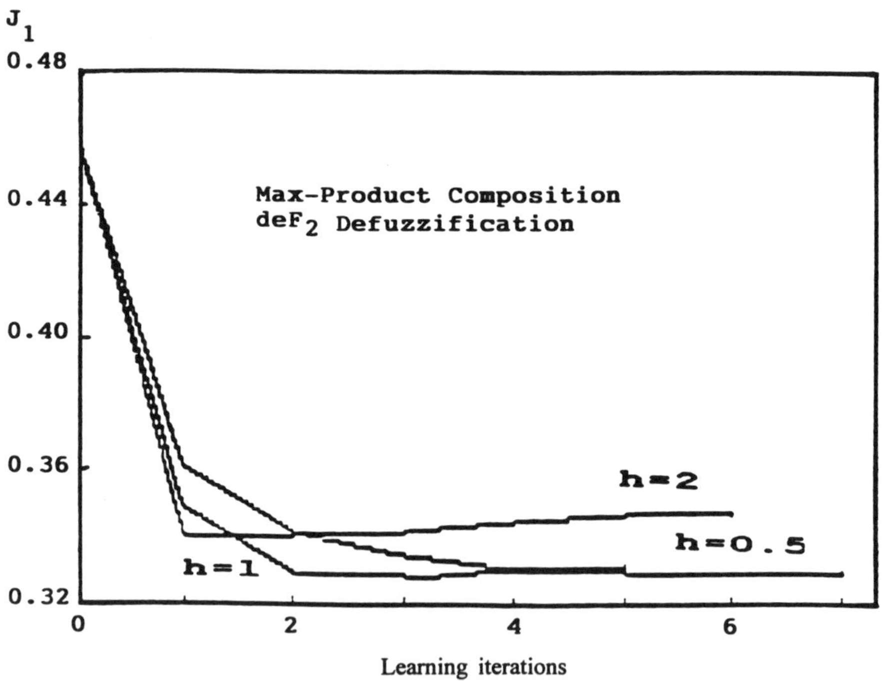

Figure 4.8 Sample simulation results showing effect of learning step (Reproduced from Xu and Lu, 1987, © 1987 IEEE)

In this example, we examine not only the effect of learning step η on J_1 (J_2), but also the effects of different composition and defuzzification methods on J_1. The defuzzification methods, denoted by deF_1, deF_2, deF_3, are maximum membership, mean-area and fuzzy mean methods respectively. The sample simulation results showing the effect of the learning step are given in Figure 4.8. We can also clearly see that deF_1 provides the worst performance and the combination of deF_3 and the 'max–product' composition provides a satisfactory performance even for data 2 with noise.

Example 4.3

This example concerns fuzzy identification for a gas furnace. In this example, 296 I/O data pairs are gathered from a gas furnace (Box and Jenkins, 1970). The input $u(k)$ is the inlet methane rate and output $y(k)$ is CO_2 concentration in the outlet gas. The data are written in a data sequence $\{y(k), u(k), k = 1, 296\}$. The fuzzy identification procedures are as follows:

(a) Determine the universes of discourse U and Y: $y(k) \in F(Y)$ and $u(k) \in F(U)$, by checking the ranges of I/O data (both mean values and standard deviation).

(b) Determine all referential fuzzy sets. Let $r = 5$, and the referential fuzzy sets $\{A_i, B_i, i = 1, 5\}$ have their memberships similar to those in Figure 4.7.

(c) Assume the structure of the rules is

$$\{y(k - \tau_1), u(k - \tau_2)\} \Rightarrow y(k)$$

where τ_1, τ_2 are to be determined.

(d) The performance index is defined as

$$J = \frac{1}{286} \sum_{k=11}^{296} [y(k) - y^p(k)]^2 \tag{4.56}$$

The task of the fuzzy identification is to determine the structure parameters τ_1, τ_2, and the fuzzy relation R. The relations between the resulting performance and τ_1, τ_2 are given in Table 4.4. It can be seen from the table that J reaches its minimum

Table 4.4 Max–min composition and deF_3 (Reproduced from Xu and Lu, 1987, © 1987 IEEE)

τ_1	τ_2				
	2	3	4	5	6
1	1.5221	1.3598	1.0689	1.4727	1.7702
2		2.1368	1.4860	1.5277	
3		3.0435	1.9839	1.8860	
4			2.5997		

Table 4.5 Combination of the composition and defuzzification (Reproduced from Xu and Lu, 1987, © 1987 IEEE)

Composition defuzzification	max–min deF_3	max–min deF_2	max–product deF_3	max–product deF_2
J	1.0689	0.6018	0.8501	0.4555

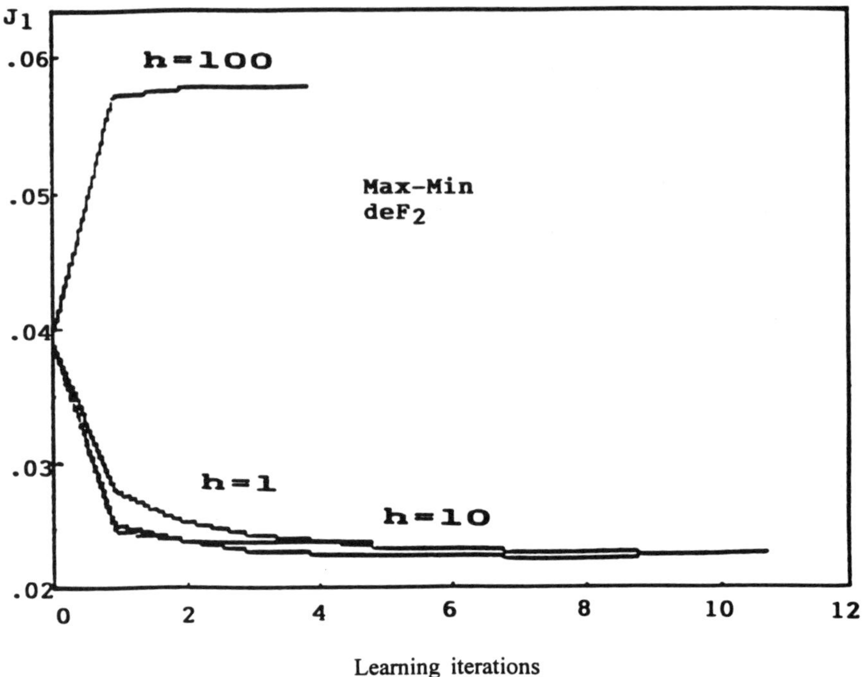

Figure 4.9 Convergent response of J with self-learning (Reproduced from Xu and Lu, 1987, © 1987 IEEE)

when $\tau_1 = 1$ and $\tau_2 = 4$. The combination of the different composition and defuzzification methods is shown in Table 4.5. It can be seen that the minimal J corresponds to the 'max–product' and deF_2. The convergent response of J with self-learning given in Figure 4.9 shows that the learning step length $\lambda = 1$ provides the best result.

4.5 APPLICATION OF FUZZY MODELING IN PREDICTION OF PRODUCT DISTRIBUTION OF FCCU

4.5.1 Problem statement

The fluidized catalytic cracking unit (FCCU) is a typical complex system with highly nonlinear, strong coupling, time-varying and significantly uncertain properties.

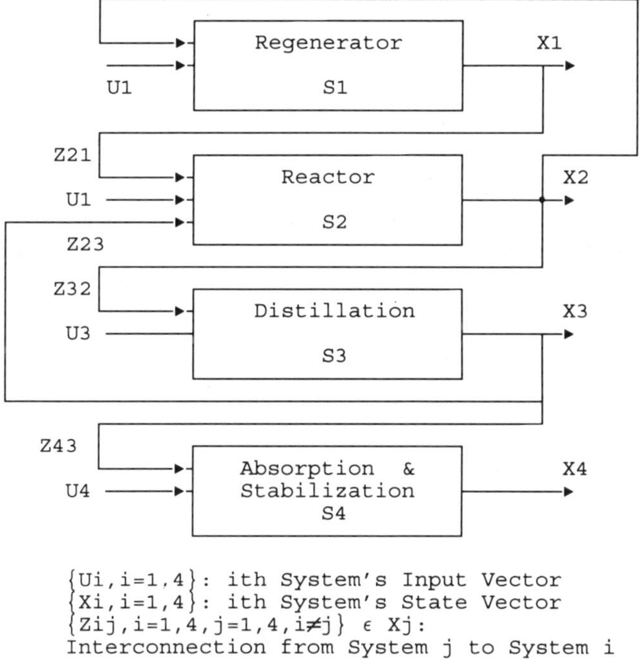

Figure 4.10 Illustration of FCCU process

During the last few decades, many efforts have been made to develop a dynamic model which can be used in advanced optimization control for an FCCU (Bromley, 1981; Brice *et al.*, 1983, etc.). However, since most systems developed are mainly based on the process principle and math-model based parameter estimation techniques, it is difficult to develop a simplified, real-time control model for practical applications.

In general, an FCCU can be divided into reactor, regenerator, distillation and absorption/stabilization: four interconnected subsystems as shown in Figure 4.10. The cracking product distribution which describes the cracking reaction degree is not available for sensory measurement at the outlet of the reactor; instead, it can only be measured by the flow rate sensors after the distillation column. The pure time delay L from the reactor to the distillation column depends on FCCU operating conditions, particularly the production throughput and system pressure. For this specific FCCU under study, the L is within 15–20 minutes. Obviously, if the optimization control actions are based on product distribution information with such a long time delay, it must be too late to control the cracking reaction conditions. This is why most of the existing FCCU optimization control is still based on the reaction temperature and an off-line mathematical model.

A fuzzy relational model with self-learning to predict the FCCU cracking product profile has been developed (Lu and He, 1991). The fuzzy model based estimator can provide the quantitative relations between the operating conditions and the rates of various cracking oil/gas products, such as gas, gasoline, light diesel, etc., from the cracking reactor.

4.5.2 Fuzzy relational model with self-learning

Based on the analysis of the FCCU principle, the operator's experience and production data, a large number of fuzzy rules can be proposed to describe the fuzzy relations between the reaction conversion rate and the operating conditions in an FCCU. To simplify the complicated representation, the fuzzy rules can be simply written in the following form:

$$C = \mathcal{F}(u_1, u_2, u_3, u_4) \tag{4.57}$$

where

$\mathcal{F}(*) =$ a fuzzy operator describing the fuzzy rule base;
$C =$ conversion rate;
$u_1 =$ reaction temperature;
$u_2 =$ reactor pressure;
$u_3 =$ recycling oil rate;
$u_4 =$ ratio between the catalyst weight and flow rate of the oil to the reactor.

We have learned that to obtain a numerical solution, a fuzzy rule base can be cast into a corresponding fuzzy relational model (or so-called 'fuzzy associative memory'). As a result, the structured knowledge can then be encoded to the relevant unstructured knowledge/numerical frame with the ability to deal with the numerical operations. Consequently, the resulting fuzzy relational model describing the numerical relations between the cracking reaction product distribution (replacing conversion rate) and the operating conditions can be expressed in the following form:

$$y_j(k) = u_1(k) \circ u_2(k) \circ u_3(k) \circ u_4(k) \circ R_j(k), \quad j = 1, 4 \tag{4.58}$$

where

$y_1 =$ gasoline rate;
$y_2 =$ light diesel rate;
$y_3 =$ cracking gas rate;
$y_4 =$ heavy diesel rate.

The initial fuzzy relations $\{R_i(0), i = 1, 4\}$ are determined by using a fuzzy identification technique as introduced in section 4.4 based on the off-line batch data $\{x_i(k), y_i^d(k - L), i = 1, 4, k = 1, K_f\}$. y_i^d denotes the measurements of the product rates at the outlet of the distillation column. Based on the initial fuzzy relations $\{R_i(0), i = 1, 4\}$ and the I/O observations, the real-time fuzzy relations $\{R_i(k), i = 1, 4\}$ can be solved through the recursive fuzzy identification algorithms, namely by upgrading the relevant fuzzy rules. As a result of applying this fuzzy relational model, the predicted product distribution can be provided based on the real-time fuzzy relations and input measurements. The measurements of

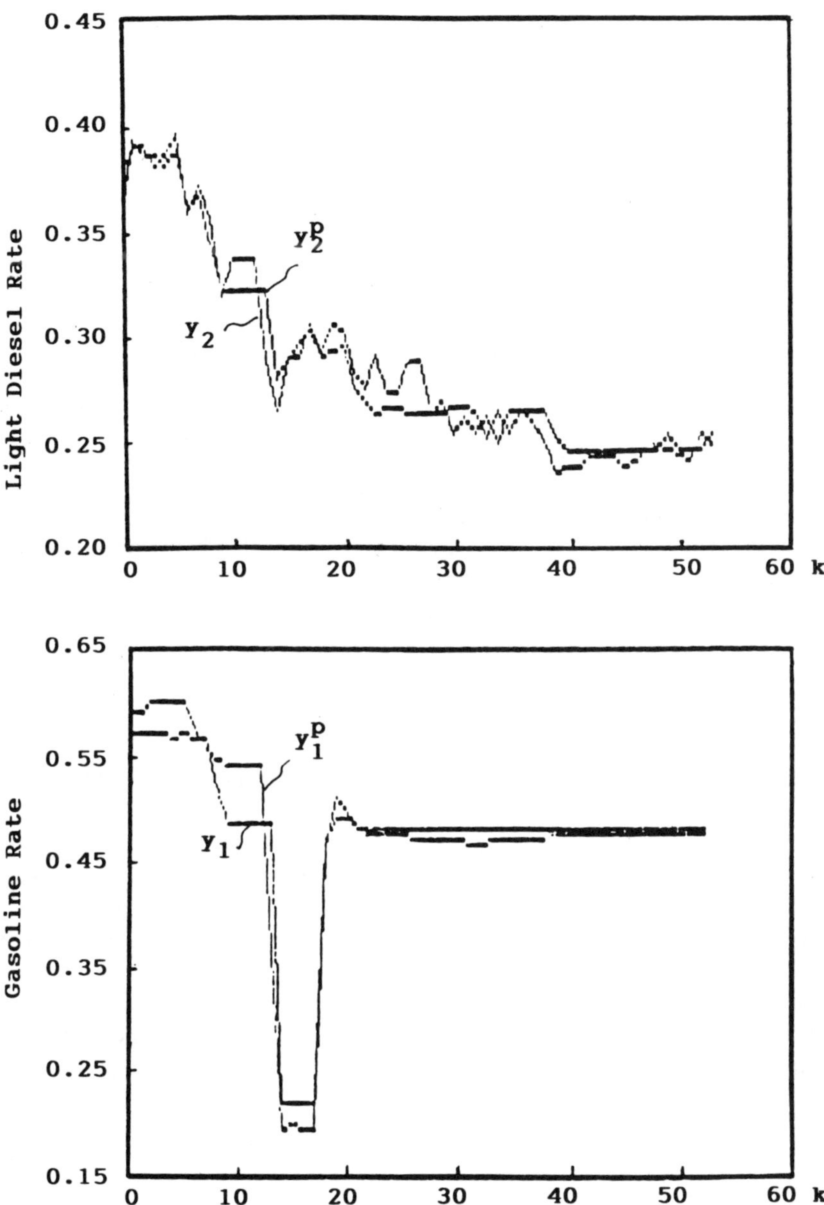

Figure 4.11 Sample comparison between actual and predicted rates of gasoline and light diesel

the product distribution at the outlet of the distillation column, $\{y_i^d(k-L), i=1,4\}$, are used as a feedback signal for self-learning. The function of the self-learning is to make the resulting estimator adapt to changes of FCCU environment.

Finally, it should be noted that the data preprocessing, fuzzification and defuzzification as we described before are certainly required in practical applications, which will not be repeated here.

4.5.3 Industrial application results

The sample comparison between the actual and the predicted rates of gasoline and light diesel are shown in Figure 4.11. We can see that the predicted errors for both gasoline and light diesel are small enough to satisfy the requirement for real-time optimization control purpose. As a result of applying this fuzzy estimator, the unmeasurable internal information can be reconstructed to serve as the sources for systems control and optimization. Due to introducing the self-learning algorithm in this estimator, the predicted results are robust and adaptive; even the production environment significantly changes, for instance, the changes of the cracking oil properties or production load.

This fuzzy model has also been applied to an FCCU expert optimization control system in which the model not only serves as an estimator but also serves as one of the key bodies in reasoning and coordination. A good performance has been provided through the increase of the rate of light oil, such as gasoline and light diesel, and optimization of the operations. The applications of the fuzzy model in FCCU expert optimization control will be introduced in detail in Section 6.3.2.

4.6 PATTERN RECOGNITION MODELING AND ESTIMATION

Since the astonishing progress on neural network technology, the potential applications of pattern recognition in dynamic system modeling and estimation techniques have been one of the most attractive techniques in industrial control. Similar to fuzzy modeling and estimation as discussed in section 4.5, pattern recognition based estimation is also a math-model free estimation technique, which is particularly applicable to a complex industrial system with rich data information, but less prior system structured knowledge. This section presents the applications of the statistical fuzzy pattern classifier (SFPC) (Zhou *et al.*, 1989) in system modeling and estimation.

4.6.1 Problem statement

Suppose we have a nonlinear MISO dynamic system governed by the following input and output pattern mapping:

$$y(k+1) = \Psi(y(k), U(k)) \tag{4.59}$$

where

$\Psi(*) =$ an unknown nonlinear stochastic operator;
$U \in R^r =$ system input vector;
$y =$ system output.

The problem of pattern recognition modeling and estimation under consideration is to provide an estimate of the real-time output, $y^p(k)$, in terms of the observations of the system inputs $\{U(k), k = k - p, k\}$, and the infrequent output measurements at the specific discrete time points, $y(mk)$, in which m is a large number. The observations of $y(mk)$ are used as a self-learning feedback signal. When applying pattern recognition techniques in solving a numerical functional estimation problem, we will find how to design a classifier with a reasonable number of classes, and satisfactory accuracy is one of the key issues in real application. Here we introduce the design of a pattern recognition estimator based on SFPC and their applications in industrial systems.

4.6.2 Architecture and algorithm

Similar to the traditional and fuzzy estimators, a generic structure of a pattern recognition estimator can be schematically shown in Figure 4.12(a). It can be seen that a pattern recognition based estimator consists of three major components, namely data preprocessing, a pattern classifier model and a learning mechanism.

We realize that the pattern recognition techniques are related to the data quality being used to establish the sample patterns, so that how to select the sample data has been a key issue in developing a pattern recognition model. We have investigated that the major concern in data preprocessing is that the data should contain rich information, in other words the data should well represent system features, have a reasonable density and less noise corruption. In general, data filtering, normalization and feature extraction as introduced in the previous chapters are the main procedures in data processing to make the normalized data uniformly distributed in an input space.

The traditional nearest neighbor classification technique is one of the most popular methods in pattern classification, since it does not require prior statistical information to perform the classification tasks. However, when applying this method to a dynamic system, to give the resulting classifier high accuracy of adaptation and make its class based output not too discrete, a high number of classes is usually required, such that the computer memory and CPU time will significantly increase. To solve these problems, a statistical fuzzy pattern classifier (SFPC) with multi-center pattern description is developed for dynamic system pattern recognition estimation.

Fuzzy membership function

There are a number of methods of choosing the fuzzy membership. Here we will introduce a method where by the membership is heuristically determined based

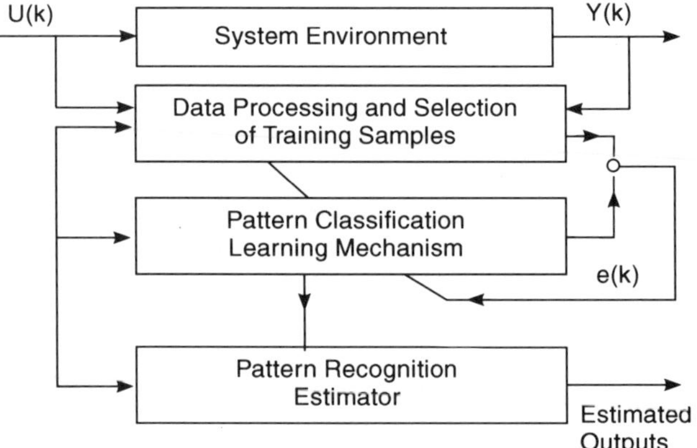

(a) Structure of pattern recognition estimator

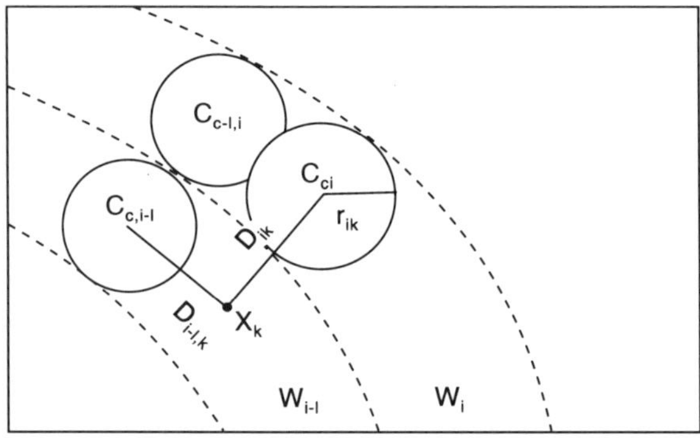

(b) Geometrical illustration of the membership function

Figure 4.12 Structure of pattern recognition of estimation

on the geometrical properties of the data set. Let the system output y be divided into N subjects, i.e. N classes denoted by $\{W_i, i = 1, N\}$, where each class may be represented by N kernels, $\{C_i, i = 1, N\}$. The fuzzy membership function for the event X, $\{A_{ij}(X), i = 1, N; k = 1, C_i\}$, can be estimated by its effective distance to the kernels, D_{ik}, being an efficient measure for membership function classification

$$D_{ik} = g(D_{ik}/r_{ik}), \quad i = 1, N; \;\; k = 1, C_i \tag{4.60}$$

where D_{ik} is the Euclidean distance between X and X_{ik} which is defined as

$$D_{ik} = d(X, X_{ik}) \tag{4.61}$$

and r_{ik} is the efficient range of the kth kernel and $g(*)$ is a nonlinear operator. Then the fuzzy membership function can be given as follows:

$$A_{ik}(X) = f(D_{ik}), \quad i = 1, N; \;\; k = 1, C_i \tag{4.62}$$

Let $m = \sum_{i=1}^{N} C_i$ be the total number of kernels. Without loss of generality, we denote ik as i $(i = 1, m)$, then $A_{ik}(X)$ can be replaced by $A_i(X)$. To make the range of $A_i(X)$ [0, 1], some additional conditions are needed:

- the membership function $A_i(X)$ is inversely proportional to the efficient distance D_i, namely

$$A_i(X) = \frac{A_i}{D_j}\frac{A_j}{D_i}, \quad 1 < i, j < m \tag{4.63}$$

  This condition means that the less the distance between the pattern and class, the higher the pattern's membership function;

- the normalization condition, namely

$$\sum_{i=1}^{m} A_i(X) = 1 \tag{4.64}$$

Using conditions 4.63 and 4.64 yields

$$A_i(X) = \prod_{i=1}^{m} D_i \Big/ \left(\prod_{i=1, i\neq 1}^{m} D_i + \prod_{i=1, i\neq 2}^{m} D_i + \cdots + \prod_{i=1, i\neq m}^{m} D_i \right) \tag{4.65}$$

Let $m = 2$, then we have

$$\begin{aligned} A_1(X) &= D_2/(D_1 + D_2) \\ A_2(X) &= D_1/(D_1 + D_2) \end{aligned} \tag{4.66}$$

The lower value of m can be used as an approximate expression of $A_i(X)$ in the real-time computation. The geometrical illustration of the membership function is given in Figure 4.12(b).

Finally, it should be pointed out that the membership function is related to the concrete geometric position of the classes in the distribution of the given set. In other words, A_i and A_j are functions of both D_i and D_j. In addition, A_i is independent of the other number of X in the data set, thus the classification of patterns will be stable.

Multi-center pattern description

In pattern description, the key problem is how to describe the real pattern space more simply and efficiently. Here we introduce an approach to a multi-center pattern description for a dynamic system based on the method of an optimal performance index classifier (Gu, 1982). Suppose the output y is divided into N

classes $\{W_i, i = 1, N\}$; the detailed procedures are as follows:

(1) Get the event set X_k that belongs to W_k from the data set, and the initial center can be produced by using the following method:

- calculate the mean vector and mean distance of the observed events in the event set, namely,

$$XM = \frac{1}{N_k} \sum_{i=1}^{N_k} X_{ki} \tag{4.67}$$

$$XD = \frac{1}{N_k} \sum_{i=1}^{N_k} \|X_{ki} - XM\| \tag{4.68}$$

where

XM = mean value vector of X_k;
XD = mean distance between X_i and XM;
X_{ki} = ith pattern feature vector;
N_k = number of events in X_k;

- find the first initial center corresponding to the event being the nearest to the total mean value vector. Suppose the distance between X_{ki} and XM is defined by

$$DM_i = \|X_{ki} - XM\| \tag{4.69}$$

the distance XD of the patterns;

- based on the least distance clustering, all observed events are classified to the clusters with the least distance between the clustering center and the event pattern. Consider C_c^N is the center of cluster c and the observed events of cluster c are X_{kc}, then

$$C_c^N = \frac{1}{N_c} \sum_{c=1}^{N_c} X_{kc} \tag{4.70}$$

where N_c is the observed event number in cluster c. If C_c^N is not equal to C_c, then consider C_c^N as a new cluster center and repeat the previous procedures. After the classification, the new clustering centers, $\{C_i^N, i = 1, N\}$, are established.

(2) Calculate the performance index M:

$$M = \frac{1}{N_c(N_c - 1)} \sum_{i=1}^{N_c} \left(\frac{N_i}{D_i^2 + D_i} \sum_{j=1}^{N_c} D_{ij} \right), \quad j \neq i \tag{4.71}$$

where

$$D_{ij} = \|C_i^N - C_j^N\|, \quad i = 1, N_c - 1, \; j = i + 1, N_c \tag{4.72}$$

$$D_i = \frac{1}{N_i} \sum_{i=1}^{N} \|X_{ji} - C_i^N\|, \quad j = 1, N_c \tag{4.73}$$

where

D_{ij} = distance between cluster i and j;
D_i = mean distance;
N_i = number of patterns which belong to center C_i^N;
X_{ij} = sample pattern which belongs to center C_i^N.

(3) Successively decrease the desired distance between the clusters to change N_c and repeat (1)–(2); denote

$$M_{\max} = \max\{M^N, M^{N-1}\} \tag{4.74}$$

(4) When the maximum number of N_c are reached, then end the iterative calculation and the clustering results are provided corresponding to the maximum M. When applying this method to a dynamic system, the system estimated output can be calculated as follows:

$$y^p(k) = \sum_{c=1, N_c, i=1, N} A_{ci}(X) y(c, i) \tag{4.75}$$

where $A_{ci}(X)$ is the fuzzy membership function and $y(C, i)$ is defined as the corresponding output of every C_i, being the mean output of the observed events that belong to W_i.

4.6.3 Industrial application – real time quality estimation and control

Problem statement

The vapor pressure is one of the most important quality criteria of the gasoline finally produced from the stabilization tower in an FCCU. However, since the vapor pressure of the gasoline cannot be measured on-line, the operator controls the vapor pressure based on the infrequent analysis data obtained from the quality control laboratory. Obviously the gasoline quality cannot be effectively controlled based on such long-term, infrequent quality information. Due to the complexities, such as nonlinear, time-varying and particularly compositional uncertainties of the raw gasoline entering the stabilization tower, it is difficult to develop an estimator based on a precise mathematical model. The problem under study is to design a math-model free state estimator which can provide on-line estimates of the unmeasurable vapor pressure of the gasoline based on the relevant measurable inputs and the infrequent, off-line measurements of the vapor pressure.

Here we will introduce an application of a pattern recognition estimator in on-line estimation of the gasoline vapor pressure in a large production scale stabilization tower in an FCCU. Based on the process principle and operation data analysis, the functional relations between system inputs and output can be expressed as follows:

$$y(k) = P(u_1(k), u_2(k), u_3(k-L)) \tag{4.76}$$

where

$P(*)$ = a nonlinear stochastic pattern mapping operator;
y = gasoline vapor pressure;
u_1 = tower bottom pressure;
u_2 = temperature of gas phase of reboiler;
L = time delay;
u_3 = heat content factor, which can be written as

$$u_3 = \frac{F_g(T_2 - T_1 + C_1)}{C_2 F_1} \tag{4.77}$$

where C_1 and C_2 are adjustable constants and F_1 and F_g are the inlet flow rate and finishing gasoline flow rate respectively.

Synthesis of pattern recognition estimator

The initial sample patterns were established based on the off-line batch I/O data pairs, $\{u_1^j(k), u_2^j(k), u_3^j(k-L), y^j(k), j = 1, K\}$. To reduce the corrupting noise in the output data, a data filtering approach in conjunction with feature space is used here. The feature space can be established based either on the input or on the output domain. As a result, in each subspace there are a number of data pairs, for instance in subspace j:

$$\{x_1^j(k), x_2^j(k), x_3^j(k-L), y^j(k), k = K_0, K_1\} \tag{4.78}$$

The average weighted recursive filtering approach can be used to provide the sample patterns, $\{x_1^j, x_2^j, x_3^j, y^j\}$, which describe the pattern mapping from a numerical input space to an output space with the classification. Based on the sample pattern established, the multi-center representation of pattern class can be further established with the algorithms introduced in the previous section. Finally, the real-time vapor pressure of the gasoline can be estimated in terms of the statistical fuzzy pattern classification and a multi-center representation of all pattern classes.

In addition, to adapt to the changes of production environment and compensate for the errors of the initial patterns, the initial sample patterns are modified on-line by using the learning algorithm with minimum performance index:

$$J = \frac{1}{2N} \sum_{k=k-N}^{k} (y(mk) - y^p(mk))^2 \tag{4.79}$$

where $y(mk)$ is the infrequent observations of the gasoline vapor pressure provided from the quality control laboratory.

4.6.4 Industrial application results

The pattern recognition estimator as developed has been successfully applied in a large production scale stabilizing tower to provide real-time estimates of the gasoline vapor pressure being used as a feedback signal in a cascade control system. The comparative results between the estimated and the actual vapor pressure are given in Figure 4.13. It can be seen that the estimated errors are satisfied for the purpose of a real-time application. In addition, the comparison of the real-time response between the conventional control and the state feedback with a pattern recognition estimator is shown in Figure 4.14, which shows that the errors between the set point and the actual vapor pressure are significantly reduced through using a real-time pattern recognition estimator in a vapor pressure control system. As a result, the gasoline quality may be improved by using this math-model free estimation technology.

Finally, it should be pointed out that the pattern recognition estimation modeling and techniques can be applied to many other industrial processes, particularly for

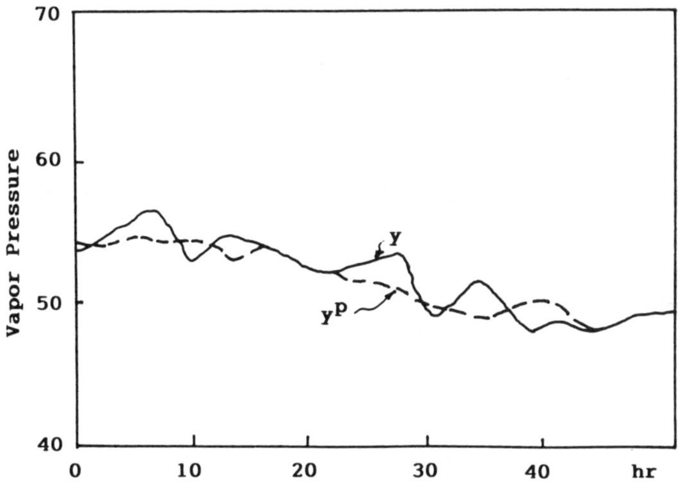

Figure 4.13 Estimation results of vapor pressure given by pattern recognition state estimation

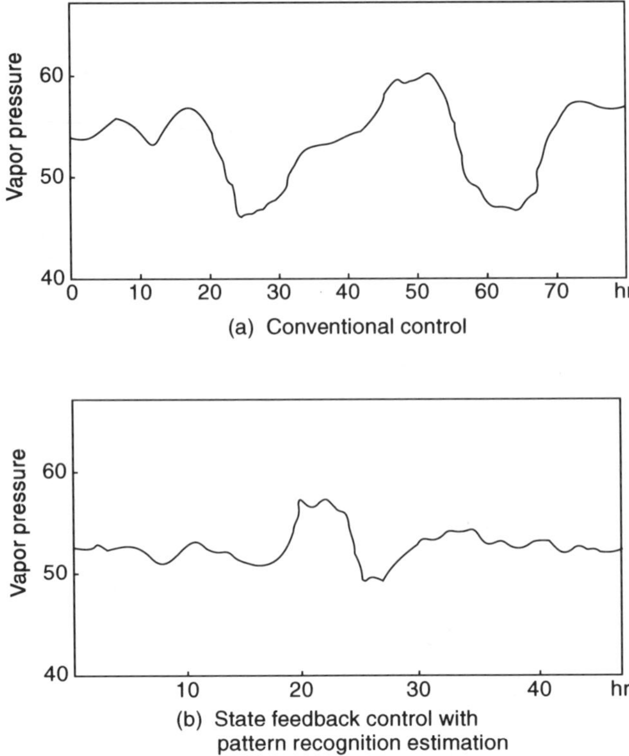

Figure 4.14 Closed loop responses of conventional and state feedback control with pattern recognition estimation

quality prediction and control. The pace of adaptation to system environment change depends on the frequency (corresponding to 'm' in 4.79) of receiving the actual measurements being used as a learning feedback signal from the quality control laboratory. In other words, the less the time interval (m), the more effective the learning activity.

4.7 NEURAL NETWORK MODELING AND ESTIMATION

We discussed in Chapters 2 and 3 that a multi-layer feedforward neural network can be used as an associative memory to perform a functional approximation through learning from samples. In comparison with other math-model free estimation techniques, the application of neural networks in estimation systems can provide quantitative numerical solutions through encoding and decoding the information/knowledge with the structured or the unstructured form. Similar to the estimation techniques introduced in the previous sections, a neural network can also serve as a math-model free estimator with the functions of filtering, estimation and prediction. In this section we focus on applications of neural networks in system modeling and estimation.

4.7.1 System modeling and estimation for static systems

Suppose we have a nonlinear steady state system governed by the following projection from its input space to output space:

$$Y_k = F(U_k) \tag{4.80}$$

where

$Y \in R^n$ = output vector;
$U \in R^r$ = input vector;
$F(*)$ = the pattern mapping operator illustrating the uncertain relations between the system inputs and outputs.

The task of neural network modeling is to establish a neural model, $\mathcal{N}: U \Rightarrow Y$, that can approximate the nonlinear function $F(*)$ just in terms of the observations of the system I/O data $\{U_k, Y_k, k = 1, K\}$, without knowing the system structure $F(*)$. In addition, the task of a neural estimator is to estimate the system outputs with the system inputs to minimize the error between the actual outputs Y and the estimated outputs Y^p, namely

$$\min \sum_{k-K}^{k} (Y_k^p - Y_k)^T (Y_k^p - Y_k) \tag{4.81}$$

To develop a neural network model and the relevant estimator, in general, the following architectures can be used.

Off-line neural network training

Suppose a set of system normalized I/O training patterns are represented as $\{U_k, Y_k, k = 1, K\}$ which serve as the neural network's inputs and the desired

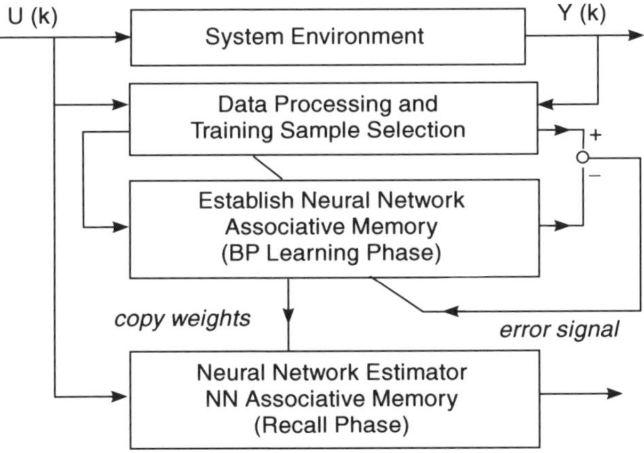

Figure 4.15 Illustration of neural network estimation

outputs, being the teaching signals respectively. A multilayered neural network produces its output patterns $\{Y_k^p, k = 1, K\}$ under its input of $\{U_k, k = 1, K\}$. The connection weights in the neural network are iteratively adjusted based on the error between the desired outputs Y_k, and neural network outputs, Y_k^p, with a BP-learning algorithm (pattern-wise or batch-wise). After a large number of pattern presentations and learning iterations, the error gradually approximates to zero when the system information or knowledge is distributed, stored in the neural network weights as an associative memory. Then the frozen weights can be loaded to another neural network that has exactly the same topology and parameters as the neural network used in the training phase. The resulting neural network can then be used as an estimator as shown in Figure 4.15.

Adaptive training of a neural network model

Based on the concept of generalization when the neural network topology is determined, the estimation accuracy of a neural network depends on the difference between the training environment and the working environment. Moreover, while a neural estimator works under a variable system environment, adaptive training with the updating of the neural associative memory is needed. Instead of batch-wise training, an adaptive neural network, e.g. Adaline network, can be used.

Neural network inverse model

A neural network inverse model means establishing an inverse pattern mapping from system output space to input space, namely, $F^{-1} : Y_k \rightarrow U_k$. A neural network based inverse model with both training and recall can be illustrated in Figure 4.16. Eventually, the functions of a neural network inverse model are the same as an inverse math-model. The application of neural network inverse models involves model predictive control, adaptive control, and system decision making.

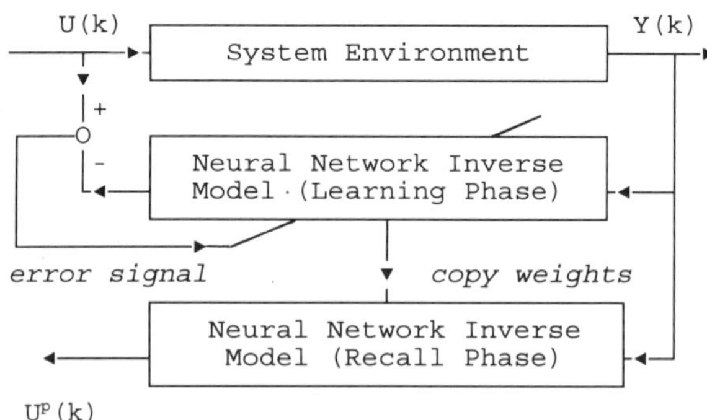

Figure 4.16 A neural network inverse model with training and recall

4.7.2 Application of dynamic systems

Architecture of neural network for dynamic modeling

When dealing with applications of neural networks in dynamic system modeling and estimation, neural network training is based not only on the current patterns, but also on the past patterns.

Many papers (Narendra and Parthasarathy, 1990; Qin *et al.*, 1992, Bhat and McAvoy, 1989) have been published to investigate neural network based dynamic system modeling. The research focuses on the structures and real-time learning algorithms of the neural networks to provide more accuracy and faster convergent results.

As noted by Narendra and Parthasarathy (1990), two major architectures, multilayered feedforward networks (FFN) and recurrent networks (RecN), have been used in solving pattern recognition problems and associative memories as well as optimization problems respectively. From a system point of view, a multilayered network represents a static nonlinear pattern mapping and, on the other hand, a recurrent network is represented by a nonlinear dynamic feedback system. Figure 4.17 (Qin *et al.*, 1992) shows the basic concepts of a regular multilayered neural network and recurrent network. It can be clearly seen that in a recurrent network

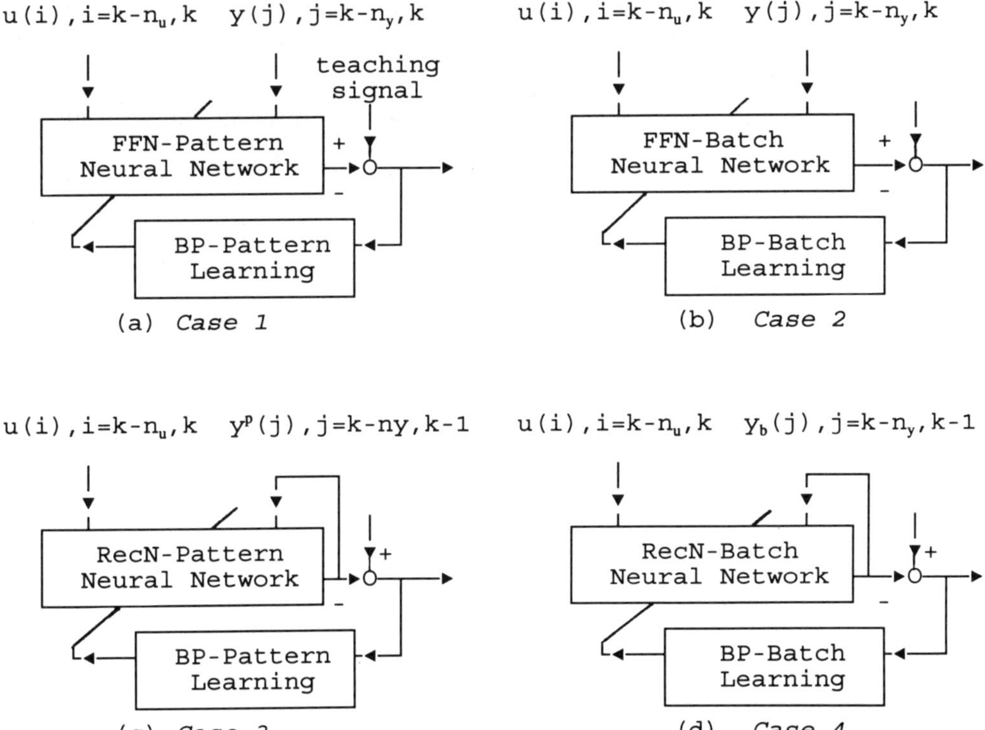

Figure 4.17 Basic concept of FFN and RecN (Reproduced from Qin *et al.*, 1992, © 1992 IEEE)

the network outputs are recurrently fed back as the neural network inputs. As a result, the recurrent learning rules are required to train a recurrent network.

Let us take the following SISO nonlinear autoregressive with exogenous inputs system as an example for the investigation of neural network applications in dynamic system modeling:

$$y(k) = F(y(k-1), \ldots, y(k-n_y), u(k), \ldots, u(k-n_u)) + v(k) \tag{4.82}$$

where

$F(*)$ = a valued nonlinear function;
u, y, v = system input, output and noise;
n_y, n_u = time delays representing the system orders for output and input respectively.

The overall I/O pattern mapping represented by a feedforward multilayered neural network can be written in the following generic form:

$$y^p(k) = \mathcal{N}(W, X) \tag{4.83}$$

where

$\mathcal{N}(*)$ = nonlinear pattern mapping or an associative memory;
W = total connected weight matrix from input layer to the output layer;
X = inputs to the network, consisting of observed system inputs and observed/estimated outputs.

The forward calculations of the four different cases are given as follows (Qin *et al.*, 1992):

<u>Case 1 (FFN-Pattern)</u>:

$$y^p(k) = \mathcal{N}(W(k), y(k-1), \ldots, y(k-n_y), u(k), \ldots, u(k-n_u)) \tag{4.84}$$

<u>Case 2 (FFN-Batch)</u>:

$$y_b(k) = \mathcal{N}(W(i), y(k-1), \ldots, y(k-n_y), u(k), \ldots, u(k-n_u)) \tag{4.85}$$

<u>Case 3 (RecN-Pattern)</u>:

$$y^p(k) = \mathcal{N}(W(k), y^p(k-1), \ldots, y^p(k-n_y), u(k), \ldots, u(k-n_u)) \tag{4.86}$$

<u>Case 4 (RecN-Batch)</u>:

$$y_b(k) = \mathcal{N}(W(i), y_b(k-1), \ldots, y_b(k-n_y), u(k), \ldots, u(k-n_u)) \tag{4.87}$$

where $y^p(k)$ and $y_b(k)$ represent the neural network outputs in the pattern-wise learning (PWL) and the batch-wise learning (BWL) respectively and i denotes the batch iteration instance. The weights $W(k)$ and $W(i)$ correspond to PWL and BWL respectively.

The following conclusions are useful in developing a neural network algorithm for real-time application:

- if the learning rate is slow enough, the pattern learning approximates the batch learning, and FFN-pattern learning is a first order approximation of FFN-batch learning;
- from an application point of view, since the learning rate can be controlled in practice, RecN-pattern learning with a small learning rate is much more simple and realistic in real-time application than RecN-batch learning;
- in comparison with FFN networks, the RecN is less sensitive to noise.

Applications in statistical prediction of digital signals

The neural network can also be applied to predict the future values of a time-correlated digital signal from the present and the past input samples. Wiener developed an optimal linear least squares filtering technique for the signal prediction (Kailath, 1981). Based on Winner's theory, an optimal prediction filter can be directly performed by using an adaptive neural network (filtering). The detailed structure of an adaptive neural network is shown in Figure 3.14.

Finally, it should be noted that the applications of neural networks in real-time state estimation can also be used in discovering unmeasurable, internal system information, if the data pairs of input, output and state variables are available for off-line training. The structure is similar to a regular Kalman filter, but its math-model is replaced by a neural network.

4.7.3 Numerical examples

Example 4.4

This example is an application of a neural network in establishing a common gas P–V–T state model. The gas state equation describing the quantitative relations between the pressure, volume and temperature (P–V–T) for a non-ideal gas is usually established with nonlinear aggregation based on the experimental laboratory data. However, it is still a difficult task to establish a gas state equation due to the lack of structure of the nonlinear state equation for a particular gas. This example presents an application of a neural network in establishing a P–V–T model based on the I/O data produced by a simulated plant.

Suppose that the Beattie–Bridgeman gas state equation expressed as follows serves as a simulated plant to produce the I/O data describing the quantitative relations (Franks, 1972):

$$V = F(T, P) = (RT + d/V + h/V^2 + q/V^3)/P \tag{4.88}$$

where

$d = RTB_o - A_o - RC/T^2$;
$h = -RTB_oB + AA_o - RB_oC/T^2$;
$q = RB_oBC/T^2$;
R = gas constant;
T = absolute temperature;
V = volume.

The values of constants A_o, B_o, A, B, C for isobutane are

$$A_o = 16.6037$$
$$B_o = 0.2354$$
$$A = 0.11171$$
$$B = 0.07697$$
$$C = 3 \times 10^6$$

A total of 400 data pairs $(V_i, T_i, P_i, i = 1, 400)$ produced by 4.88 serve as the training patterns for neural network training, and V and P both corrupted by random noise are used as inputs and T is defined as an output. A neural network

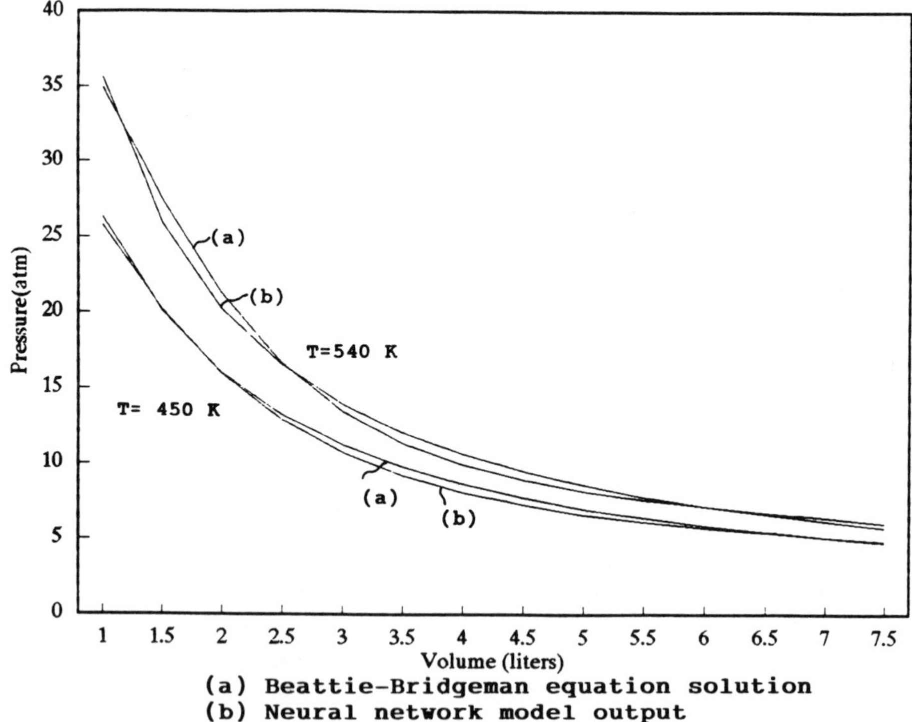

Figure 4.18 Numerical relations between P–V–T given by a neural gas state model

with two input nodes related to T and V, ten hidden nodes and one output node related to P is designed for P–V–T neural network modeling. The batch-wise conjugate gradient descent BP-learning with an adaptive learning rate is used in this example. After 30 000 batch presentations of the training patterns, the error between the neural network output and actual teaching signal is less than 0.04%, and the neural network weights are frozen.

The application of the resulting neural network gas state model with the well trained frozen weights can precisely provide the relations between the inputs (V and P) and output T as shown in Figure 4.18. It can be seen that the error for the working patterns being not trained is also small enough to be applied in practice. Similarly, in variables P, V and T, we can select any two as the inputs for neural network training; as a result, instead of a gas state equation, a neural network model can be readily established to describe the quantitative relations between P, V and T based just on the laboratory experimental data.

Example 4.5

This example concerns the application of a neural network in modeling a continuous optical strip centering system. Strip centering control is one of the most popular problems in a steel mill to wind a steel coil precisely. The key of the winding control system is an optical/digital position transducer which monitors the position of the steel strip to the mill centerline. The transducer uses a two-line scan camera and a microprocessor based system to derive a position feedback signal which is processed to activate a hydraulic actuator. The system under investigation

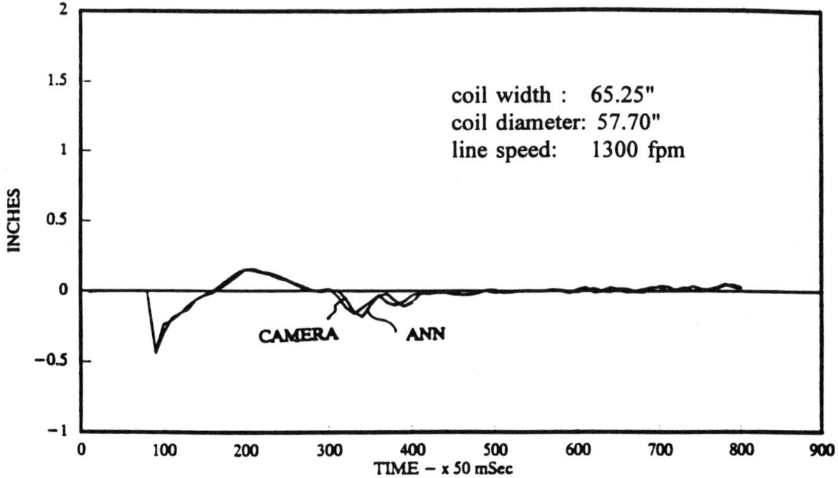

ANN: predicted by a neural network model
CAMERA: measured by a camera sensor

Figure 4.19 Position prediction given an adaptive neural network

has the following major characteristics:

- the system dynamics depend on the mill operating conditions, such as the physical dimensions and weight of the coil, the line speed, etc.;
- the system is sensitive to sensor measurement delay and noise.

A sensor based feedback controller with fixed parameters can not provide robust control performance, particularly when the production operating conditions significantly change, so that adaptive predictive control is required to produce precise centering in the winding. An adaptive predictive control is generally based on a real-time dynamic model or its inverse model. To avoid the time consuming real-time parameter estimation and adaptive mechanism, here an adaptive prediction model with a neural network is established for further application in the adaptive predictive control system. The comparison between the actual camera sensor measurements and the predicted output provided by the adaptive neural network is shown in Figure 4.19. It can be seen that the neural network provides excellent results even when the production environment significantly changes.

From the discussions of this section, we realize that a neural network is a powerful tool in system modeling and estimation, because neural networks have excellent behavior in nonlinear function approximation, learning and associative memories. Neural networks are particularly applicable to those systems which have rich data, but poor knowledge.of principles.

The evaluation of system observability is an important issue in designing a state estimator. In math-model free estimation, the physical concept and definition of system observability can still be used. From an application point of view, if the qualitative correlations between the system output and state vectors are studied with pattern feature, quantization and/or resulting model sensitivity analysis either in the design phase or in simulation studies, a realistic math-model free estimator can still be designed even without having a state space model which is required to verify the observability. The detailed investigation of the system observability for a math-model free estimation system is beyond the scope of this book.

Finally, it should be noted that the techniques of system modeling and estimation may also be applied in many other industrial control subjects, for instance dynamic feedback control, quality control (or statistical process control), fault detection and diagnosis, etc. These subjects will be investigated in the following chapters.

5
Dynamic controls

One of the major tasks for an intelligent control system is to provide a desired or satisfactory system behavior under an unknown and/or uncertain controlled environment. In this chapter, we focus on the applications of intelligent system methodologies in the design of a 'non-math-knowledge' oriented or so-called 'math-model-free' dynamic control system which has robust and adaptive behavior. Understanding the general concepts and methods of math-model based adaptive control will help in learning intelligent control. This is because both intelligent control and adaptive control expect to solve the same problem, but they have different problem backgrounds and thus use different technology. This chapter starts with a brief review of math-model based adaptive control, and then investigates the system architectures, strategies, design and industrial applications of neural networks, fuzzy logic, and rule based dynamic controls.

5.1 REVIEW OF ADAPTIVE CONTROL

5.1.1 Introduction

The classic dictionary meaning of 'adapt' or 'adaptation' is usually expressed in terms of biological adaptation to environmental changes. However, in the system control field, the concept of 'adaptive control' is related to the match of two major components in a control system, namely, 'controlled environment' and 'controller'. In other words, the goal of an adaptive control system is to provide robust behavior through automatically modifying controller parameters and/or structure, when the controlled environment changes within a specific region. In math-model based methodology, both environment and controller are described by relevant mathematical representations, i.e. differential or difference (matrix) equations in the time domain or transfer functions in the frequency domain. As a result, the dynamic response of a control system is also governed by a mathematical representation derived in terms of the signal flow between system environment and controller. The fundamentals of adaptive control are based on general theories of system stability, optimal control, stochastic approximation and dual control.

In the past three decades, many efforts (Astrom and Wittenmark, 1989; Astrom, 1983; Gupta, 1986) have been made to explore novel adaptive identification and control algorithms for complex systems under significantly uncertain environments. However, the development of adaptive control systems for an unknown, nonlinear and time varying environment has not been completely solved in both theoretical and practical aspects. From an industrial application viewpoint, the needs of adaptive control are to design a control system which can effectively work under a set of operating conditions, but with very limited knowledge of system dynamics. In general, the controlled environment in industrial control can be divided into the following two categories:

(1) *Continuous Production Systems.* In process industry, many systems, for instance distillation columns, chemical reactors and thermal processing processes, etc., belong to this family. System dynamics usually continuously and slowly change with time. For instance, any changes of raw material properties, production planning and scheduling will result in significant changes of system dynamics, e.g. static gain, time constants and time delay. In many cases, the controlled plant is nonlinear and time-varying, but with slower time-response and unvarying structure. Many R&D activities of adaptive controls are based on this kind of industrial environment.

(2) *Batch and Discrete-Event Driven Production Systems.* Batch and discrete event driven production processes have been popularly applied in process and manufacturing industries respectively. For instance, batch chemical reactors are widely used in biochemical and fine chemical production. In comparison with continuous production systems, from a system viewpoint, each batch can be represented as a specific point in an entire time (batch sequence) coordinate. Consequently, the system dynamics (both parameters and structure) can be a discrete distribution with respect to the batch events. In other words, both system structure and/or parameters may be quite different from batch to batch. As a result, the design of an adaptive control system will be more difficult, if the sequence of the batch operations is governed by a set of system models with variable structure as shown as follows:

$$X_{k+1} = \Phi_i(X_k, U_k, \theta_i), \quad i = 1, N \tag{5.1}$$

where i denotes the batch index in the batch operation sequence.

The batch systems are even more popular in many manufacturing systems, for instance, a steel rolling mill. However, in manufacturing production, a batch operating system is usually called a 'discrete-event driven system'. In general, the batch processing time in a manufacturing system is much less than in process industry; an adaptive control system thus has to make a 'rapid transition' from one job to another, otherwise the products made during the transition may be not qualified. In fact, some time-consuming adaptive control algorithms obviously are not applicable to many manufacturing systems.

5.1.2 Problem statement

Consider a nonlinear system with the following nonlinear state space model:

$$X_{k+1} = \Phi(X_k, U_k, \theta) \tag{5.2}$$

where both model structure, $\Phi(*)$, and parameter vector, θ, may be perturbed within the specific areas, $\Phi \in \Sigma_\Phi$ and $\theta \in \Sigma_\theta$. The design of an adaptive control system is to develop an adaptive control law

$$U_k = \Psi(X_k, \xi) \tag{5.3}$$

which results in a dynamic response of the closed loop system as

$$X_{k+1} = \Phi(X_k, \Psi(X_k, \xi), \theta) \tag{5.4}$$

where control law Ψ and/or parameter vector ξ are automatically modified in terms of either a given performance criterion or an on-line identified model 5.2. They correspond to the 'direct' or 'indirect' approach to adaptive control.

Problem 1: Direct adaptive control

The direct adaptive control can be represented as an on-line optimization problem shown as follows:

$$\min_{\Psi, \xi} \sum_k \| X_k^d - X_k \|^2 \tag{5.5}$$

under an unknown system environment represented by 5.2, where $\Phi \in \Sigma_\Phi, \theta \in \Sigma_\theta$, X_k^d denotes a desired trajectory of the state vector.

In principle, various optimization approaches may be used to solve problem 5.5, namely determining both the control law and its parameters. However, in many practical applications, the problem can be reduced to only solving the optimized parameters under a control law with fixed structure.

Problem 2: Indirect adaptive control

In an indirect adaptive control scheme, the implementation of an adaptive control is divided into two phases: (1) identifying the system dynamic model (time-domain or frequency-domain), and (2) solving the system control law and its parameters based on the identified system model, desired control criteria and constraints. The identification method being used in adaptive control can be from simple step response or frequency response methods to more complicated approaches to recursive least squares, stochastic approximation, maximum likelihood, extended Kalman filtering, etc. However, it should be noted that whatever kind of on-line identification (or parameter estimation) method is used, how to add the test signals to a real-time working production system with minimal effective on the normal operation and quality of products is one of the major concerns in the practical applications, particularly for those systems which are highly sensitive to disturbance. The model based real-time tuning or creation of a control law in phase (2) can be designed with classical control or modern control techniques.

Finally, it should be emphasized that even though rapid progress in adaptive control has been made in both academic and industrial aspects, structured adaptive control is still in a very early phase. Now we introduce some major schemes of parameter adaptive control (Astrom, 1983), which will be beneficial in learning about dynamic controls.

5.1.3 Schemes of parameter adaptive control

An ordinary feedback control system consists of a controlled system environment and a controller (or so-called 'regulator') with adjustable parameters. The problem in parameter adaptive control is to find a realistic way of modifying the controller parameters in response to the changes of system environment dynamics. The detailed goal of a parameter adaptive control system is to minimize the error between the set point and the actual response of the controlled variables or the sensitivities from system dynamics to the controlled criteria under an uncertain (deterministic or stochastic) system environment. The framework of parameter adaptive control can be described with three most popular schemes: 'gain schedule', 'model reference control' and 'self-tuning control' (Astrom, 1983). Now we will highlight their structures and functions.

Gain schedule

In many industrial systems, we find that the system dynamics, e.g. static gain, time constants and time delay, are dependent on the production operating conditions. For instance, the system dynamics of a distillation column are governed by its production scheme, production throughput, properties of raw material, etc., and in a gauge control system for a steel rolling mill, its dynamics are related to the steel grade (chemical composition), steel temperature, line speed, gauge reduction pace, etc. Obviously, if we can extract the key feature (auxiliary) variables representing the system operation condition and its related dynamics, a suitable coincidence between system dynamics and controller parameters may then be established and further used for controller tuning. Since originally only process gain was considered as the auxiliary variable, the approach is called 'gain schedule'. The major advantage of using gain the schedule approach in performing an adaptive system is that no on-line identification or optimization search is needed; consequently, the tuning of the controller parameters is straightforward, and the dynamic response is much faster than is given by other methods. However, on the other hand, to perform the gain schedule for adaptive control, it is actually, a time consuming job to find the quantitative relations between process auxiliary variables, controller parameters and the corresponding closed loop stability and performance with many simulation runs, particularly for those multivariable, nonlinear systems. Moreover, we can see that the gain schedule is only an open loop compensation without any feedback information related to the actions taken with the gain schedule; then the resulting system performance and stability can be managed if and only if the working domain of the system environment is fully recognized in the system design phase.

Model reference adaptive system (MRAS)

The model reference adaptive system (MRAS) was originally proposed by Whitaker and co-workers (1958) to solve servo problems. The system consists of a reference model and a closed control loop. The goal of MRAS is to force the controlled variable, y, to track (or duplicate) the output of the reference model, y_m through adjusting the controller parameters. In other words, MRAS can be governed by the following on-line optimization problem:

$$\min_{\xi} \sum_{k} \|y_m(k) - y(k)\|^2 \tag{5.6}$$

where ξ denotes the adjustable parameter vector of the controller, and as a result, the error response should converge to zero:

$$\lim_{k \to \infty} (y_m(k) - y(k)) \leq \epsilon \tag{5.7}$$

where ϵ denotes a specified small positive number. Obviously, the key in performing MRAS is to design the adjustment mechanism. The following 'MIT-rule' was originally used as parameter adjustment mechanism:

$$(\xi_{k+1} - \xi_k)/\Delta t = \eta e \nabla_{\xi} e \tag{5.8}$$

where

$\Delta t =$ sampling time interval;
$e =$ model error;
$\nabla e =$ vector of sensitivity derivatives of the error with respect to the adjustable parameters in vector ξ;
$\eta =$ learning rate, being a small constant, to determine the rate of adaptation.

In comparison with the gain schedule approach, MRAS is a feedback system, and its error response (stability and adaptation rate) can be controlled by adjusting η.

Self-tuning regulator (STR)

The 'self-tuning regulator' (STR) control generally consists of recursive parameter estimator and regulator design, the latter involves the determination of both the regulator law and its parameters. The STR was proposed by Astrom *et al.* (1977) for the stochastic minimum variance control problem. To apply STR in industrial systems, it is critical to select the on-line parameter estimation method, the perturbed test signals and the approach to the regulator design. The system response of an STR is dependent upon both the performance of parameter estimation and the match between the designed control law and the identified process model.

To apply STR in practice. The following issues must be investigated in detail (Isermann, 1982):

- parameter estimation in a closed loop;
- selection of the criterion in the design of a control algorithm;
- selection of control algorithms for adaptive control which satisfy:
  - closed-loop identifiability;
  - small computation expense and storage memory for the controller parameter calculation;
  - applicability to many processes and signals.

In comparing MRAS with STR, we find that they are similar in that each system has two loops, i.e., an 'internal regular feedback control loop' and an 'outer loop' being used for adjusting the regulator's parameters. However, on the other hand, their main difference is that in MRAS, the parameters are directly adjusted in terms of the error performance; instead, in STR the regulator is designed with the knowledge of the identified model. In addition, as pointed out by Astrom (1983), the MRAS was obtained by considering a deterministic servo-problem and the STR by considering a stochastic regulation problem.

Finally, it should be emphasized that the concepts and applications of adaptive control have been widely extended to a variety of control systems, such as feed-forward, decoupling (Smith predictor), dead-time, optimal and large scale system controls.

5.2 STRATEGIC SCHEMES OF MATH-MODEL FREE DYNAMIC CONTROL

5.2.1 Problem statement

From section 5.1 we can realize that in an adaptive control system both process model and control algorithms are represented by the relevant mathematical descriptions. Moreover, modern control techniques, e.g. parameter estimation, stochastic control and optimal control, are used in either model identification or design of a matchable control law. However, some industrial processes are too complicated to be modeled and/or controlled by math-algorithms, because they are highly nonlinear and significantly uncertain with unknown structure and imprecise information. With the recently rapid progress of intelligent learning system techniques, e.g. neural network, fuzzy logic and rule based expert systems, the intelligent dynamic (or adaptive) control has become a hot subject in both theoretical and practical aspects. Because modern control theory for nonlinear systems has not been fully developed, it will be more difficult to establish a systematic discipline for intelligent control systems. However, even if the theoretical system and foundation of intelligent control have not been established, studies on its practical methodologies and applications should not be influenced by this fact. This chapter addresses the concepts, architectures and algorithms of intelligent dynamic controls with the framework of math-model based adaptive control.

What is an intelligent dynamic control system? Following the concepts, principle and methodology of intelligent systems as we discussed in the previous chapters, we can find that an intelligent dynamic control system should be able to effectively adapt to system environment changes to improve its control performance through learning, associative memory and/or self-organization. In comparison with adaptive control, there are the following key features of intelligent dynamic control:

(1) Both process behavior and control law are represented by a specific knowledge base or an associative memory, which could be fuzzy logic, neural network, rule base or some hybrid form.

(2) Instead of numerical computation algorithms being used in math-model based control, in an intelligent control system the control law can be performed with parallel information processing, learning, associative memory or reasoning, etc.

(3) The system design is mainly based on analytical principle of process, knowledge of human expertise and historical production data.

(4) The system I/O signals are processed through encoding and decoding, e.g. fuzzification, defuzzification, vector quantization, feature extraction, etc.

The problem of intelligent dynamic controls is to design an intelligent control system with similar schemes of 'gain schedule', 'direct' and 'indirect' adaptive control in terms of intelligent system technologies. The problem of an intelligent control system can be generally described as: suppose we have an uncertain (deterministic, stochastic or fuzzy) controlled environment, $\{\Phi : X_k, U_k \Rightarrow X_{k+1}, \Phi \in \Sigma_\Phi\}$, and the relevant control criteria associated with the system constraints, where Φ can be a neural network, fuzzy associative or production rule base, etc. The design of an intelligent control system is to synthesize a knowledge based control law with learning, associative memory or self-organization to satisfy both desired control criteria and constraints. The control law, $\{\Psi : X_k \Rightarrow U_k, \Psi \in \Sigma_\Psi\}$, can also be performed by a selected form of knowledge representation.

5.2.2 Schemes of intelligent dynamic control

The geometrical description of an intelligent control system can be illustrated in Figure 5.1. Similar to adaptive control, the intelligent control system also constructs a suitable coincidence between variable controlled environment and control law to produce a satisfactory control performance. However, instead a of math-algorithm, this coincidence is established with some human-like knowledge and associative

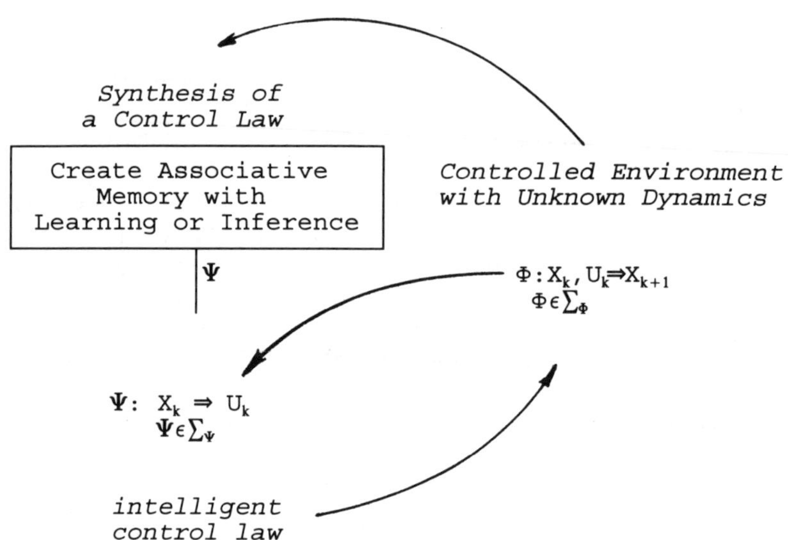

Figure 5.1 Geometrical description of intelligent control systems

memory representations. With this general strategic concept, intelligent control can be further represented in the following detailed forms.

'Gain schedule' type of intelligent controls

Implementation of the 'gain schedule' type of intelligent controls usually consists of the following two phases:

(1) Establishing pattern mapping between the system environment E and the auxiliary variable, AV, the latter can usually be determined with feature extraction in a multi-dimensional space of system characteristic parameters, $C \in R^p$. For instance, in industrial control, the auxiliary variable is related to the parameters which represent the production operating conditions, such as production rate, product grade, production scheme, raw material property, etc. The process of identifying AV can be represented as $\{E(C_k \in R^p) \Rightarrow AV_k\}$. In fact, the AV_k is a scalar with either an event-driven discrete code or a continuous number. If the physical mechanism in finding the auxiliary variable is not explored, the techniques of pattern classification or self-organized learning can be used in solving this problem.

(2) Establishing pattern mapping between the auxiliary variable and the control law (or controller parameters), namely, $\{\mathcal{IC} : AV_k \Rightarrow \Psi_k, \xi_k\}$. This mapping can be either a simple rule base or a functional hetero-associative memory. Obviously, knowledge of control expertise and computer simulation are required in performing phase (2).

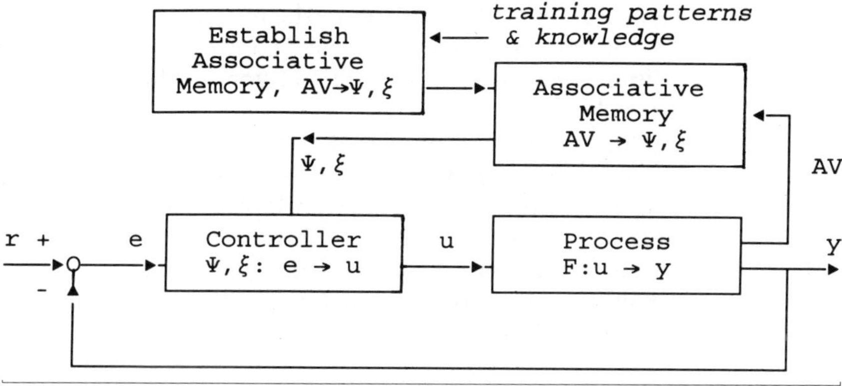

Figure 5.2 Architecture of 'gain schedule' type of intelligent control

The architecture of the gain schedule type of intelligent control is schematically shown in Figure 5.2.

'Direct' type of intelligent control

To perform the 'direct' type of intelligent dynamic control, the manipulation actions provided by the control law optimize the given criterion. The problem can be described as

$$\min_{\Psi,\xi} \sum_{k} \| y_m(k) - y(k) \|^2 \tag{5.9}$$

subject to an unknown process model with $\{\Phi : X_k, U_k \Rightarrow X_{k+1}\}$.

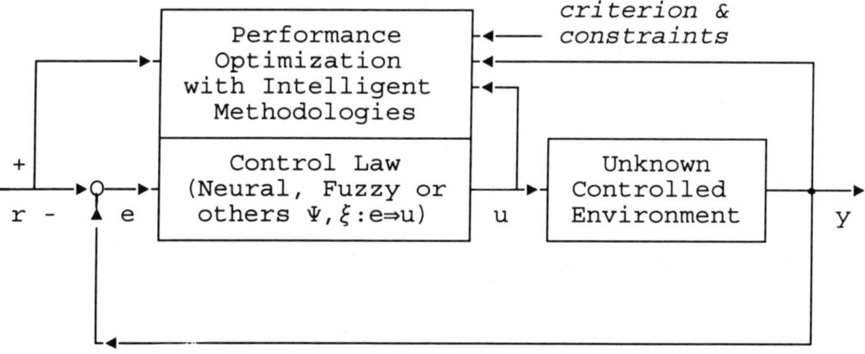

Figure 5.3 Architecture of 'direct' type of intelligent control

Similar to MRAS, y_m and y denote the outputs of the reference model and process respectively, and Ψ, ξ represent the knowledge based control law and its adjustable parameters, e.g. the adjustable weights in a neural network or the fuzzy relations in a fuzzy model, respectively.

The architecture of a 'direct' type of intelligent control is shown in Figure 5.3. It can be seen that the error, $e = y_m - y$, between the reference model and the closed control loop is used as a 'learning' or so-called 'adaptive' feedback signal. The solution to the optimization problem 5.9 is a pattern mapping or associative memory from the error, e, to the control manipulation variable, u, i.e. $\{\Psi, \xi : e_k \Rightarrow u_k\}$. In general, a direct optimization strategy can be used to find a control sequence which may directly minimize the error without having an on-line process model.

'Indirect' type of intelligent control

The architecture of the 'indirect' type of intelligent control is given in Figure 5.4. Similar to the STR type of adaptive control, the working process of this system is also divided into two interactive phases, namely, identifying the process knowledge base and determining the control law, or simply the controller parameters. These two procedures can be further represented as $\{U_k, X_k \Rightarrow \Phi \text{ and/or } \theta\}$ and $\{\Phi, \theta \Rightarrow \Psi \text{ and/or } \xi\}$, where Φ and Ψ can be any form of knowledge representation.

Finally, it should be noted that the 'gain schedule', 'direct' and 'indirect' schemes are three basic forms commonly used in intelligent controls. Some other higher level control strategies, such as intelligent coordinator and supervision, are also important in making a number of control loops work together, or coordinating dynamic controls and supervisory optimization. It is clear that this kind of intelligent capability can be viewed as meta-knowledge in artificial intelligence.

Now we will introduce some intelligent control strategies with neural, fuzzy and expert systems methodologies. Since the detailed fundamentals of intelligent systems have been addressed in the previous chapters, here we will only present their applications in the design of a dynamic control system.

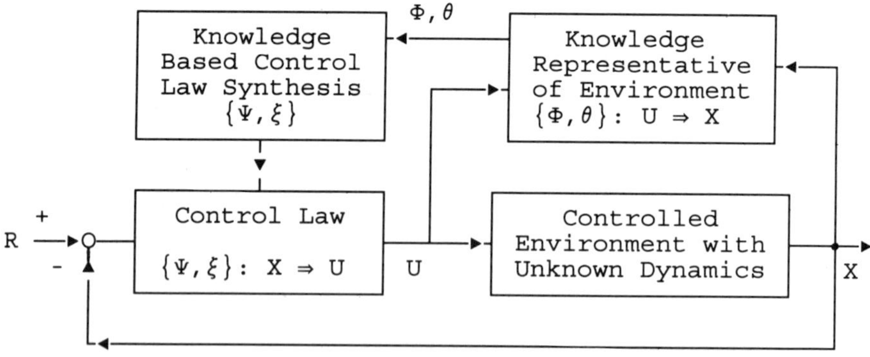

Figure 5.4 Architecture of 'indirect' type of intelligent control

5.3 NEURAL NETWORK ADAPTIVE CONTROLS

5.3.1 Introduction

As we investigated in the previous chapters, a neural network can be applied to nonlinear function approximation (mapping) or modeling for static and dynamic systems with supervised or recursive learning. The analytical relation between system identification and neural network modeling is an attractive problem for both control and neural network communities. It should be kept in mind that the applications of nonlinear identification are still limited to particular input–output models. Moreover, of course, the structure identification for nonlinear systems is a more difficult problem than that of parameter estimation. However, on the other hand, it has been proved that any continuous function can be arbitrarily well approximated by a multilayer feedforward network with only a single hidden layer, where each neuron has a continuous sigmoid function. On the basis of this conclusion, clearly the structure determination, i.e. selection of nonlinear signal function and the neuron number in the hidden layer, is not as difficult as in nonlinear system identification, particularly when the structure learning of the neural network is introduced.

The applications of a variety of neural networks for system control have been extensively studied in the past few years (Hunt *et al.*, 1992; Willis *et al.*, 1992; Psaltis *et al.*, 1988; Kung and Hwang, 1989; Qin and McAvoy, 1992a, Sastry, 1994, etc.). The neural model based dynamic controls, such as neural model predictive control, neural model identification based control, neural-self-tuning control, etc., have been widely investigated. If we examine the neural model based control systems, we will find that the math-model is replaced by a neural network with forward or inverse forms. Here we will only investigate the design and applications of the neural controllers with direct feedback; namely a simple neural adaptive controller can be designed and implemented without having off-line training and consequently the application of a neural adaptive controller will be as easy as a conventional feedback controller. The following sections will address the development of a neural adaptive controller (with a direct feedback), simulation studies and real-time industrial applications.

5.3.2 Problem statement

In general, the neural adaptive control can also be represented by a dynamic optimization problem. The problem is to find a neural control law with both the neural structure parameters, θ, and the synaptic connection weights, W, which minimize the desired criterion. The neural network adaptive control can be formulated as follows:

$$\min_{\theta_k, W_k} J(X_k, U_k) \tag{5.10}$$

subject to the system model, e.g.

$$X_{k+1} = \Phi(X_k, U_k) \tag{5.11}$$

$$Y_k = G(X_k) + v_k \tag{5.12}$$

and the structure of the designed neural network control law:

$$U_k = \mathcal{NN}_k(X_k, R_k, \theta_k, W_k) \tag{5.13}$$

where

$X \in R^n$ = n-dimensional system state vector;
$R \in R^n$ = n-dimensional system reference state vector;
$U \in R^r$ = r-dimensional control vector;
$Y \in R^m$ = m-dimensional output vector, i.e. $Y \in X$;
$\Phi(*) \in \Sigma_\Phi$ = an uncertain nonlinear discrete-time state space model;
$G(X_k)$ = output equation;
$J(*)$ = objective function;
v = n-dimensional measurement noise vector.

The state vector X can be estimated by the state estimation (Kalman filter, a neural network, pattern recognition or fuzzy model based estimator) with the observed I/O sequence $\{U_k, Y_k\}$. The neural adaptive control law with a direct state feedback can be simply described as follows:

$$\mathcal{NN}_k(\theta_k, W_k) : X_k, R_k \Rightarrow U_k \tag{5.14}$$

The utilization of the neural networks for adaptive control has the following functional structures:

(1) The neural network is designed to learn process dynamics (or the dynamics of the inverse model), and then be used as a model based adaptive control law through the retrieval process. It should be noted that the control system behavior is heavily dependent on the 'generalization' performance, in other words the difference between the training and working environments. In some complex industrial systems, particularly for discrete-event driven production systems, e.g. steel rolling mills and other manufacturing systems, it is very difficult to extract the feature test signals to be used in the neural network training that can cover all working conditions. Consequently, the resulting generalization performance may be poor in some working areas.

(2) The neural network is designed as a self-tuning mechanism. The task of the neural network is to provide optimal tuning parameters for a math-control algorithm to minimize the error between the reference and actual values of the system output (or state) vector. Obviously, the system behavior, e.g., adaptation and robustness, are mainly governed by the math-control algorithm used.

(3) The neural network is designed as a direct feedback adaptive controller (Ichikawa *et al.*, 1992, Lu, *et al.*, 1992, Lu 1994), namely, the neural network controller receives the system observed output and/or the estimated state variables, and then provides its control actions to the controlled system

environment to optimize the control criterion through real-time information processing. In comparison with the above two forms, the neural network serves as an adaptive controller with self-adjustment capability, such that network off-line training is not needed, and then the generalization performance only depends on the neural network structure and real-time learning.

5.3.3 Neural adaptive control system with direct feedback

Architecture

Figure 5.5 illustrates the architecture of a neural adaptive control with direct feedback. The state variables are estimated by a neural network estimator, namely, $NNE : \{U_k, Y_k \Rightarrow X_k^p\}$, and the weights of the neural network controller are real-time upgraded by a specific learning algorithm. The detailed structure of the neural network controller with adaptive learning is schematically shown in Figure 5.6. The development of the neural network algorithm for adaptive control is dependent on the control goals and knowledge of the system environment. The general steps of the development can be described as follows.

(1) Determine the system input, output and state variables, control criterion, constraints and qualitative knowledge of the controlled system environment.

(2) Based on the problem background, select the proper architecture for neural network adaptive control, namely the network topology and signal activation function. In this study a single-layer feedforward network with adaptive back-propagation (BP) learning is used to construct a neural adaptive controller.

(3) Based on the selected neural network architecture and I/O signals, determine signal coding/decoding policies. In general, 'min–max', 'mean–variance', '0, 1' or 'nonlinear' approaches can be selected to match the neural network I/O and other related properties.

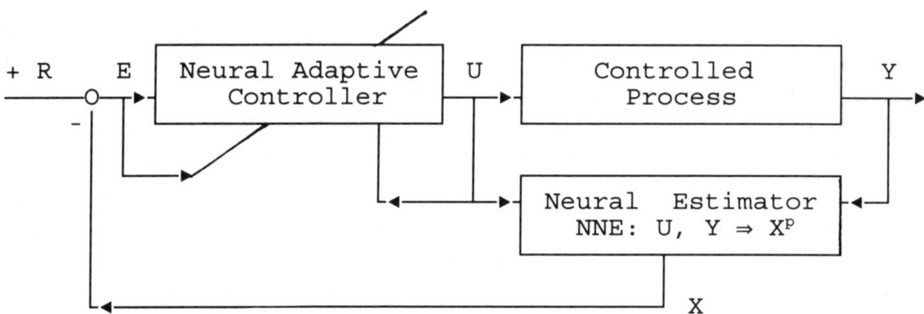

Figure 5.5 Neural adaptive control system with direct feedback

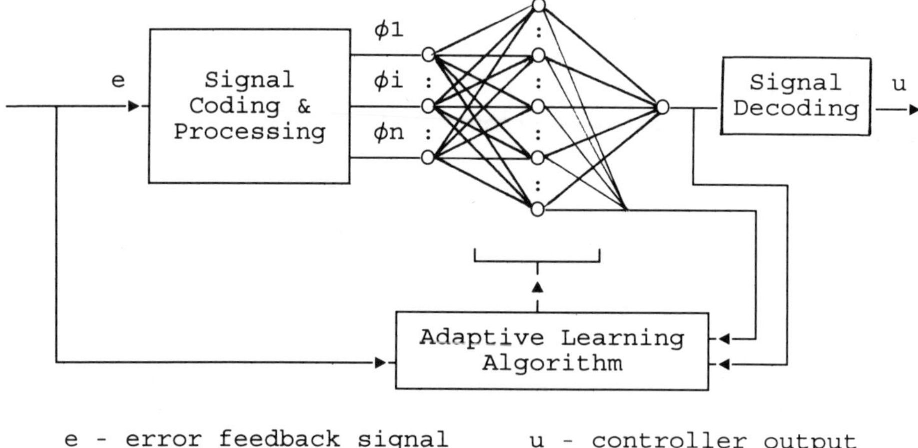

Figure 5.6 Structure of a neural adaptive controller

(4) Design the signal preprocessing algorithm, depending on the signals and/or signal functions that are desired as the neural network controller's inputs. From a knowledge gathering and information processing point of view, this step is important to make the neural network communicate with the system environment. In general, the dynamic and/or nonlinear functions of the error ($e(k) = r(k) - y(k)$, or $E(k) = R(k) - Y(k)$), i.e. $\{\phi_i(e(k)), i = 1, p\}$, are selected to be the neural adaptive controller's input signals.

(5) Based on the selected neural network architecture, design a proper learning algorithm to minimize the desired criterion subject to the system constraints.

Learning algorithm

When a neural network is used as an adaptive controller, in contrast to neural network modeling, the development of the learning algorithm is based on a cascade system with both the neural network and the controlled plant. In fact, the latter becomes part of the network, but is not adjustable. Psaltis *et al.* (1988) proposed a so-called *'specialized learning architecture'* for the neural network controller, in which the plant gain, i.e. the partial derivatives ($\partial y/\partial u$), around the operating point should be calculated in real-time. However, with the noise in many industrial control systems and the limit of the level one control computer, it is hardly possible to determine the real-time plant gain without adding special test signals. In this study, the controlled plant is considered as an additional part of the network with structure and/or parameter uncertainties. These uncertainties can be detected and compensated through the neural feedback control. Without loss of generality, suppose the neural network structure parameter vector θ is fixed, then only the adjustable weights are to be upgraded through real-time learning. Consider a single-input single-output

(SISO) system with following scalar criterion:

$$\min_{W_k} J(W) = \frac{1}{2}\sum_k \| r_k - y_k \|^2 \tag{5.15}$$

subject to

$$u_k = \mathcal{NN}(e(k), \phi_i(e(k)), \lambda, W_k) \tag{5.16}$$

and the unknown plant

$$\mathcal{P} : u_k \rightarrow y_k \tag{5.17}$$

As we mentioned above, if $\mathcal{P}$ is part of the network with non-adjustable structure and parameters, then the learning algorithm can be developed on the basis of BP-learning with a sigmoid activation function. When the gradient method is applied in solving the optimization problem as shown in equations 5.15–5.17, the network weights are upgraded with the following delta rule:

$$\frac{\partial J(W)}{\partial W} = (r - y)\left(\frac{\partial y}{\partial u}\right)\left(\frac{\partial S(\beta)}{\partial \beta}\right)\left(\frac{\partial \beta}{\partial W}\right) \tag{5.18}$$

then

$$\Delta W(i) = \eta(r - y)S'(\beta)u(i) \tag{5.19}$$

$$\beta = \sum_i^p W(i)x(i) \tag{5.20}$$

where

$x(i) = \phi_i(e(k)), i = 1, p$ are the neural controller's inputs;
$S(\beta)$ is the neuron's signal activation function;
$u = S(\beta)$ is the neural controller's output;
η is the learning rate.

As a result, the transfer function of the neural network for adaptive control can be written as

$$u_k = S(\beta_k) \tag{5.21}$$

where

$$\beta_k = \sum_{i=1}^p W_k(i)X_k(i) \tag{5.22}$$

$$W_{k+1}(i) = W_k(i) + \Delta W_k(i) \tag{5.23}$$

where k is the time index.

In comparing 5.17 with 5.18, we can see that the unknown plant gain $(\partial y/\partial u)$ is included when we tune the network learning rate η. The plant uncertainty (dynamic and steady state) will be first detected by neural network inputs and the learning feedback signal, and then further compensated by an on-line learning action.

System property analysis

Let us consider a system represented by the nonlinear I/O model

$$y_{k+1} = F(y_k, \ldots, y_{k-n}, u_k, \ldots, u_{k-r}, \theta) \tag{5.24}$$

and suppose the math-representation of a neural adaptive controller is as follows:

$$u_k = N(\phi_i(e(k)), W_k(i), \eta), \quad i = 1, p \tag{5.25}$$

Substituting 5.25 to 5.24, the closed-loop dynamics yield

$$y_{k+1} = F(y_k, \ldots, y_{k-n}, N(k), \ldots, N(k-r), \theta) \tag{5.26}$$

where $N(k)$ denotes the neural controller's output at time index k.

It can be seen from 5.25 and 5.26 that the closed loop performance is governed by the designed $\phi_i(e(k))$ and η. The term η controls the system stability and convergent responses. Finally, it should be noted that the system uncertainties of the structure $F(*)$ and parameters θ are compensated by adapting the synaptic connection weights $W(i)$ to minimize the desired control performance. The following simulation studies with a benchmark plant and the industrial applications show the excellent performance in both robustness and adaptation of the neural adaptive controller as developed in this study.

Simulation results

A benchmark plant (Graebe, 1993) proposed for the IFAC '93 (Sydney) World Congress is applied in the simulation studies of the neural adaptive controller. This section presents the simulation results comparing the robust and/or adaptive controls (Zhou and Kimura, 1993) based on the Sydney benchmark plant.

Description of Benchmark Problem The Sydney benchmark problem is to design a controller for a time-varying (SISO) plant. The plant operates at three stress levels, which represent the different levels of the plant uncertainty. The aim of the benchmark plant control is to design a robust and/or adaptive controller which satisfies the following desired dynamic performances under the setpoint square-wave changes between $+1.0$ to -1.0 with a period of 20 seconds for all three stress levels:

- plant output $y(k)$ satisfies $-1.5 \leq y(k) \leq +1.5, \forall k;$

- undershoot/overshoot of $y(k)$ is less than 0.2 most of the time;
- fast setting and short rising time;
- zero steady state tracking error.

To have a representative picture of the plant dynamic responses, it is required to simulate the plant with at least 15 runs over a 20-second period for each stress level. The nominal transfer function of the Sydney benchmark plant with the parameter perturbation for the stress levels 1, 2 and 3 can be written as

$$G(s) = \frac{Y(s)}{U(s)} = \frac{K\omega_0^2(-T_2 s + 1)}{(s^2 + 2\xi s + \omega_0^2)(T_1 s + 1)} \tag{5.27}$$

At stress level 1:

$$T_1 \in [4.8, 5.2], \quad T_2 \in [0.35, 0.45],$$
$$\omega_0 \in [3.5, 6.5], \quad \xi \in [0.20, 0.40],$$
$$K = 1.0.$$

At stress level 2:

$$T_1 \in [4.7, 5.3], \quad T_2 \in [0.30, 0.50],$$
$$\omega_0 \in [2.5, 7.5], \quad \xi \in [0.15, 0.45]$$
$$K \in [0.85, 1.15];$$

At stress level 3:

$$T_1 \in [4.7, 5.3], \quad T_2 \in [0.25, 0.55],$$
$$\omega_0 \in [2.0, 8.0], \quad \xi \in [0.15, 0.45],$$
$$K \in [0.50, 1.5].$$

Obviously, the higher the stress level, the more significant the uncertainty of the controlled plant.

Simulation results for benchmark plant (Lu, 1994) In order to test the control performances of the neural adaptive controller and to compare with the robust and/or adaptive control strategies, the Sydney benchmark problem is used for our simulation tests. The simulation results of the neural adaptive controller with BP-learning for the Sydney benchmark plant under three different stress levels are given in Figures 5.7, 5.8 and 5.9. In making the simulation runs, the benchmark plant is supposed to be unknown, and the 15 runs for each stress level are on the basis of the random changes of the plant parameters within the corresponding 'min–max' ranges. During the simulation tests, only the learning rate η

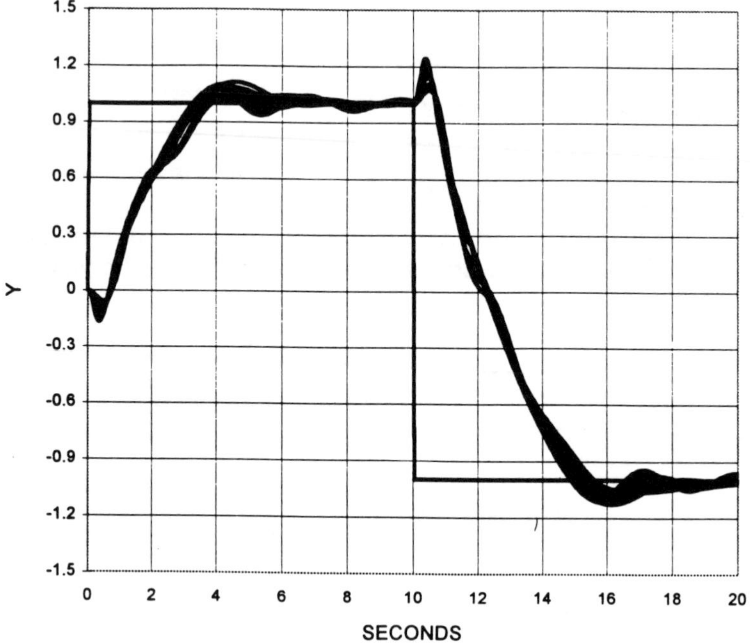

Figure 5.7 Neural adaptive control for stress level 1 of Sydney benchmark plant

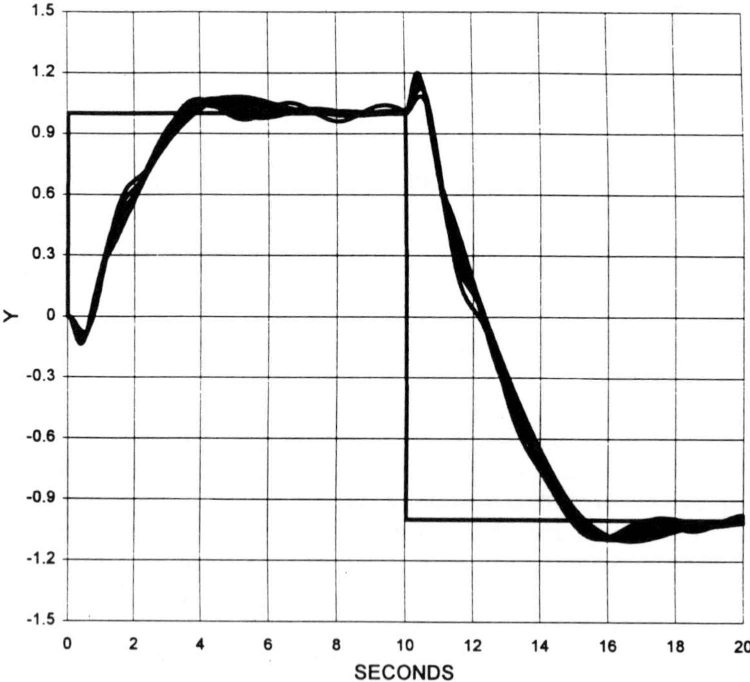

Figure 5.8 Neural adaptive control for stress level 2 of Sydney benchmark plant

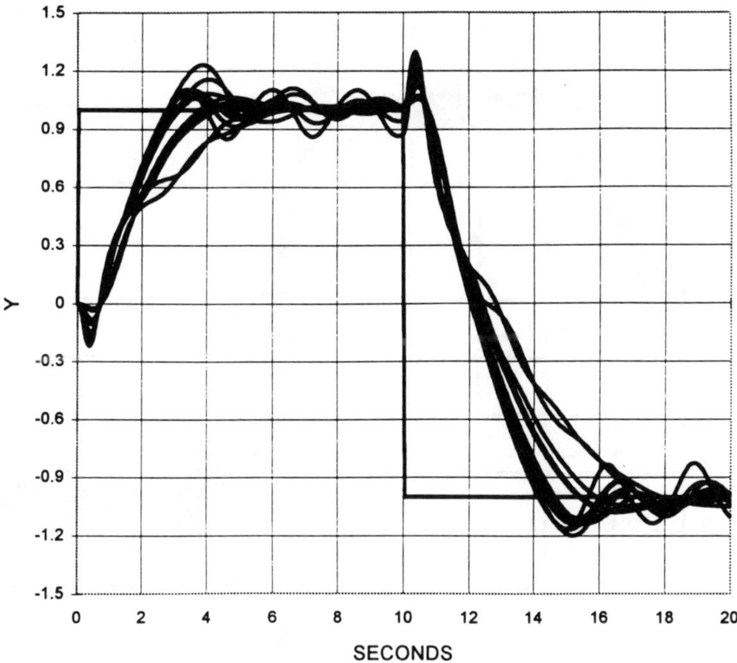

Figure 5.9 Neural adaptive control for stress level 3 of Sydney benchmark plant

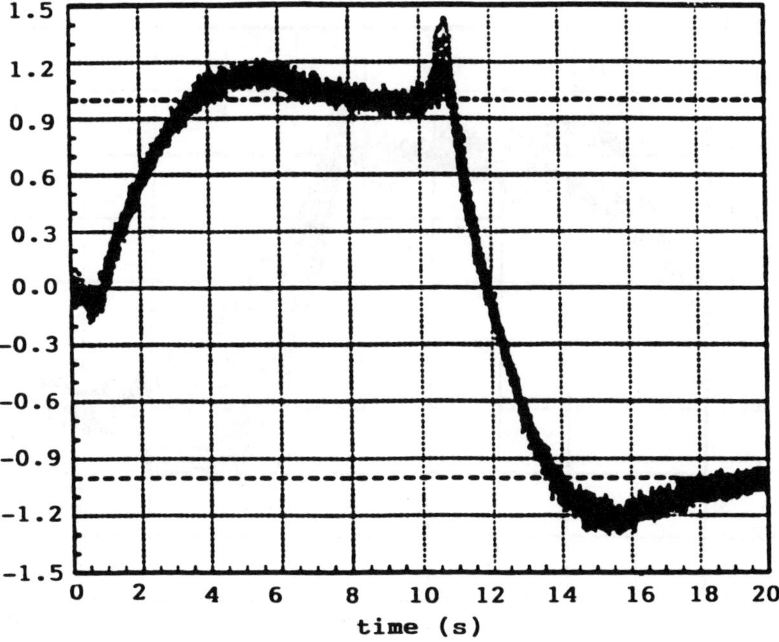

Figure 5.10 Robust control for stress level 1 of Sydney benchmark plant (Zhou and Kimura, 1993)

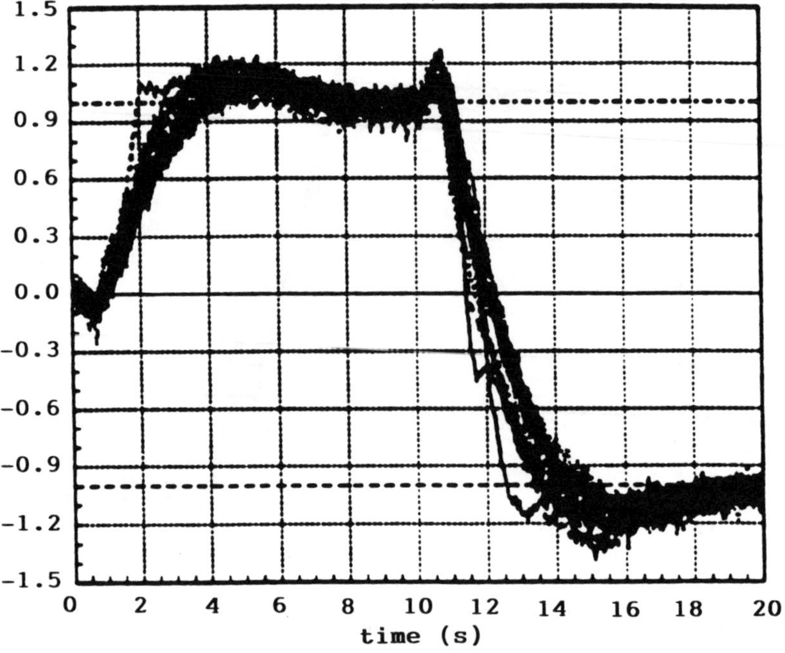

Figure 5.11 Robust control for stress level 2 of Sydney benchmark plant (Zhou and Kimura, 1993)

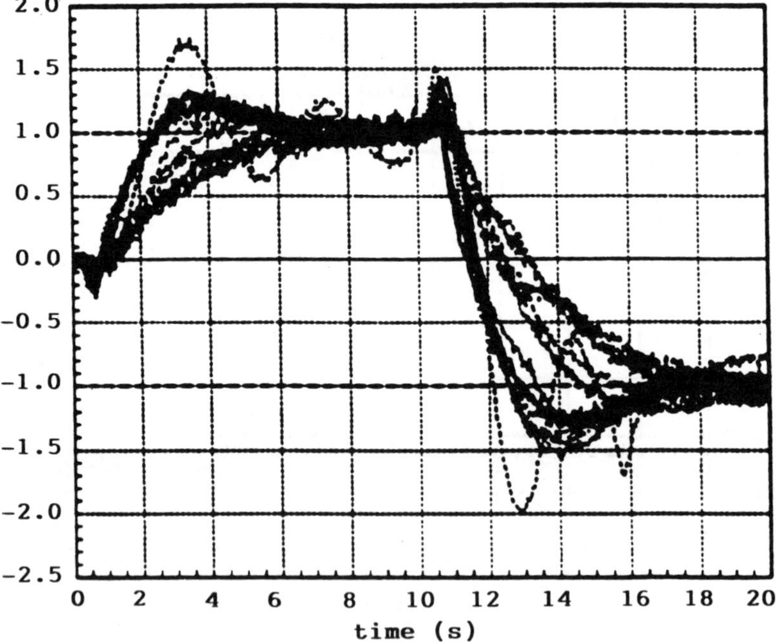

Figure 5.12 Robust control for stress level 3 of Sydney benchmark plant (Zhou and Kimura, 1993)

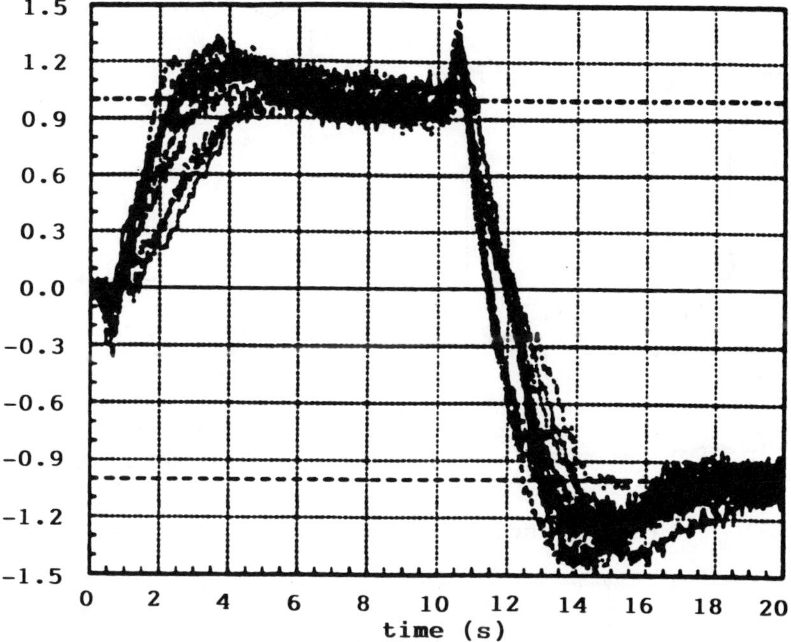

Figure 5.13 Robust and gain adaptive control for stress level 3 of Sydney benchmark plant (Zhou and Kimura, 1993)

is tuned, and finally a proper fixed tuning of the learning rate is used for all three stress levels.

From the simulation results given in Figures 5.7 to 5.9, we can see that the dynamic responses with the neural adaptive controller are comparable to many results published in the IFAC 12th World Congress (IFAC, 1993). To make a clear comparison, the simulation results with an H_∞ robust control and a combination of robust and adaptive control (Zhou and Kimura, 1993) are given in Figures 5.10 to 5.13. These simulation results show that for stress levels 1 and 2 a fixed robust controller can be designed to achieve the required performance; however, for stress level 3, an adaptive gain with a robust stability degree assignment should be included in order to provide a better dynamic performance. More simulation results with various robust and adaptive control strategies can be found in (IFAC, 1993).

5.4 AN INTEGRATED NEURAL SYSTEM FOR COATING WEIGHT PREDICTION AND CONTROL

5.4.1 Introduction

A modern hot dip galvanize/galvanneal production line is capable of producing galvanized and galvanneal products that meet the tight automotive market

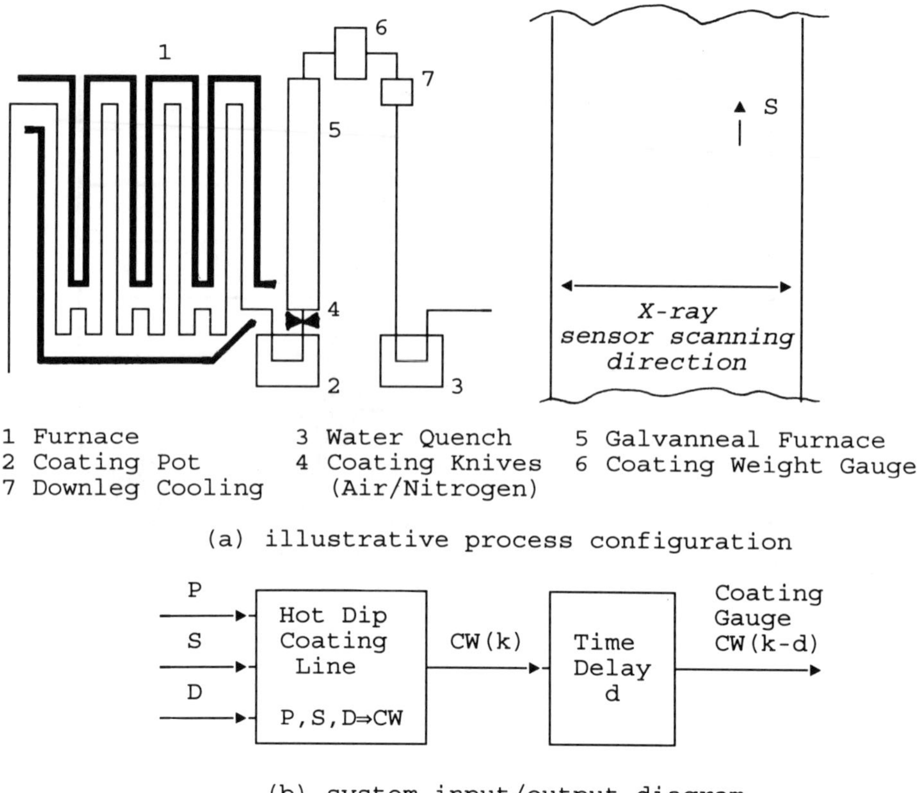

Figure 5.14 Illustration of a hot dip coating line in a steel mill

tolerances. The line is made up of three main sections, i.e. the entry, processing and delivery sections. A general overview of a hot dip galvanize/galvanneal line is illustrated given in Figure 5.14. The processing section will be the focus of our study. This section will introduce development and industrial application of an integrated neural system for coating weight prediction and control at a hot dip coating line (Markward and Lu, 1995).

The process section consists of an annealing furnace, a zinc pot, an air knife, a zinc iron thickness measuring sensor with scanning and a temper mill/tension leveler. The line speed, pressure and knife gap are the dominant factors in determining the coating weight. The pressure at the air knife is the primary control action of the zinc thickness. Several other factors that affect the coating weight are strip cleanliness, steel grade, strip roughness, strip temperature and knife height; however, these variables are difficult to measure and are masked by the first three inputs of speed, pressure and gap.

The control of the coating weight is very important in producing qualified products with the low operating cost. In other words, an automatic coating weight control system that delivers small variation in the machine direction can have tighter coating weight aim specifications, lower zinc consumption rates and less head and tail loss on coating weight aim changes. The goal for the coating weight control system is to have a single standard deviation of less than 3%. The regression model and conventional PID algorithm are commonly used for coating weight prediction, static feedforward control and feedback control respectively.

This section presents development and application of an integrated neural system in coating weight prediction and control; namely, a neural network with batch-wise learning is used to perform the zinc pot coating weight prediction, another neural network is used as a feedforward control and, correspondingly, a neural adaptive control with real-time learning is used as an adaptive coating weight feedback control. The detailed comparison between the regression model based prediction and the neural network based prediction is investigated through both simulation and actual mill data. Some sample closed-loop responses with the neural system are given to show the excellent results for the real-time applications. The problem formulation, system configuration and industrial tests and analysis will be addressed in the following sections.

5.4.2 Problem formulation

General introduction

To precisely control the coating weight, the following problems should be considered.

Real-time prediction of the zinc pot coating weight The zinc pot coating weight can be represented as the following steady-state nonlinear function:

$$CW_p(k) = \Phi(P(k), S(k), D(k), \theta) \tag{5.28}$$

where

Φ, θ = nonlinear function and parameter vector respectively;
CW_p = zinc pot coating weight;
P = pressure;
S = line speed;
D = air knife gap;
k = index of the scanning times.

In fact, the actual zinc pot coating weight is not available for sensor measurement, so that the prediction of the zinc pot coating weight is important for coating weight control. The problem of the coating weight prediction is to find both Φ and θ which minimize the error between the actual and predicted zinc pot coating weight in terms of the mill's historical data; this can be represented as the following optimization problem:

$$\min_{\Phi,\theta} \sum_k (CW_p(k) - CW_p^p(k))^2 \tag{5.29}$$

subject to the I/O data sequence, where CW_p^p denotes the predicted zinc pot coating weight which can be predicted through the sensor's observed coating gauge and other related mill operating parameters.

Coating weight control Both the steady-state and dynamic behaviors of the production line heavily depend on the operating conditions, e.g. the level of the desired coating weight, line speed, zinc coating type status, etc. To meet customer order and other production environment changes, the setpoint of the coating weight and/or the line speed may be frequently and significantly upgraded in the operation. As a result, the coating weight prediction and control system should be designed with highly robust and adaptive behavior. The goal of designing a coating weight control system is to develop an adaptive control law which makes the actual zinc pot coating weight track the desired coating weight precisely under a variable production environment; this can be further written in the following mathematical form:

$$\min_{P(k)} \sum_k (R_p(k) - CW_p(k))^2 \tag{5.30}$$

subject to

$$P = C(R_p, CW_p, S, D, \xi) \tag{5.31}$$

and other production related constraints, where C, ξ denote the FBC/FFC law and the corresponding parameter vector respectively, and R_p represents the desired zinc pot coating weight. The main aims of the coating weight control are to reduce the standard deviation of the coding weight error and the transition footage from one coil to the another under various operating conditions. Now we will address both the original and the neural system configurations for coating weight prediction and control.

Original system configuration

Regression model for coating weight prediction and FFC The original system consists of the regression model based coating weight prediction and steady-state feedforward control and a PID type of feedback control. A block diagram of the original system is shown in Figure 5.15. The line speed is controlled by the operator input or a level 2 thermal model. The knife gap is adjusted by the operator. The feedforward control responds to the line speed, desired coating weight and air knife gap changes. The feedback control responds to the coating weight gage feedback changes. The output of both feedforward and main (coating weight) feedback controllers is serves as a setpoint of the secondary (pressure) controller in the FFC/FBC (cascade) system. The regression model based feedforward control algorithm FFC can be represented as the following steady-state relation:

$$FFC : R_p, S, D \rightarrow P_{ffc} \tag{5.32}$$

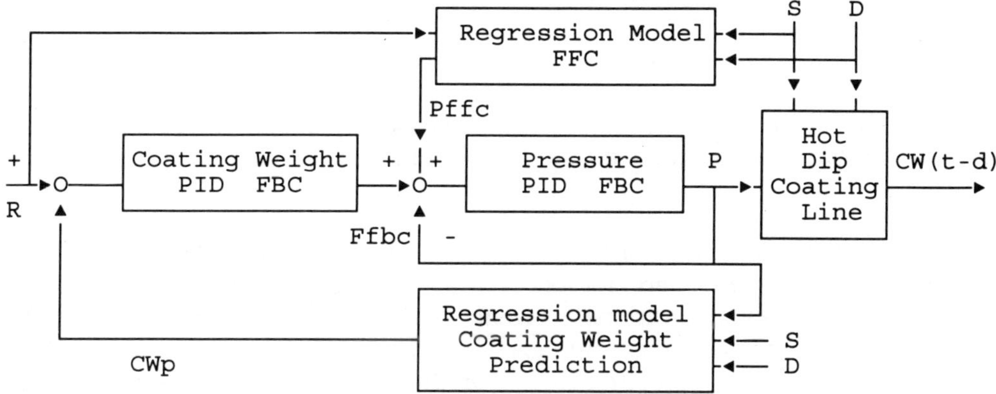

Figure 5.15 Regression model based coating weight prediction and control

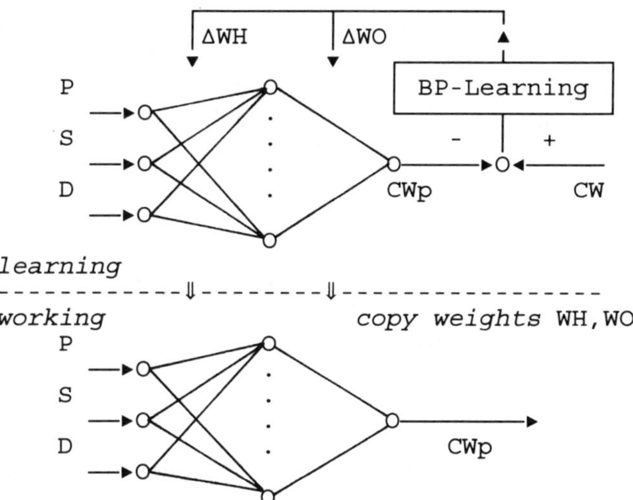

Figure 5.16 Neural network based coating weight prediction

where P_{ffc} serves as a reference of the pressure controller. In fact, the FFC algorithm 5.32 may be developed with a regression coating weight prediction model with the following linear form:

$$CW_p^p = C_s f(S) + C_p f(P) + C_d f(D) + \beta \tag{5.33}$$

where $f(*)$ is a simple nonlinear function and the constant coefficients $\{C_s, C_p, C_d, \beta\}$ are determined with a linear regression algorithm. The solution to the regression model is done off-line with batch data. The coefficients are then downloaded to the system for real-time applications. To improve the accuracy of the regression model a gain term α was added, then equation (5.33) becomes

$$CW_p^p = \alpha(C_s f(S) + C_p f(P) + C_d f(D) + \beta) \tag{5.34}$$

where α as a learning factor is a slow trim gain based on the difference between the regression model based predicted coating weight and the gage feedback coating weight.

Coating weight feedback control The feedback control with its inputs of both the desired and the predicted zinc pot coating weight is designed to provide an output P_{fbc} serving as an additional reference of the pressure controller. In fact, the functions of FBC are to compensate those disturbances not being included in FFC and the inaccuracies of the coating weight prediction and FFC models as well. The FBC can be expressed as follows:

$$FBC : \{R_p, CW_p^p\} \Rightarrow P_{fbc} \tag{5.35}$$

where FBC is a popular PID type of control law. Then a pressure setpoint with weighted sum of P_{ffc} and P_{fbc} is applied to the pressure controller. To make the system more stable, the running of FFC and FBC is logically managed by a set of simple rules: for instance, when the coating weight setpoint or line speed significantly changes, the system will first run FFC, and then FBC will run when the error of the coating weight has entered into a given area, etc. Finally, it should be pointed out that the FBC is only a cascade control system without any robust and/or adaptive policies. Since the coating production line is a highly nonlinear system with both static gain and dynamics, and the aim coating weight and line speed significantly change under the various operating conditions, e.g. the production scheduling and furnace status, the resulting coating weight may not be satisfactory, or sometimes the system may be unstable in some operating areas.

5.4.3 Integrated neural system for coating weight prediction and control

An integrated neural system has been developed and successfully applied in coating weight prediction and control at a modern hot dip coating line (HDCL) (Markward and Lu, 1995, Lu and Markward, 1995) Similar to the original system, the neural system also consists of the coating weight prediction, FFC and FBC. Now we will introduce the details.

Neural network model for coating weight prediction Physically, the hot dip coating line has significant uncertainties due to producing various galvanized and galvanneal products according to customer orders and with an uncertain production environment. A linear regression model with a fixed structure can hardly match the entire working conditions. Based on the universal approximation of multilayer neural network in pattern mapping and associative memory, in this study a neural network state estimator has been developed and applied in the real-time prediction of the zinc pot coating weight. The general form of this neural estimator can be represented as

$$\mathcal{NN} : S, P, D \Rightarrow CW_p^p \tag{5.36}$$

The neural estimator developed consists of the following major components:

(a) Signal preprocessing and training sample selection

This step includes the signal filtering, I/O data normalization, and the training pattern selection with least-distance classification. It should be emphasized that the performance of the 'generalization' or the 'robustness' not only is dependent on the network's topology and the neuron activation function designed, but also heavily relies on the learning environment, namely the characteristics of the training samples. A program to select the training sample selection under a real-time environment has been applied; as a result, the size of the training data set and the required training time are significantly reduced, but on the other hand the training environment created still contains rich information with uniform data distribution. The mill test data have shown that the application of this software may make the training data set more informative day after day through gathering the updated information from the working environment.

(b) Multilayer feedforward neural network with BP-learning

A feedforward network with only one hidden layer is applied in the state estimator. A batch-wise, conjugated gradient back propagation (BP) learning algorithm is used to establish the neural network associative memory with synaptic weights. To make the learning more efficient, a one-dimensional search method is used to optimize the learning rate.

(c) Multilayer feedforward neural network for retrieval

A multilayer feedforward neural network with only forward path is used as a real-time associative memory. After the completion of the learning, the resulting synaptic weights are loaded as a data file for realtime applications. The prediction model runs in real-time through recall of the relevant weights files. The architecture of a neural estimator for this application is illustrated in Figure 5.17, from which we can see that the signal processing, learning and retrieval jointly work together to establish an associative memory with periodic weight upgrading.

Neural network for coating weight FFC A similar multilayer neural network is established for the coating weight FFC, i.e.

$$\mathcal{NN} : R_p^d, S, D \Rightarrow P_{ffc} \tag{5.37}$$

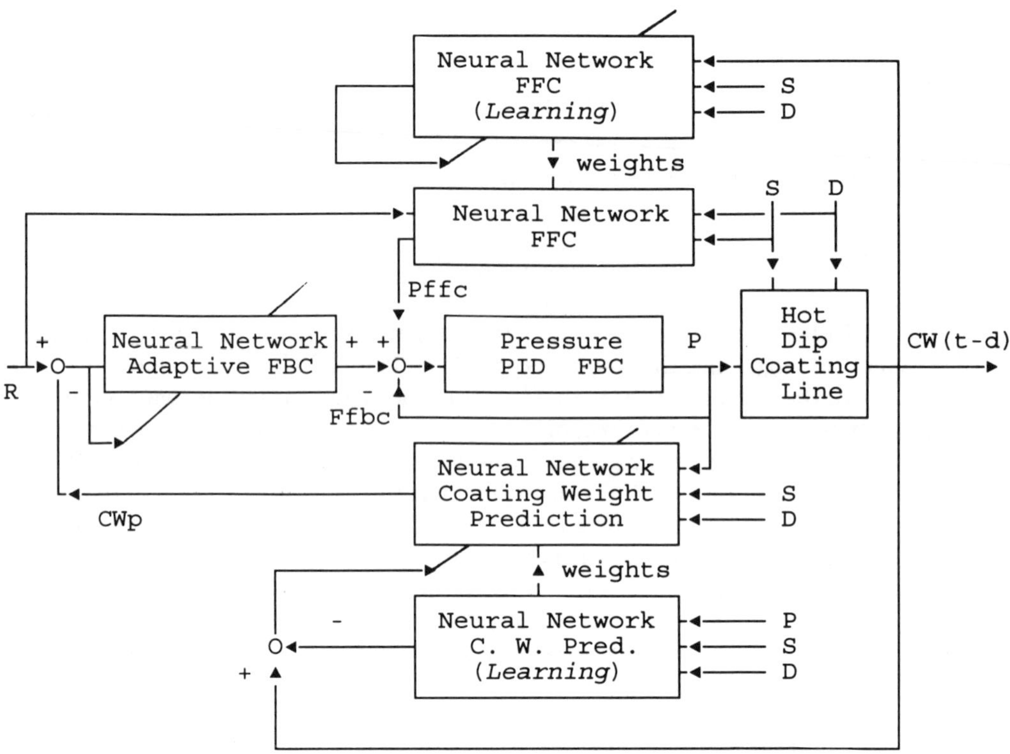

Figure 5.17 An integrated neural system for coating weight prediction and FFC/FBC control

It can be seen that the pressure P_{ffc} being the neural FFC output is dependent on the aim coating weight R_p^d, line speed S, and gap distance D. The signal processing, sample selection, training and retrieval are the same as we described above.

Neural Network for Coating Weight Adaptive Control The PID controller in the original system is replaced by the neural adaptive controller as developed in section 5.3. The design of a neural adaptive controller is to find a control law which can make the zinc pot coating weight track its set point with minimum error, namely:

$$\min_{W(k)} J(W(k)) = \sum_k (R^d(k) - CW_p^p(k))^2 \tag{5.38}$$

The neural control law can be represented as

$$P_{fbc}(k) = \mathcal{NC}(e(k), x_i(k), W(k), \eta) \tag{5.39}$$

where

$R^d(k)$ = desired level of the coating weight;
$CW_p^p(k)$ = predicted zinc pot coating weight;
η = learning rate;
$W(k)$ = synaptic weights;
$e(k) = R^d(k) - CW_p^p(k)$;
$\{x_i(k), i = 1, r\} = \phi_i(e(k))$, is an r-dimensional network input vector.

From equation 5.39, we can see that the neural adaptive control law performs a pattern mapping from its multiple inputs to a single output through a nonlinear function vector, namely $\mathcal{NC} : \phi_i(e(k))$. The closed-loop system stability is controlled by adjusting the learning rate η.

By combining the neural adaptive control with neural coating weight prediction and feedforward control, an integrated neural system can be as shown in Figure 5.17. It should be noted that the associative memory for coating weight prediction and feedforward control is established through batch-wise BP-learning with periodic upgrading, and the neural network for adaptive control is an adaptive network, i.e. its weights are upgraded at the end of each scan.

Practical aspects in neural networks applications It should be noted that a neural network can be used either as an associative memory for steady state modeling, or as a dynamic component for adaptive control. Moreover, just like a math-model based system with both estimation and control, a number of neural networks with different functions can also work together to form an integrated neural system. The following practical aspects must be considered in developing such an integrated neural system.

(a) While a neural network is developed for a static-prediction or static-FFC model, the generalization performance is an important factor for real applications. In general, the generalization depends on the network topology, activation function, learning performance and the difference between the learning and working environments. In this study, to follow the mill production environment changes, the training data set is updated periodically. On the other hand, the pace of the learning is controlled by a proper learning rate which will affect the associative memory established. The illustration of the learning pace is given by the stability–plasticity dilemma as we noted in section 3.4.5. From this dilemma we can realize that the learning pace should not be too fast or too slow. In this study, an adaptive learning rate is used during the learning phase.

(b) The coordination between a number of neural networks in an integrated system is another important issue to make the system robust and stable. This is particularly critical while a number of dynamic neural networks work together. In this study, an Adaline network (Widrow and Winter, 1988) with least-squares learning was also developed for adaptive (dynamic) coating weight prediction. The dynamic response of the coating weight prediction is given in Figure 5.18. It can be seen that the prediction results are excellent. However, when both Adaline prediction and FFC/FBC work together, in

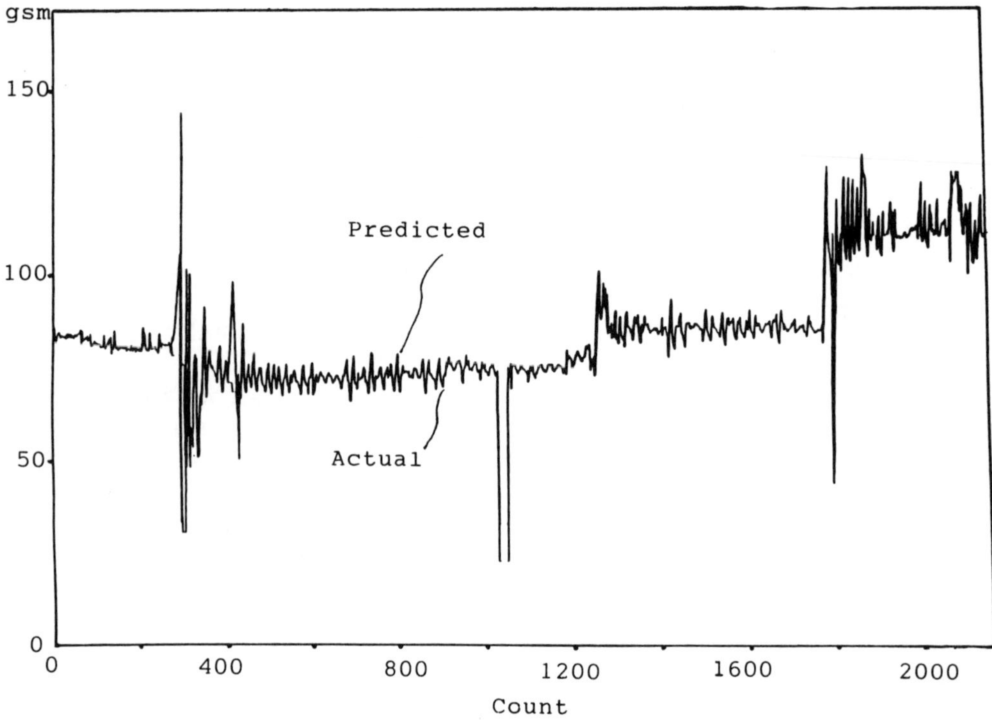

Figure 5.18 Coating weight prediction given by an Adaline neural network

some working conditions the entire system may be unstable. It has been recognized that more studies on neural system stability must be made with two or more dynamic neural networks working together.

Mill application test results The neural software developed for the coating weight prediction, FFC and adaptive FBC has been integrated and loaded into a real-time computer control system. A database on a workstation consists of the mill's real-time data, historical data, training samples and the network synaptic weights. The real-time system with the integrated neural system communicates with the local control loops at level 1 and with the production line through the computer network communication. The integrated neural system has been running successfully. In comparison with the original system, 5% zinc savings have been provided through reducing the aim coating weight additive and the transition footage has also been reduced by 8%. Here we give some test results and the comparison with the original system.

Regression model vs. neural model for coating weight predictioin

Figures 5.19 and 5.20 show the quantitative relationships between the air knife pressure and the coating weight for both the galvanized and the galvanneal

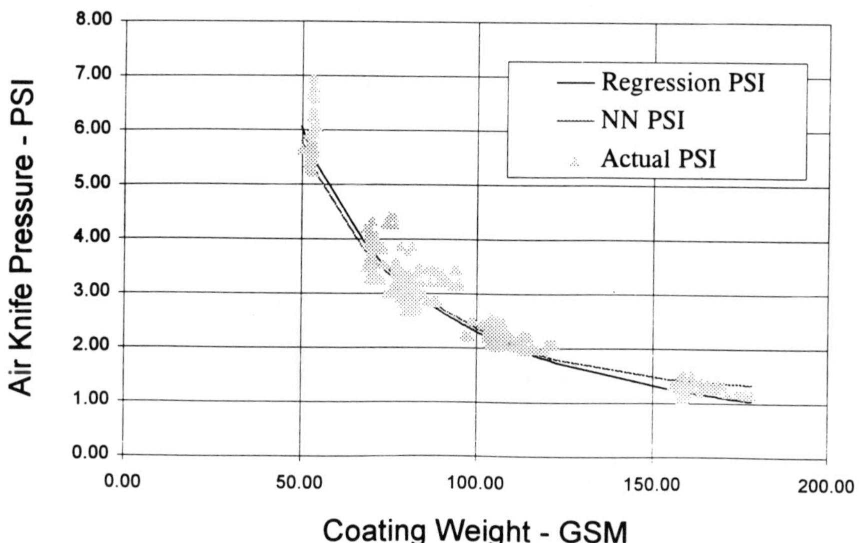

Figure 5.19 Comparison of the prediction results between regression and neural models for galvanneal production

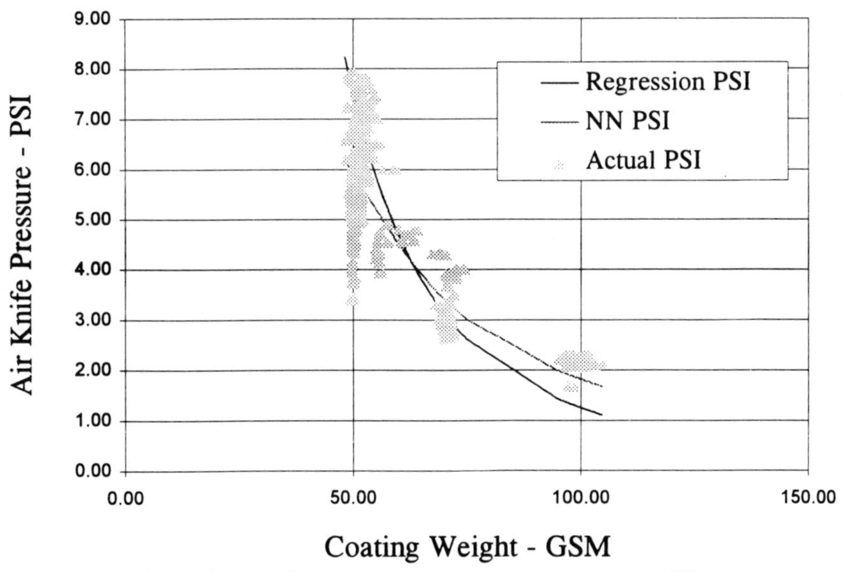

Figure 5.20 Comparison of the prediction results between regression and neural models for galvanized production

production. In the figures, the solid line and the dashed line represent the results given by the regression model and the neural network model respectively, and the dots are the actual production data. From the data shown in the figures, we can clearly see that the production data are closer to the neural network model.

Closed loop responses of the coating weight control

The sample closed-loop dynamic responses of the coating weight with both neural prediction and neural FFC/FBC are given in Figures 5.21 and 5.22. They are related to the changes of the reference coating weight and the line speed. It can be seen that the integrated neural system can provide satisfactory performance in real-time applications, even when the production environment, e.g. the reference coating weight, line speed or air knife gap, significantly changes.

Concluding Remarks This study has shown how to apply neural network technology to a real-world industrial process. Here we highlight the following points as our concluding remarks.

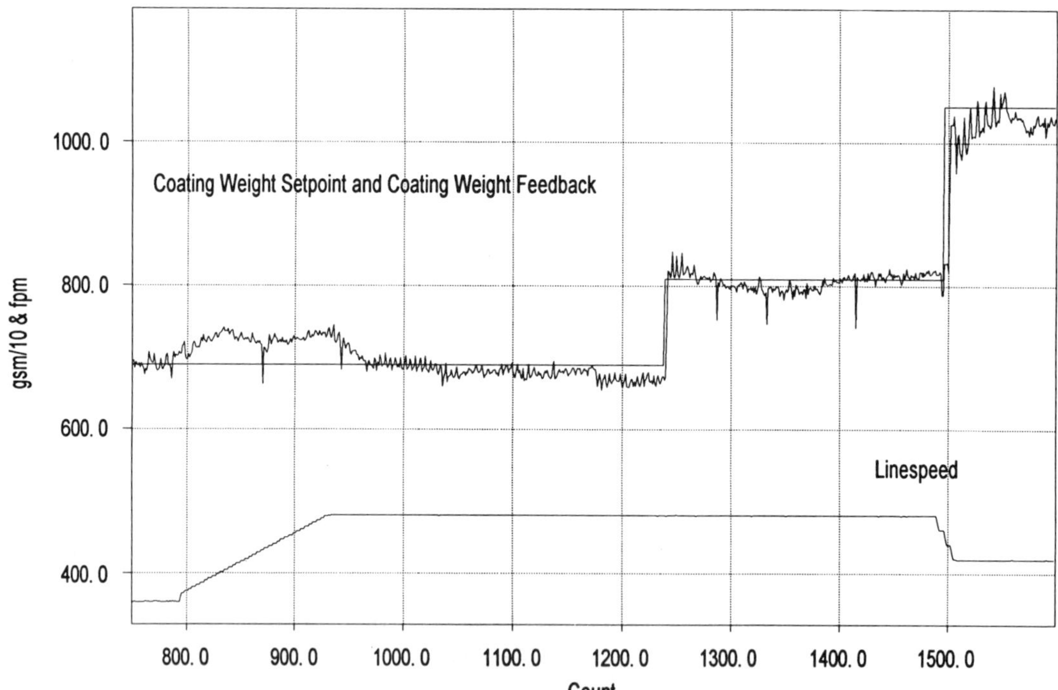

Figure 5.21 Closed-loop responses of coating weight control for galvanized production

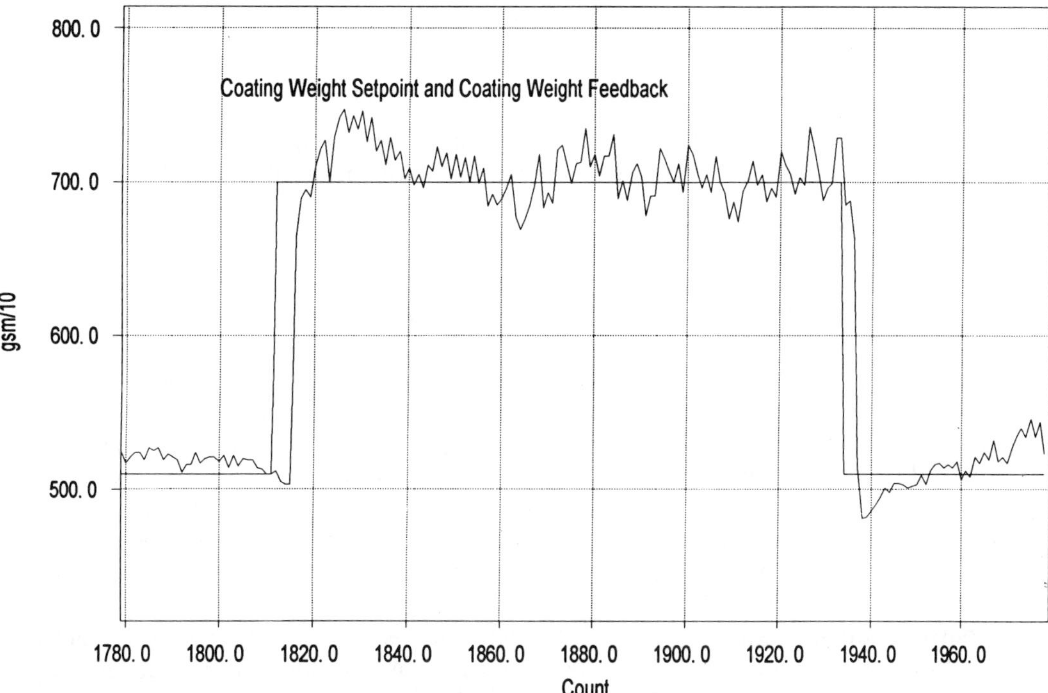

Figure 5.22 Closed-loop responses of coating weight control for galvanneal production

Regression vs. neural approximation

It is clear that both the regression and the neural approximation have a similar purpose, namely, to establish a functional relation (or pattern mapping) between system inputs and outputs through training, but they use a different architecture and algorithm. In general, the regression method is good for linear systems, or a nonlinear system with some knowledge of the structure. It should be noted that with advanced nonlinear regression techniques, the system structure may also be identified; however, this will be a difficult job for practicing engineers. In contrast, a neural network as a simple tool can be used as a universal approximation with more robust results for various nonlinear systems without knowing the system structure. On the other hand, the solution to both regression and neural network come from the same optimization problem. However, the solution to the regression is obtained directly by a least-squares oriented matrix manipulation; instead, the solution to the neural network approximation is provided by searching in an error space with some optimization method, e.g. a descent gradient method. The detailed analysis of this issue can be found elsewhere (Qin and Rajagopal, 1993). From this study, we can see that the neural network provides better coating weight prediction than the regression model for both galvanized and galvanneal production. The major merits of the neural network prediction are:

- powerful learning and associative memory capability;
- easier application applied in practice, namely, a universal structure can be used to approximate various nonlinear systems with different properties;
- less sensitivity to the input correlation and noise.

Neural network model based control vs. neural control with a direct feedback

While reviewing the applications of neural networks in advanced control, we will find that many neural controllers are formed through off-line training under some special test signals. Obviously, this is a time consuming and difficult job for many industrial processes with significant uncertainty. From this study, we can see that the application of an adaptive neural controller in coating weight control only requires very limited knowledge of the process, and no off-line training is needed. The neural controller with only a single layer can be performed even at level 1 of a hierarchical computer system.

Finally, from this study we can also see that an integrated coating weight prediction and FFC / FBC system can be developed and implemented through using a number of neural networks. Obviously, the hybrid system consisting of both math-algorithms and neural networks should also be feasible and applicable in real industrial application.

5.5 FUZZY LOGIC CONTROLS

5.5.1 Introduction

Fuzzy logic control (FLC) has been recognized as one of the most effective intelligent control strategies for industrial plants which have highly nonlinear, unknown and/ or uncertain dynamics. As we have learned, that a fuzzy knowledge based governed by fuzzy 'If-Then' rules represents human-like knowledge. However, it should be emphasized again that fuzzy rules are under a structured knowledge paradigm, but can be manipulated with a numerical frame through fuzzy reasoning or fuzzy modeling. On the other hand, fuzzy learning strategies with self-adjusting fuzzy membership functions provide a remarkable capability in designing an adaptive system working under an uncertain environment. These features make fuzzy systems more realistic, manageable and attractive in applications of systems dynamic control than regular expert systems which are poor and difficult in quantitative operation and learning.

In more detail, the major advantages of FLC can be summarized as follows:

- FLC is a math-model free control strategy; as a result, the math-model development, being a most difficult task, can be avoided;

- FLC is able to provide a 'universal approximation' (Wang, 1992) or so-called 'nonlinear–associative memory' and 'area to area pattern mapping' from its input signals to control actions, and the nonlinearity can be designed through selecting the type of membership function and adjusting the related parameters;
- actually, FLC combines human qualitative knowledge (fuzzy rules) with process I/O quantitative data; consequently, the resulting fuzzy control algorithm works with both linguistic fuzzy information and sensory I/O data. This function cannot be easily performed by either an expert system or a mathematical algorithm;
- the applications of neural network learning in updating fuzzy membership functions makes it simpler to implement an adaptive FLC;
- finally, from an application point of view, in comparing with math-model based control, FLC is much easier to design and implement through either programming or fuzzy VLSI chips.

However, on the other hand, the design of an FLC starts with a collection of linguistic fuzzy rules, e.g. the fuzzy relations between fuzzy inputs and fuzzy outputs in an FLC, hence if this kind of knowledge is not available it will be not realistic to design an FLC. Even the fuzzy knowledge might be directly extracted from system I/O data through unsupervised learning as we introduced before; however, the problem turns out to be much more complicated. In this section, we will focus on FLC architectures, design issues and applications.

5.5.2 Architecture of FLC

The general principles and structures of FLC have been extensively studied in the last few years (Buckley, 1992; Lee, 1990). Similar to any other feedback controller, a FLC can also be represented as an optimization problem: find a fuzzy feedback control law which minimizes the error between the reference set point and the actual value of the controlled variable under a variable controlled environment. A fuzzy control law can be described as following the general form:

$$\mathcal{F} : \{\Psi_i(e(k)), i = 1, r\} \Rightarrow \Psi u(k) \tag{5.40}$$

where $\mathcal{F}$ denotes a fuzzy rule set or fuzzy associative memory of an FLC, and $\{\Psi_i(e(k), i = 1, r\}$ and $\Psi u(k)$ denote the FLC's inputs and output. It can be seen that an FLC can have multiple inputs which are the desired functions of error, $e(k)$, and the fuzzy control law is performed by a fuzzy inference engine.

Figure 5.23 shows a general structure of a typical FLC with or without self-tuning functions. From Figure 5.23 we can see that any FLC consists of following the three components:

- Input Signal Coding: An FLC may have a number of fuzzy inputs which are obtained through taking the functions of error feedback signal, (e.g. error,

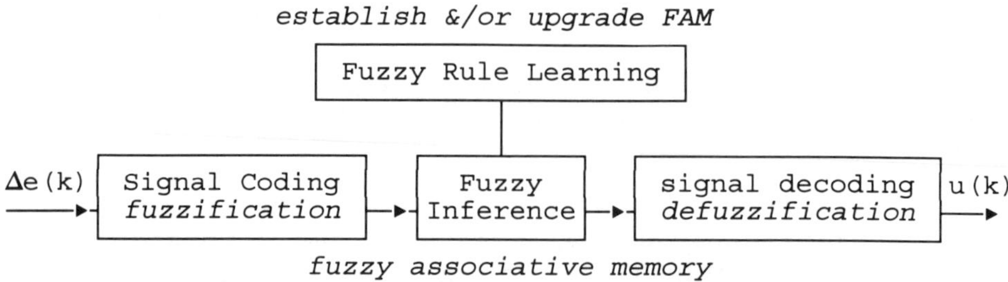

Figure 5.23 General structure of a fuzzy logic controller

$e(k)$, and change in error, $\Delta e = e(k) - e(k-1)$, and fuzzification of the error functions. As a result, an FLC can directly receive non-fuzzy information from the controlled environment through sensory measurement and/or any other data communications.

- Fuzzy Inference Engine: The task of fuzzy inference is similar to a math-control algorithm; however, the fuzzy inference only results in providing a fuzzy output (e.g. change in control action) in terms of designed fuzzy rules and fuzzy inputs.
- Defuzzification of Fuzzy Output: The fuzzy output provided by an FLC can serve as a control manipulator with non-fuzzy numerical values, only if the fuzzy variable is decoded through defuzzification.
- Fuzzy Rule Learning: To design an adaptive FLC, the fuzzy rules (rule structure and/or membership function) should be capable of being upgraded based on the real-time responses and given control criteria. However, in many applications only parametric fuzzy relations are modified.

The fuzzy rules for a FLC may be designed case by case; however, some FLC has been widely investigated and successfully applied in practice. For example, if taking the fuzzified error, E, and fuzzified change in error, ΔE, as the FLC's inputs and the change in control action (fuzzy variable), ΔU, as FLC's output, the general fuzzy control rules can be represented as follows:

$$\text{if } E \text{ is } A_i \text{ and } \Delta E \text{ is } A_j \text{ than } \Delta U \text{ is } B_k$$

where A_i, A_j and B_k are fuzzy variables which can be selected from a set of {NL, NM, NS, ZO, PS, PM, PL}, where NL, NM, PS, ZO, etc., are the abbreviations for the commonly used qualitative terms 'negative large', 'negative medium', 'positive small', 'zero', etc. Finally, a fuzzy matrix type of table can be established to describe the fuzzy relations between error related signals and the control actions. For many real-world problems, fuzzy control rules can have multiple inputs and multiple outputs, and sometimes related specific functions can be introduced in conjunction with either inputs or control actions. In addition, it should be kept in mind that the 'min' and 'max' operators correspond to 'and' and 'or' respectively in making the fuzzy inference.

5.5.3 Design of fuzzy-neural adaptive control

Similar to the regular and neural adaptive controls, the task of an adaptive FLC is to maintain satisfactory control behavior through making the fuzzy control law adapt to the changes of the controlled environment. What are adaptive fuzzy systems? Kosko (1992) noted that *adaptive fuzzy systems learn to control a complex process very much as we do. They begin with a few crude rules of thumb that describe the process. Experts may give them the rules. Or they may abstract the rules from observed expert behavior. Successive experience refines the rules and, usually, improves performance.* The core of an FLC is fuzzy associative memories (FAMs) which map from fuzzy set to fuzzy set from the input space to the output space. Based on the general concept of adaptive fuzzy system, the main function of an adaptive FLC is to upgrade FAMs with learning. Many fuzzy learning and self-tuning algorithms (Nomura *et al*, 1992; Xu and Lu, 1987; Wang, 1993; Qin, 1994) have been developed in the past years. However, it should be noted that combining fuzzy logic with neural networks is one of the most effective strategies either in designing a fuzzy system or in making a fuzzy system adaptive and robust. In this section, we will focus on how to apply neural network learning in fuzzy control systems to construct NN-driven fuzzy reasoning systems. Takagi (1994) investigated how neural networks and fuzzy logic have been jointly applied to Japanese consumer products, and indicated that as of September, 1990, there were 14 consumer products which have been furnished with neural and fuzzy technologies.

In this section, we mainly address the concepts and strategies of adaptive fuzzy systems with the neural network oriented fuzzy membership adaptation. In general, neural networks can serve as development tools for determining the fuzzy membership functions in the design phase, or as adaptive components for real-time upgrading of the membership functions for fuzzy systems. Fuzzy logic can encode expert knowledge directly using fuzzy rules. However, in developing a fuzzy system, how to determine the fuzzy membership function which can encode structured knowledge into a numerical frame is a critical issue in establishing a fuzzy associative memory. The self-learning ability of neural networks can automatically determine fuzzy membership in terms of the initial designed membership function and off-line data. For instance, the membership function for a washing machine with its inputs (cloth mass, impurity of water, time differential of impurity) and its outputs (water quantity, water flow rate, washing time, rising time and spinning time) can be determined by a neural learning system tool.

Moreover, fuzzy membership can also be upgraded in real time through neural learning in an adaptive manner; as a result, an adaptive fuzzy system with neural network learning can be developed. Suppose an FLC is governed by a fuzzy relation, $Y = X \circ R$, where X and Y represent fuzzy input and output vectors respectively, '$\circ$' denotes 'max–min composition', and R is defined as a fuzzy matrix representing a set of fuzzy associative rules. The task of a fuzzy adaptive system is to calculate R on-line with the fuzzy system inputs, and outputs and the resulting system performance. A general architecture of an adaptive FLC is shown in Figure 5.24. In principle, the neural network learning can upgrade either the membership functions or the fuzzy rules. However, in most real-time applications, an optimization oriented supervised learning algorithm is used to upgrade the fuzzy relation R.

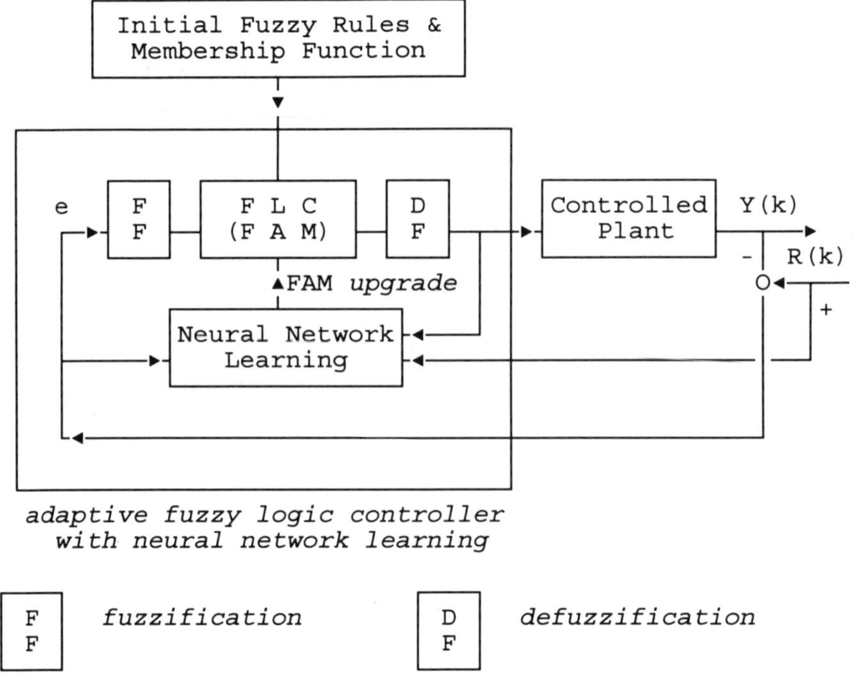

Figure 5.24 General structure of an adaptive fuzzy logic controller

A more sophisticated fuzzy–neural system can be designed in terms of unsupervised learning; as a result, a set of fuzzy rules, or so-called fuzzy associative memories, FAMs, can be extracted based on system I/O data.

In addition, the applications of neural networks in FLC can also be designed as a secondary corrector as used in Japanese consumer products. In this kind of scheme, the neural network receives the sensory information and produces the additional correction of the FLC's output. For instance, in a Hitachi washing machine, a parallel neural network provides a set of correcting values for washing time, rinsing time and spinning time, in terms of fuzzy inputs of cloth mass and quality, and electrical conductivity. Finally, it should be noted that the fuzzy–neural system is able to deal with both structured knowledge and numerical knowledge in terms of fuzzy logic and neural network respectively. Consequently, the fuzzy–neural system can process not only the sensor oriented information, but also the human preference factors; obviously, the marriage between fuzzy and neural technologies results in a high level of intelligence.

5.6 EXPERT CONTROL

5.6.1 General architecture and functions

In the design of a math-model based control system, the task is to develop a control strategy and algorithm which should meet the desired control criteria and the relevant constraints under a controlled environment with either explicit or uncertain and unknown dynamics. However, due to the complexity existing in many real-world control problems, not all control problems can be well represented, solved and/or implemented by the traditional mathematical methodologies and tools. Consequently, the applications of a human-like knowledge system in design and performance of real-time dynamic control has been an attractive field in both academic and industrial application aspects. Astrom *et al.* (1986) indicated that the objective of expert control is to encode knowledge representations and decision capabilities to allow intelligent decisions and recommendations to be made automatically rather than by preprogrammed logic which treats each case explicitly. Based on the characteristics of rule based expert systems and the demands of a dynamic control for a complex systems, expert control can be viewed as an application of expert systems in solving various sophisticated problems in systems control. In other words, an expert control system with knowledge representation and inference engine emulates human expert behavior in solving the control problems within a particular domain. The general scheme of an expert control system is illustrated in Figure 5.25, from which we can see that an expert control system functionally works as follows:

(1) The system receives information through the sensory measurements, man–machine interface and/or data communication from other systems. This information can be numerical (quantitative) or linguistic (qualitative).

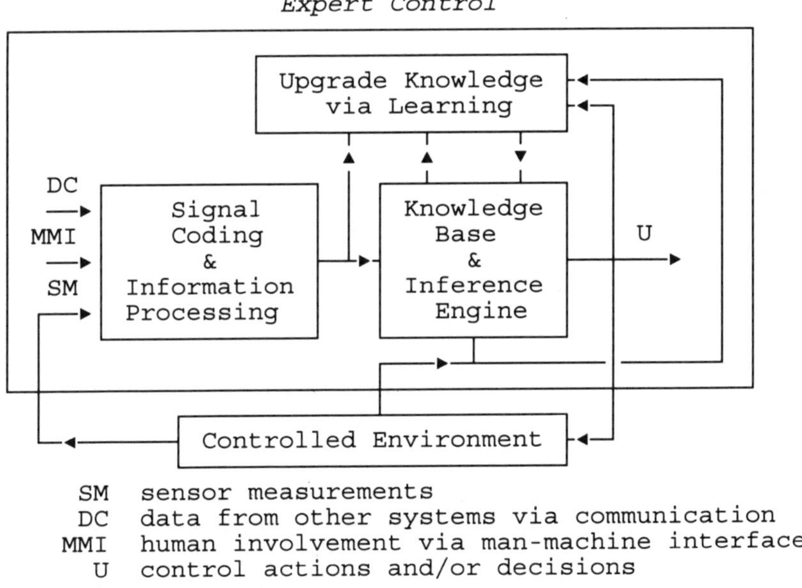

Figure 5.25 General scheme of an expert control system

(2) Based on the real-time information received, a knowledge base and inference engine provide the decision making to maintain the control system's performance stable and satisfactory under a significant uncertain environment, such as changes of disturbances, setpoint, system dynamics, etc. The decision making can be just tuning of the controller's parameters, to change the control algorithm and/or structure, or to send a command to other con trol systems (located at level 1 or level 2) to perform system coordination between a number of control systems.

(3) Rule learning and deep reasoning perform higher level intelligence in an expert control system. Functionally, the applications of rule learning and deep reasoning can make a system more flexible and adaptive through extending or changing the search space. Consequently, we may avoid the 'empty of solution' outcome sometimes occurring in a heuristic search process. On the other hand, any new rules explored through rule learning and deep reasoning should be automatically embedded and reorganized into the original knowledge base.

Since expert control works in a real-time manner, how to design a knowledge base and inference engine is a critical issue. Moore (1991) provided the details of G2, a real-time expert system tool.

5.6.2 Aspects in design of an expert control system

Knowledge acquisition and representation

Similar to many other expert systems, both knowledge acquisition and knowledge representation are the key in the design of an expert control system. In general, the knowledge can be obtained through the following channels:

- process engineers' knowledge, e.g. process characteristics and behavior;
- control experts' knowledge, e.g. the controller parameter tuning, nonlinearity, stability, controllability, observability, etc.;
- operators' experiences in the process response time, gain, interconnections, etc.;
- data and related information, e.g. the steady-state, statistical and dynamic production data can be used to extract the system knowledge with the supervised and/or unsupervised learning.

Eventually, the knowledge in an expert control system can be represented in different forms, for instance production rules, framework, math-algorithms and other associative memories with pattern recognition, fuzzy rules and neural networks. Moreover, a hybrid knowledge representation may be established based on the problem background.

Functions of expert control

Since expert control can be used in solving a variety of industrial control problems, it is important to be aware of the reasons for using expert control, and how to interface with current computer systems. Generally speaking, expert control can be applied in the following areas:

- self-tuning for conventional controllers;
- control algorithm and/or structure adaptation;
- coordination of interconnected control systems or between dedicate and supervisory levels;
- process monitoring;
- control for batch and discrete-event driven systems;
- on-line quality control, etc.

The system design depends on the desired functions and the real-time information available. It will be difficult to give a generic methodology for the design of expert control. In the following section, we will give a working example to describe the applications of expert control in industrial control.

5.6.3 *Working example*

An expert optimization control system (Lu *et al.*, 1995) was developed and successfully applied in a production scale fluidized catalytic cracking unit (FCCU). The entire system consists of the rule based dynamic control and coordination, fuzzy model and prediction, and expert supervisory optimization. In this section, we only present the topics of rule based dynamic control and coordination; the other topics will be introduced in Chapter 6. The setpoints of some dedicated dynamic control loops are governed by the on-line optimization results. Consequently, the coupling and stability of the dedicated control systems become a critical problem. In this expert optimization control system, both 'control rule' and 'coordination rule' are used to make the global system stable and satisfactory. Now we briefly address the structures of control rule and coordination rule.

Control rule base

It should be emphasized that the control loops in level 1 are the foundation for real-time optimization. Because the optimization may result in significantly changes of the controllers' setpoints, it is critical to design the controllers as the manipulators of optimization with good servo-tracking performance. The major functions of the control rules are to automatically select the control algorithms and their tuning parameters for the control loops in level 1 based on the optimization results and/or other operating conditions.

To illustrate the function of the control rule base, here we take distillation column level control as an example. From analysis of production data and principles, we find that after furnishing FCCU with a real-time optimization control the level control of the distillation column becomes extremely difficult due to the serious uncertainty between the level and the optimized parameters, e.g. the reactor temperature. The level control system with the conventional control algorithm sometimes turns unstable. To solve this problem, a hybrid intelligent controller with a combination of fuzzy logic control and human expertise knowledge is applied to update both the control law and its parameters. As a result of this application, the performance of level control becomes stable and satisfactory (He and Lu, 1988).

Coordination rule base

The coordination between the supervisory optimization control (level 2) and the dedicated control (level 1) in a hierarchical computer control system has been one of the major issues in practice. If we did not make good coordination, the system would be seriously disturbed or unstable, particularly for systems with nonlinear, strongly coupling and/or multiple time-scale behavior. To avoid this kind of situation, in addition to designing the dedicated control loops with robust or adaptive characteristics, the coordination between optimization actions and dynamic responses of the local control loops is an important measure. This kind of coordination makes a compromise between the optimization requirement and the capability of accepting the desired setpoint changes in the local control loops. For instance, the optimization of light oil rate will result in significant changes of the distillation column level. If the distillation column can not accept the optimization actions, the system will automatically adjust the optimization results to maintain a normal (or at least stable) operating status.

Rule structure

The rule base is designed with a modular structure, independent of the inference mechanism, such that both the knowledge representation and the inference may be flexible and efficient. To unify and simplify the inference process, each rule in the rule base is defined as a standard set with the following four elements:

$$[\mathrm{RN}, \mathrm{RI}, (\mathrm{A1}, \ldots, \mathrm{An}), \mathrm{B}]$$

where

RN is the rule name, usually presented by the number of the rule;
RI is the rule catalog, representing the rule base for control, optimization and coordination;
A1, ..., An and B denote the fuzzy condition and action respectively.

As an example, the rule describing the relations between the level of the distillation column, gasoline rate, light diesel rates and conversion rate can be represented as follows:

Rule 41: If level of distillation column is high (Lc > Lcm)
and gasoline rate is low (y1 < y1m)
and light diesel rate is high (y2 > y2m)
then conversion rate is low (B41)

This rule can be further expressed in the following format:

[41, 3, (Lc > Lcm), (y1 < y1m), (y2 > y2m), B41]

where the label '3' illustrates category of the rule set. Here '3' means that the rule '41' belongs to the *'coordination rule base'*.

Inference mechanism

The event (or condition) driven forward inference engine is used in this real-time expert control system. The control and coordination rule base are controlled by the relevant real-time clock. The entire inference process that includes the set of successful rules, intermediate and final inference results is stored to complete a search tree, that will be further used to explain the inference process and/or to be supplied for modification of the rule base. The industrial application results will be given in Chapter 6.

Finally, it should be emphasized that similar to the math-model based dynamic control, the structural property analysis, such as stability, observability and controllability, for intelligent control systems are also important in successfully designing a real working system. However, these topics are beyond the scope of this text. Narendra and Parthasarathy (1990) have investigated these issues in depth.

6 Optimization control techniques

6.1 INTRODUCTION

In industrial control, optimization control can be divided into two categories: 'dynamic' and 'steady state', which correspond to 'optimal control' and 'supervisory control' respectively. An optimal control creates a time oriented dynamic control trajectory to optimize a desired criterion subject to a set of constraints. The optimal control can be achieved by feedback, feedforward, adaptive or robust controls, etc. The main approaches to optimal control are 'maximum principle' (Brogen, 1985), and 'dynamic programming' (Bellman and Dreyfus, 1962), etc. The mathematical solutions of optimal control are based on a dynamic model, e.g. differential, difference or state space equations. However, the applications of optimal control to a discrete event dynamic system (DEDS) in manufacturing or communication systems are much more complicated, since it is difficult to describe the behavior of a DEDS with conventional mathematical tools.

In contrast, supervisory optimization control provides a steady state solution which is represented by a vector in a multi-dimensional space. Correspondingly, both criteria and constraints are governed by algebraic equations. The most popular approaches to steady state optimization are 'linear programming', 'nonlinear programming', 'dynamic programming', 'heuristic search', etc. However, since many industrial systems are high dimensional and highly nonlinear, application of the regular operations research approaches may not be applicable to providing a reasonable solution. This chapter starts with a review of on-line computer optimization control, and then focuses on intelligent supervisory control and production scheduling. The following topics with industrial working examples will be addressed:

- fuzzy model based optimization control;
- dynamic model based expert optimization control;
- intelligent production scheduling;

- computational evolution optimization;
- Hopfield network and simulated annealing optimization.

6.2 REVIEW OF ON-LINE OPTIMIZATION CONTROL

6.2.1 Introduction

Functionally, optimization can be applied in the following industrial control areas (Latour, 1979).

Optimizing operating conditions

The function of supervisory optimization control is to determine operating conditions (e.g. setpoints of local control loops) which can make the specified criteria be optimal (or satisfactory) subject to the given system constraints, under a variable or uncertain controlled environment. Eventually, the optimal operating conditions are based on process physical principles, e.g. mass and energy balance, chemical reaction, etc. For instance, the cracking product distribution of a fluidized catalytic cracking unit (FCCU) depends on reaction temperature, recycle oil rate, system pressure, catalyst activity, etc. In a hierarchical computer control system, the supervisory optimization control is usually placed in level 2.

Optimizing allocation

The proper allocation of a limited resource among parallel processes is another important application of optimization in industrial control. Typical examples are optimum load distribution in parallel steam generators as economic dispatch, loading electric generating plants, optimal energy management for multiple fuel gases, blending of gasoline or other materials, balancing multiple pass furnaces, etc.

Optimizing scheduling

Optimizing scheduling can be applied in continuous, semi-batch and batch production processes. The function of production scheduling is to provide time-sequenced job and machine decisions, for instance, scheduling of production operation from basic oxygen furnaces to continuous casters in the steel industry; scheduling for coke oven operation sequence; production sequence for batch production in fine chemical and food industry; scheduling for entire catalyst replacement in FCCU, cleaning heat exchanger bundles and steam de-coke of olefin furnace, etc.

When designing an optimization system, we should follow the following steps:

(1) Define the size of the optimized environment.

(2) Determine the purposes of optimization, and then select an objective function which depends on process independent variables (controllable and uncontrollable). If there are multiple criteria, the problem can be converted into a single criterion one through introducing penalty terms in the objective function or placing some lower priority criteria as additional constraints.

(3) Specify process constraints, which could be 'min–max' ranges for the values of variables and/or their differential rates, or some specific functional relations among the variables.

(4) Formulate a process model to describe the explicit math-functional relations or an associative memory (pattern mapping) between the variables.

(5) Design an optimization strategy and algorithm to deal with the problem under consideration.

(6) Carry out computer simulation studies.

(7) Develop real-time computer software for on-line optimization.

6.2.2 Math-programming based optimization

Linear programming

Many practical optimization problems in engineering can be formulated as a linear programming (LP) problem. The general form of LP can be described as follows:

$$\max P = \sum_{j=1}^{n} c_j x_j \tag{6.1}$$

subject to

$$\begin{aligned} \sum_{j=1}^{n} a_{ij} x_j = b_i, \quad i = 1, \ldots, m \\ x_j \geq 0, \quad j = 1, \ldots, m \end{aligned} \tag{6.2}$$

where P and $\{x_i, i = 1, \ldots, m\}$ denote the criterion and independent variables respectively, and c_j, a_{ij} and b_i are constants determined by the system environment.

In 6.1–6.2, if the solution of x_i can only be integers, then the LP problem becomes an 'integer programming' one. Moreover, if the solution of x_i can only be '0' or '1', then the problem becomes a so-called '0–1 programming' one.

Nonlinear programming

If the objective is a nonlinear function of the independent variables and/or if constraint equality or inequality is nonlinear, then the problem is defined as a nonlinear programming (NP) one and can be generally described as follows:

$$\min F(X) \tag{6.3}$$

subject to

$$\begin{aligned} h_i(X) &= 0, \quad i = 1, \ldots, m \\ g_j(X) &\geq 0, \quad j = 1, \ldots, r \end{aligned} \tag{6.4}$$

where F is a nonlinear objective function, both h_i and g_j are constraints and $X = [x_1, \ldots, x_n]^T$, being an n-dimensional vector, denotes the independent variables.

Dynamic programming

'Dynamic programming' (DP) originated by Bellman is another popular optimization technique, which mainly deals with the multi-stage (space or time) oriented optimization problem and provides an efficient mechanism for sequential decision making. The DP solution of optimization is carried out stage by stage. The mathematical basis of DP is the 'principle of optimality' (Bellman and Drefus, 1962): *An optimal policy has the property that whatever the initial state and initial decision are, the remaining decisions must constitute an optimal policy with regard to the state resulting from the first decision. In this classical statement, the 'decision' means a choice of control at a particular time and a 'policy' is an entire control sequence or control action* (Brogen, 1985).

Large scale system optimization

To simplify the computation of optimization, a large scale system (i.e. high dimensional and/or seriously interconnected system) can be decomposed into a number of interconnected subsystems; consequently, a hierarchical structure with subsystems (local control stations) and coordinator for optimization can be established. The optimization solution of a global system may be provided through the iterations between the subsystems and coordinator. In other words, we first solve the optimization problem for each individual subsystem, and then send the subsystems' solutions to the coordinator and evaluate the global system's performance. If the global system is at its local optimum region and satisfies the constraints, then end the iterations and provide the local optimal solutions, otherwise the coordinator will provide and transfer the coordination variables to the subsystems. The subsystems will upgrade their solutions with the information (coordination variables) from the coordinator level. The reader can refer to Singh and Titli (1978), Lin and Lu (1988), Zheng and Lu (1988) and Wu and Lu (1987) for the detailed algorithms.

The physical concept behind large scale systems (decomposition and coordination) optimization methodology can be described as follows: even if all interconnected production units (or facilities) work at their optimum conditions, the global performance of a complex will not be necessarily optimal due to the interconnections among the production units through mass, energy and information transfer. The philosophy and strategies of large scale systems optimization may be widely used not only in math-programming based optimization, but also in intelligent optimization, particularly for corporate wide business planning, production planning and scheduling, energy management, etc.

Comments

It should be noted that although many algorithms for NP have been developed, it is still difficult to solve a complex nonlinear optimization problem by using NP due to the following factors:

- if the system has multiple optimum points, it will be quite possible to obtain a local optimum instead of the global optimum by using the gradient search oriented NP. For many nonlinear system, the NP-solution is heavily dependent on the initial point;
- it is difficult to establish a nonlinear math-model for some complex industrial systems;
- it is difficult to deal with 'qualitative', 'imprecise' and/or 'incomplete' information and knowledge when applying NP in practice.

Finally, it should be emphasized that the math-programming based optimization framework and strategies are important in developing an intelligent optimiza-tion system. In the following sections, we will introduce some intelligent optimization techniques associated with working examples.

6.3 FUZZY OPTIMIZATION CONTROL

This section focuses on the strategies, algorithms and industrial applications of fuzzy optimization control.

6.3.1 Fuzzy optimization algorithms

Without loss of generality, the problem of fuzzy model based optimization can be represented with the following example:

$$\min_{x_1,x_2} y_1 = x_1 \circ x_2 \circ \ldots \circ x_n \circ R_1 \tag{6.5}$$

subject to

$$y_2 = x_1 \circ x_2 \circ \ldots \circ x_n \circ R_2 > y_{2m} \tag{6.6}$$

and other related constraints, where $x_i, i = 1, n$ and $y_j, j = 1, 2$ denote fuzzy input and output variables respectively, and $R_j, j = 1, 2$ represent fuzzy relations.

Here we introduce an algorithm for fuzzy optimization control with fuzzy rule match (Lu *et al.*, 1995). A fuzzy relational model can be decomposed into a large number of fuzzy rules with corresponding confidence factors, for example

$$\begin{aligned} &Rule\ i(s_1, s_2, s_3, s_4, s):\\ &\quad If\ x_1\ is\ A_{1s1},\ and, \ldots, and,\ x_4\ is\ A_{4s4}\\ &\quad Then\ y_i\ is\ B_{is}\ with\ Confidence\ (cf = R_i(s_1, s_2, s_3, s_4, s)) \end{aligned} \tag{6.7}$$

where $s_i, i = 1, 4, s$ denote the relevant confidence factors for both conditions and conclusions respectively. According to the features of the fuzzy relational model, a heuristic search method can be used as follows:

Step 1: Construct a 'rule table' that consists of all the rules covered by equation 6.6;

$$\begin{array}{l} s_1^{11}, s_2^{11}, s_3^{11}, s_4^{11}, r\\ \vdots \qquad\qquad \vdots \Rightarrow y_1 = B_{1r}\\ s_1^{1r}, s_2^{1r}, s_3^{1r}, s_4^{1r}, r\\ s_1^{21}, s_2^{21}, s_3^{21}, s_4^{21}, r-1\\ \vdots \qquad\qquad \vdots \Rightarrow y_1 = B_{1r-1}\\ s_1^{2r}, s_2^{2r}, s_3^{2r}, s_4^{2r}, r-1\\ \qquad\qquad \vdots\\ s_1^{r1}, s_2^{r1}, s_3^{r1}, s_4^{r1}, 1\\ \vdots \qquad\qquad \vdots \Rightarrow y_1 = B_{11}\\ s_1^{rr}, s_2^{rr}, s_3^{rr}, s_4^{rr}, 1 \end{array}$$

Step 2: A 'proper rule table' can be established through eliminating those rules which have lower confidence factors and/or do not satisfy the given constraints:

$$\begin{array}{l} s_1^{11}, s_2^{11}, s_3^{11}, s_4^{11}, r\\ \vdots \qquad\qquad \vdots\\ s_1^{1nr}, s_2^{1nr}, s_3^{1nr}, s_4^{1nr}, r\\ \qquad\qquad \vdots\\ s_1^{r1}, s_2^{r1}, s_3^{r1}, s_4^{r1}, r\\ \vdots \qquad\qquad \vdots\\ s_1^{rnr}, s_2^{rnr}, s_3^{rnr}, s_4^{rnr}, r \end{array}$$

All rules in the 'proper rule table' must satisfy the following two conditions:

- $R1(s1, s2, s3, s4, s) > e_0$ (a pre-selected threshold);
- without loss of generality, let $y_{2m} = B_{2t}$ and define

$$p_1 = \sum_{k=1}^{t-1} R_2(s_1, s_2, s_3, s_4, k) \tag{6.8}$$

$$p_2 = \sum_{k=t+1}^{r} R_2(s_1, s_2, s_3, s_4, k) \tag{6.9}$$

$$p_1 < p_2 \tag{6.10}$$

The condition 6.10 is equivalent to the constraint 6.6.

Step 3: Comparing the measurements of the random inputs (x_3, x_4) with the conditions of each rule in the 'proper rule table', and cutting those rules that match unsuccessfully, the remainder are written into a so-called 'matched proper rule table'.

$$\begin{array}{l} s_1^1, s_2^1, s_3^1, s_4^1, s \\ \quad \vdots \\ s_1^j, s_2^j, s_3^j, s_4^j, s \\ s_1^{j+1}, s_2^{j+1}, s_3^{j+1}, s_4^{j+1}, s \\ \quad \vdots \\ s_1^n, s_2^n, s_3^n, s_4^n, s, s < r \end{array}$$

The matching process can be represented by the following inequality:

$$\begin{array}{ll} \min[poss(x_3 A_{s3}), poss(x_4 A_{s4})] > P_0, & \textit{successful} \\ \textit{otherwise}, & \textit{unsuccessful} \end{array} \tag{6.11}$$

Step 4: Select a number of rules from the 'Matched Proper Rule Table' which can steer the objective function 6.5 to be minimum. These rules are constructed as a 'maximum matched proper rule table' as follows:

$$\begin{array}{l} s_1^1, s_2^1, s_3^1, s_4^1, s \\ \quad \vdots \\ s_1^m, s_2^m, s_3^m, s_4^m, s \end{array}$$

Then further select one rule from the 'maximum matched proper rule table' that may make the objective function reach the peak value most probably, i.e.

$$de_1 > de_i, \quad i = 1, j \tag{6.12}$$

where

$$de_i = R_1(s_1^i, s_2^i, s_3^i, s_4^i, s) - R_2(s_1^i, s_2^i, s_3^i, s_4^i, s-1) \tag{6.13}$$

Step 5: Finally, the solution of the approximate optimization can be given through defuzzification of the optimized fuzzy variables (x_1^*, x_2^*),

In fact, the fuzzy optimization strategy as described above consists of fuzzy rule match, cutting and search; all these actions are performed in terms of a specified criterion, a set of constraints and a fuzzy rule based model. Now we will further address a real industrial application of fuzzy expert optimization control in an FCCU.

6.3.2 Fuzzy expert optimization control for FCCU

Problem statement

A fluidized catalytic cracking unit (FCCU) is a typical complex system in process industry. An FCCU consists of multiple sophisticated chemical and physical processes. The production goal of an FCCU is to produce the cracking light petro-oil products, e.g. gasoline, light diesel, etc., through the FFC chemical reaction of heavy oil. From a system point of view, FCCU can be viewed as a high-dimensional, highly nonlinear and seriously interconnected complex system. Consequently, it is very difficult to design a real-time optimization control with a sophisticated FCCU mathematical model. In practice, there are various optimization opportunities for an FCCU. The primary variable is 'reactor temperature', which determines the conversion of the feedstock to the lighter products. However, it should be emphasized that fixed quantitative relations between reactor temperature and the reaction conversion (or the cracking product profile) exist only if the FCCU stands under a fixed operating status. This means that the major operating condition, such as heavy oil properties, production load, catalyst weight and activities and system pressure profile, are almost constant. However, due to the complexity of FCCU, this kind of ideal conditions does not always exist. In other words, a fixed (optimized) reactor temperature will not be necessary to provide an optimal operation for all operating conditions. The selection of optimization criterion has been an important issue for FCCU optimization control. For instance, most FCCUs in North America are operated for maximum gasoline production, but some northern refineries and European refineries tend to be run for higher middle distillate yield, varying with the season (Latour, 1979).

In this study, the cracking product distribution is selected as an optimization criterion. However, this distribution is not available for sensory measurement until after the distillation column. The time-delay from reactor to distillation column depends on the processing load and system pressure; for a large scale FCCU, the time-delay can be as high as about 15–20 minutes. Obviously, if the optimization control actions are based on such delayed information, it must be too late to manipulate the reaction conditions. To design a real-time fuzzy expert optimization control, we should first examine the following two problems.

Problem 1: To develop a fuzzy estimator with self-learning to predict the FCCU cracking product profile, e.g. cracking gas, gasoline, light diesel, etc. This problem can be represented as a fuzzy associative memory

$$FAM : U \in R^n \rightarrow Y \in R^m \tag{6.14}$$

where U and Y denote the operating conditions and the cracking product distribution respectively. This problem has been investigated in section 4.5.

Problem 2: To develop a real-time fuzzy model based expert optimization control (EOC) system to approximately optimize the cracking product distribution under various operating conditions and uncertainties. This problem involves design and implementation of an expert optimization control with the combination of fuzzy model, knowledge base and reasoning. This problem can be simply represented as follows:

$$EOC : P, C_{st}, E \rightarrow U^* \tag{6.15}$$

where P and C_{st} denote the criterion and constraints respectively; E presents the system environment related parameter set, and U^* denotes an optimization solution of the supervisory control vector.

Based on the analysis of process principle, human knowledge and production data, a large number of fuzzy rules were proposed to describe the fuzzy relations between the reaction conversion rate and main operating conditions. However, to simplify the complicated representation, the fuzzy rules can be simply written in the following form:

$$C = F(u_1, u_2, u_3, u_4) \tag{6.16}$$

where

$F(*)$ = a fuzzy operator describing the fuzzy rule base;
C = reaction conversion rate;
u_1 = reaction temperature;
u_2 = reaction pressure;
u_3 = recycling oil rate;
u_4 = ratio between catalyst weight and oil flow rate to reactor.

To obtain a numerical solution, the fuzzy rule base (with structured, symbolic knowledge representation) should be converted into its corresponding fuzzy relational model (with numerical framework). As a result, the fuzzy model is able to deal with numerical operations. The resulting fuzzy relational model describing the numerical relations between the reaction product profile (replacing conversion rate) and operating conditions can be expressed in the following form:

$$y_j(k) = u_1(k) \circ u_2(k) \circ u_3(k) \circ u_4(k) \circ R_j(k), \quad j = 1,4 \tag{6.17}$$

where

y_1 = gasoline rate;
y_2 = light diesel rate;
y_3 = heavy diesel rate;
y_4 = cracking gas rate.

The initial fuzzy relations $R_j(0), j = 1,4$ are determined by using the fuzzy identification technique in terms of off-line batch data $\{x_i(k), y_j^d(k-L), i = 1,4, k = 1, K_f\}; y_j^d$ denotes the measurements of product rates at the outlet of the distillation column. Based on the initial fuzzy relations $\{R_j(0), j = 1,4\}$ and the real-time I/O measurements, the real-time fuzzy relations $\{R_j(k), j = 1,4\}$ can be solved through a recursive fuzzy identification algorithm, namely, upgrading the relevant fuzzy rules. As a result of applying the fuzzy relational model, the predicted product profile can be provided with the fuzzy relations and observation of the inputs, $\{u_i, i = 1,4\}$. The observation of the product profile at the outlet of the distillation column, $\{y_j^d(k-L), j = 1,4\}$, is used as a feedback signal for fuzzy self-learning. The function of self-learning is to make the fuzzy estimator robust and adaptive. In fact, this fuzzy model has also been applied in expert optimization control, in which the fuzzy model is not only used as an estimator, but also serves as one of the key bodies in the fuzzy reasoning and coordination.

Fuzzy expert optimization control

In general, the criterion can be selected from the following three options:

(a) maximum gasoline rate;

(b) maximum light diesel rate;

(c) maximum light product (gasoline plus diesel) rate.

The constraints are the 'min–max' ranges of the operating parameters for normal operation and safety, for example

1. level of distillation column;
2. level of recycling oil tank;
3. temperatures of 'dense-phase' and 'dilute-phase' in the regenerator;
4. production system pressure profile;
5. specified product qualities, etc.

Without loss of generality, here we take only criterion (a) as an example; thus the real-time optimization control problem can be described as follows:

$$\max_{u_1,u_2,u_3,u_4} y_1 \tag{6.18}$$

subject to

$$\begin{aligned} y_1(k) &= u_1(k) \circ u_2(k) \circ u_3(k) \circ u_4(k) \circ R_1 \\ y_2(k) &= u_1(k) \circ u_2(k) \circ u_3(k) \circ u_4(k) \circ R_2 \\ u_{i\min} &\le u_i \le u_{i\max}, \quad i = 1,4 \\ y_2 &\ge y_{2\min} \end{aligned} \tag{6.19}$$

and other related constraints, where $u_{i\min}$ and $u_{i\max}$ are defined as the 'min' and 'max' boundaries for u_i respectively. Here only u_1 and u_2 serve as on-line optimized variables, and u_3 and u_4, being provided by the expert optimization control system, serve as off-line operating guides, because it is difficult to automatically regulate them for this specific facility. The algorithm of fuzzy model based optimization control has been illustrated in section 6.3.1. This section only focuses on some application oriented issues, particularly the construction of knowledge base and reasoning with different levels. Moreover, this system consists of three databases, i.e. real-time dynamic database, steady-state database and historical database. Now we summarize both knowledge base and inference.

Control rule base The control loop at level 1 is the foundation for real-time optimization, because the optimization may result in significant changes of the local controllers' setpoints. It is critical to design a controller to be the manipulator of optimization with good servo-tracking performance. Just like an expert control system, the functions of the control rules are to automatically select the control algorithms and their tuning parameters for control loops at level 1, $\{u_i, i = 1,4\}$, based on optimization results and/or production environment. For instance, after furnishing FCCU with real-time optimization control, level control of the distillation column becomes extremely difficult due to serious coupling between the level and optimized variables. As a result, a conventional control algorithm may make the level control unstable. A set of control rules are used to select a proper control law (e.g. fuzzy logic control or PID control) and its relevant tuning parameters to make the system stable.

Optimization rule base Optimization rules representing the qualitative relations between criterion and optimized variables can be gathered through process principle, data analysis and human knowledge. For instance, many optimization rules can be directly extracted from Figure 6.1 which describes the qualitative relations between the light oil product rates and on-line optimized variables, i.e. both reactor temperature and recycle oil rate under a constant catalyst activity. It can be seen from Figure 6.1 that the optimum reactor temperature is moving with the

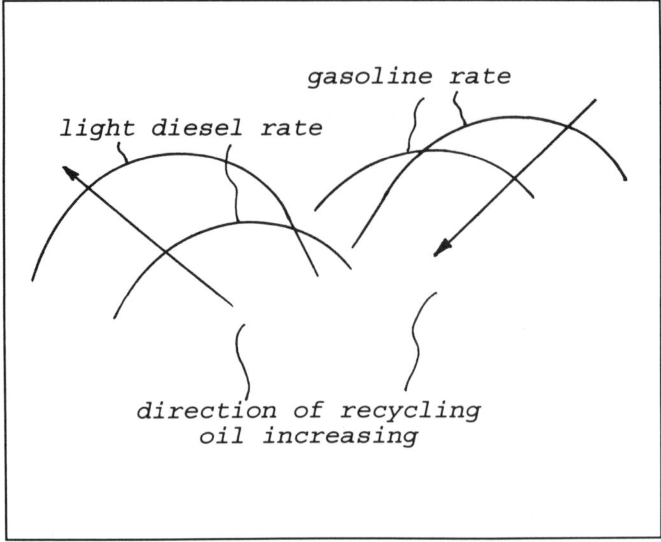

Figure 6.1 Qualitative relations between light oil product rates and on-line optimized variables

changes of the recycling oil rate. On the other hand, combining fuzzy model based prediction with optimization rules may significantly reduce the search cost. This is one of the major merits in carrying out real-time expert optimization control.

Coordination rule base The real-time optimization may result in a stability problem for local control loops. In addition to designing the local control loops with robust or adaptive features, the coordination between optimization actions and dynamic responses of local controls is an additional important measure. In fact, this kind of coordination makes a compromise between the optimization results and capability of accepting the supervised setpoints changes for the local controls. For instance, the optimization of light oil rate may result in significant changes of the distillation column level; if the distillation column can not accept the optimization actions, the system will automatically adjust the optimization results to maintain a normal (or at least stable) operating status.

On the other hand, the coordination between the optimization rules and fuzzy model based optimization search in FCCU optimization control provides a joint utilization of both qualitative and multivalued knowledge. The potential optimized actions can be widely extended to those variables which are not included in the fuzzy model. Consequently, a hybrid system is certainly more effective and better than if we use only either, optimization rules or fuzzy model based optimization search.

Structure of rule base To unify and simplify the inference process, each rule in the rule base is defined as a standard set with the following four elements:

$$[RN, RI, (A1, \ldots, An), B]$$

where

RN = rule name presented by the number of the rule;
RI = rule catalog label, indicating the rule base for control, optimization and coordination;
$A1, \ldots, An$ and B denote conditions and actions respectively.

Application of the rules' structure as noted above may provide higher search efficiency. An example rule describing the relations between distillation column level, gasoline rate, light diesel rate and conversion rate has been represented in Section 5.6.3.

Inference mechanism An event-driven forward inference engine is used in this real-time expert optimization control system. The inference tree is illustrated in Figure 6.2. The rules of control, optimization, coordination, and the fuzzy model, are controlled by a real-time clock. All nodes in Figure 6.2 are connected with a dynamic database and the lines between nodes represent the calling of the corresponding

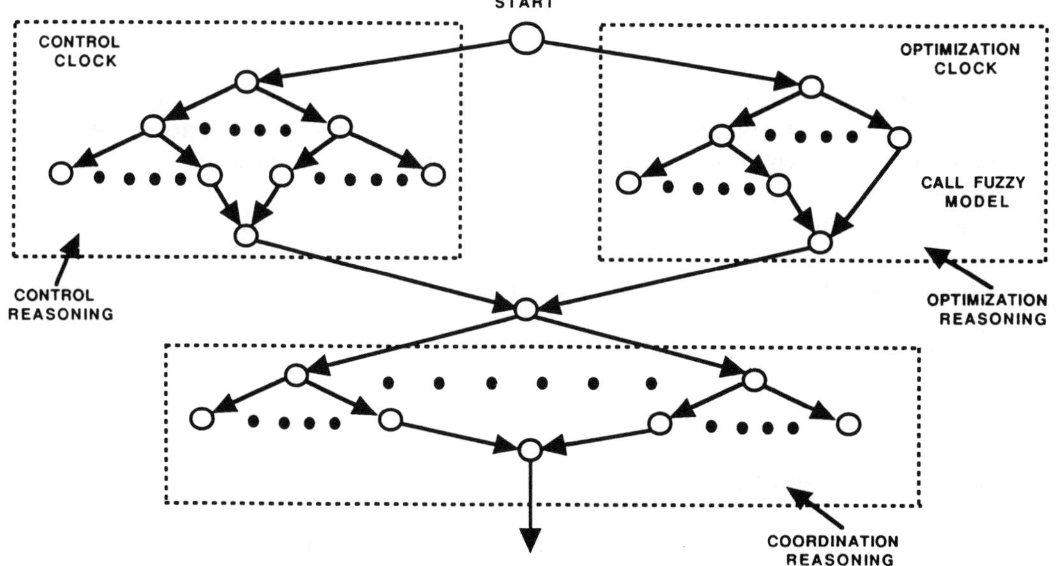

Figure 6.2 Inference tree in expert optimization control system for FCCU

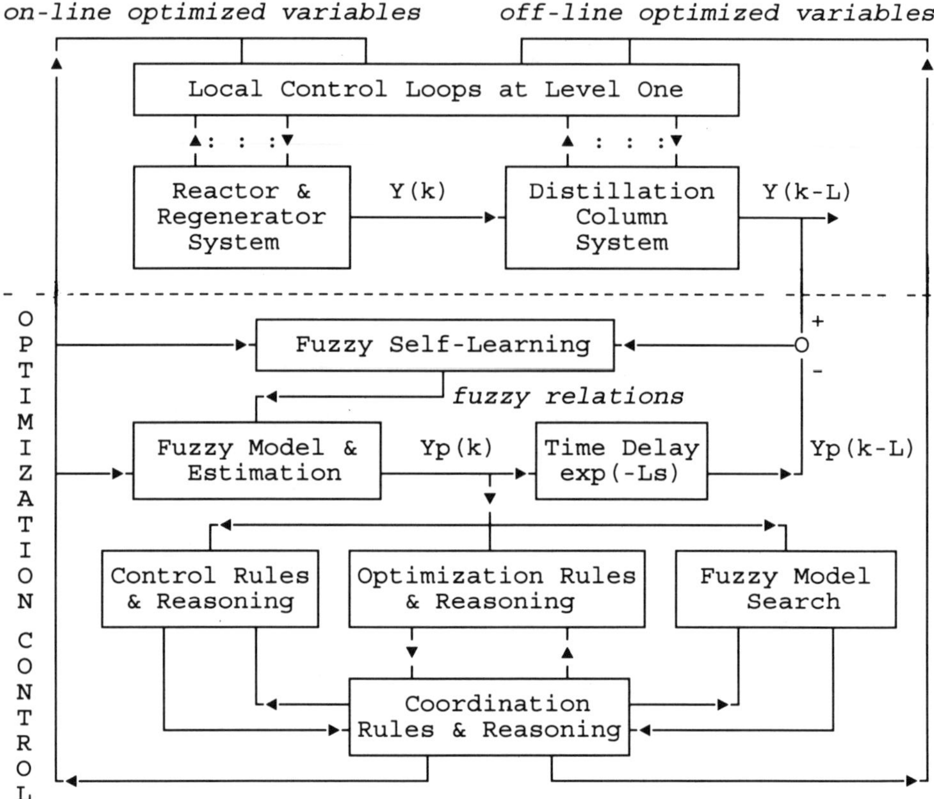

Figure 6.3 System configuration of fuzzy expert optimization control for FCCU

rules. The entire inference process, including a set of successful rules and the intermediate and final inference results, is stored as a complete search tree to be used to explain the inference process and/or to be supplied for the modification of the rule base.

The entire system configuration is shown in Figure 6.3. It can be seen that the system consists of knowledge base and reasoning with 'control', 'optimization' and 'coordination' rule sets which jointly work with fuzzy model based optimization search. In fact, the coordination rules do perform a function of 'meta-knowledge' in this system.

Industrial application results

This expert optimization control system has been successfully applied in a large scale FCCU. The optimization expert control is installed at level 2 of a hierarchical computer control system. Similar to many on-line optimization control systems, the

Table 6.1 Comparison between original system and expert optimization control in product rates

Product name	Product rate (%) Trad. control	Ex. Control	Product rate change (ton/year)
Gasoline	52.10	52.61	4590
Light diesel	27.14	27.40	2340
Heavy diesel	3.27	3.60	2970
Cracking gas	12.81	11.72	−9810
Others	4.68	4.67	—

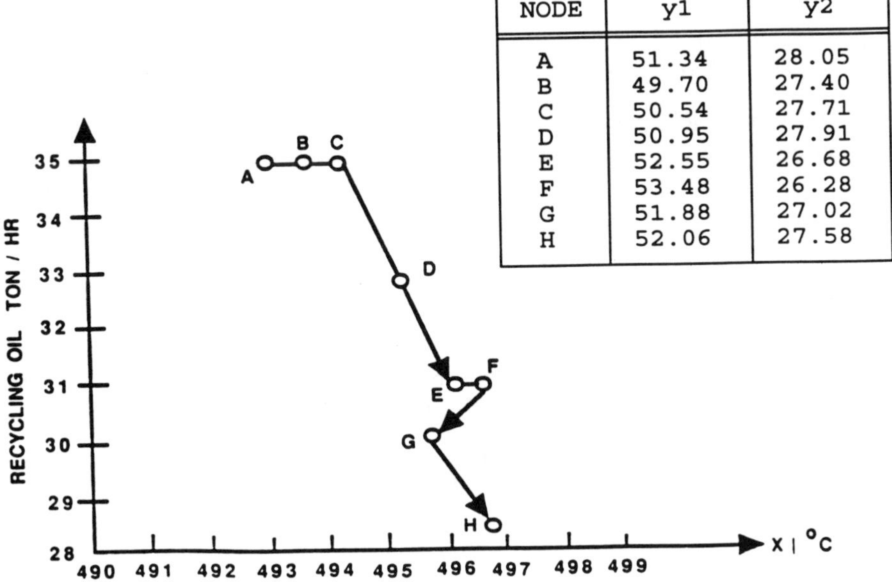

NODE	y1	y2
A	51.34	28.05
B	49.70	27.40
C	50.54	27.71
D	50.95	27.91
E	52.55	26.68
F	53.48	26.28
G	51.88	27.02
H	52.06	27.58

Figure 6.4 Sample results of fuzzy model based optimization search path for FCCU

optimization control actions are sent to the local controls only if the optimized set point change is greater than a specified threshold value.

The quantitative comparison results between the original control (with human operator's setpoint setting) and fuzzy expert optimization control are given in Table 6.1. It can be seen that considerable economic benefits may be provided through increasing both gasoline and diesel rates and decreasing cracking gas rate. The sample optimization search path is shown in Figure 6.4. This optimization control system is satisfied for real-time applications in various production environments, such as changing heavy oil properties and/or production throughput, etc.

6.4 DYNAMIC MODEL BASED EXPERT OPTIMIZATION CONTROL

This section presents a development and industrial application of dynamic math-model based expert optimization control. The idea behind this kind of expert optimization control is to take the respective advantages in math-model and rule based heuristic frames. We will start with a general introduction to the system architecture and algorithm, and then describe a real-time application for steel reheating furnaces.

6.4.1 System architecture and algorithm

A general scheme of a multi-zone reheating furnace is illustrated in Figure 6.5. Functionally, a dynamic state space model may perform the following two roles in real-time optimization for reheating furnace:

- real-time estimation or prediction for process internal (unmeasurable) information to be used as resources for real-time optimization control;
- instead of 'on-line' optimization search under a real production environment, the search is carried out 'off-line' with a state space model being viewed as a simulated environment. Consequently, much less disturbance will affect the normal operation.

On the other hand, the major merits of applying rule base and inference in a real-time optimization control system can be summarized as follows:

- the qualitative and/or human factors that cannot be included in dynamic math-model may be managed with an AI heuristic approach;
- to avoid the difficulties in applying math-programming optimization techniques for many complex industrial systems.

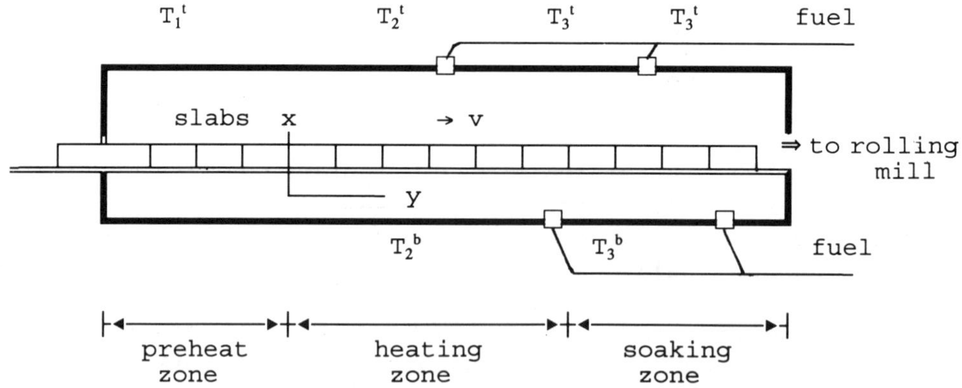

Figure 6.5 Scheme of a multi-zone reheating furnace

An example of dynamic model based optimization control can be represented as follows: select a control vector, $U(k) = [u_1(k), \ldots, u_r(k)], k = 1, K-1$, such that the performance index

$$J = \sum_{k=0}^{K-1} P(U(k)) \tag{6.20}$$

is minimized, while subject to

$$\begin{aligned} X(k+1) &= F(X(k), U(k)) \\ X(0) &= X_0 \\ X &\in \Omega_x \\ U &\in \Omega_u \end{aligned} \tag{6.21}$$

and other qualitative constraints, where U represents a control vector for process optimization, which could be the setpoints of the local controls, and $X = [x_1, \ldots, x_n]^T$ denotes a state vector and is controlled by the control vector U. The approach to the optimization problem as given is to iteratively improve the performance index, satisfying all constraints (numerical and qualitative) by heuristic rule search on the state space model 6.21–6.22. The algorithm can be summarized as follows:

Step 1: Set up stationary conditions for optimization from the normal operating data.

Step 2: Let the iterative number $i = 0$, and select the initial control vector (setpoints) $U(i)$ in terms of the stationary condition.

Step 3: Calculate the estimated state variables $X(i)$ with 6.21 and the real-time $U(i)$.

Step 4: Check if $X(i)$ and $U(i)$ meet the optimal criterion (compare $J(i)$ with $J(i-1)$) and the constraints 6.22; if yes, transfer the optimization results to the local controls, otherwise go to step 5;

Step 5: Based on the information obtained in the step 4, make further search on the search space through calling the proper heuristic search rules from the rule base, let $i = i + 1$, then return to step 3, until the optimization is solved.

The optimization control algorithm was designed as an event-driven one. This means that the optimization control algorithm is active, only if the changes of operating conditions (e.g. changes of production load, material properties and/or production scheduling, etc.) are beyond of a pre-specified threshold, or the statistical process control (SPC) data show that the process has been moving out of the optimal region. It should be emphasized that the self-learning is introduced to make the dynamic model more accurate. Finally, the key in heuristic optimization search is to develop a set of production rules which are practically reasonable. Similar to fuzzy model based optimization, process principle and data analysis, math-model based simulation results and human knowledge are commonly used to establish an optimization search rule base. Now we will present a real industrial application of dynamic model based expert optimization control in a steel slab reheating furnace.

6.4.2 Optimization control for a reheating furnace

This section addresses an industrial application of dynamic model based expert optimization control for a slab reheating furnace developed by Yang and Lu (1988).

Introduction

Slab reheating furnaces popularly used in the plate steel mills are schematically shown in Figure 6.5. The furnace can be divided into preheating, heating and, soaking zones and in general, a number of parallel furnaces sequentially supply the heated slabs for a single rolling mill. The optimization goals of the reheating furnace operation are to deliver the heated slabs 'in time' with the required (metallurgical and mechanical) uniform temperature profile, while minimizing both fuel consumption and over-heating steel losses. To perform these goals, a hierarchical computer control system with the following three levels is designed:

Level 1: 'Temperature and combustion control' for each heating or soak zone. The purpose of this level is to control zone temperatures under the most efficient combustion status, i.e. to manipulate the ratio between combustion air and fuel gas based on the oxygen analysis in the flue gas.

Level 2: 'Supervisory optimization control' for the setpoints of the zone temperatures. The purpose of this level is to optimize each zone temperature to meet the mill running pace and to minimize fuel consumption and over-heating steel losses.

Level 3: 'Production scheduling' for 'furnace-mill' system. The task of the scheduling is to assign a job-shop (slab-furnace) time-sequence, which makes the whole system run smoothly.

However, in this section we will only deal with the supervisory optimization control at level 2.

Problem formulation

To design an optimization control system, a 'slab tracking thermal model' with discrete-time state space form was developed by Yang and Lu (1986, 1988). With this model, we can track any particular slab that moves through a temperature field being a function of the furnace length dimension. As a result, the real-time temperature profile responses of any slab can be estimated through running a slab tracking thermal model. Based on the process background and the optimization requirement, the problem formulation for furnace optimization control can be considered as follows:

- select the minimum sum of the weighed square zone temperatures as a criterion; consequently, both fuel consumption and steel losses will be approximately minimized;
- the temperature setpoints of heating and soak zones are selected as the optimized control variables;
- the state space dynamic model, desired slab temperature profiles and other operating conditions are used as the constraints;
- the variations in slab size, steel grade (related to the thermal properties), slab moving speed, etc., are considered as uncontrollable disturbances, and are applied in the model calculation.

Correspondingly, the problem can be represented in the following mathematical form:

$$\min J(U) = \sum_{k=0}^{K-1} \|U(k)\|^2_{R(k)} \tag{6.22}$$

subject to

$$X(k+1) = A[X(k),k]X(k) + B[X(k),k]U(k) \tag{6.23}$$

$$X(0) = X_0 \tag{6.24}$$

$$X(K) \in [X^* - \delta, X^* + \delta] \tag{6.25}$$

$$X(k) < X_{max} \tag{6.26}$$

$$U_{min} \leq U(k) \leq U_{max} \tag{6.27}$$

where

k = discrete time instance;
K = travelling time of the slab from the furnace entrance to exit;
X = temperature profile in the dimension of the slab thickness;
X_0 = slab temperature profile at the furnace entrance;
X^* = expected slab temperature profile at the furnace exit;
δ = admissible tolerance error of the slab temperature profile at the furnace exit;
U = zone temperature distribution.
R = weighting factor

U_{min} and U_{max} are the 'min' and 'max' limits of the zone temperatures respectively. All X, X_0, X^*, δ have the same dimension, being equal to the node number used in the slab's spatial discretization. A and B are defined as system matrices, whose elements are functions of state vector X. In practice, the furnace temperature distribution can be simplified as a one-dimensional (length dimension) piecewise linear function in terms of the measured zone temperatures. For instance, the furnace temperature distribution on the continuous 'time' and 'spatial' space can be precisely represented as

$$U = [u_b(y_s, t), u_t(y_s, t)] \tag{6.28}$$

where the subscripts t and b represent the top and bottom respectively, and the notations y_s and t denote the slab position in the furnace and the time coordinate respectively. Suppose a furnace has three bottom zones and four top zones; the simplified form of the furnace temperature distribution can be expressed as follows:

$$u_t(y_s, t) = f_1(T_1^t(t), T_2^t(t), T_3^t(t), T_4^t(t), y_s) \tag{6.29}$$

$$u_b(y_s, t) = f_2(T_1^b(t), T_2^b(t), T_3^b(t), y_s) \tag{6.30}$$

where f_1 and f_2 are defined as linear piecewise functions. With equations 6.29–6.30, now criterion 6.22 can be simplified as

$$\min \sum_{i=1}^{N} [T_i^b T_i^t] R_i [T_i^b T_i^t]^T \tag{6.31}$$

where N denotes the pairs of zones, and $R_i, i = 1, N$, are defined as positive definite weighing matrices. However, in practice, only the temperatures of the heating and soaking zones, T_2^t, T_2^b, T_3^t and T_3^b, are controllable and optimized by real-time optimization.

System Configuration

A computer program flowchart of real-time optimization control for heating furnaces is given in Figure 6.6. The system working procedures just follow the steps as we described in section 6.4.1. The heuristic production rules are important in performing the search on the performance space with respect to the control variables. However, it should be kept in mind that this search is carried out only in a simulated furnace, i.e., a state space model. Here we list some sample shallow knowledge based rules used in the heuristic optimization search:

- if the average slab moving rate is increasing (i.e., mill running pace increasing), then the setpoints of heating zone temperatures should be correspondingly increasing;
- If the average thickness of the charged slabs is decreasing, then the setpoints of heating zone temperatures should be correspondingly decreasing;
- If the slabs require specific heating trajectory (due to the metallurgical reasons), the set points of the heating and soaking zone temperatures should meet those requirements.

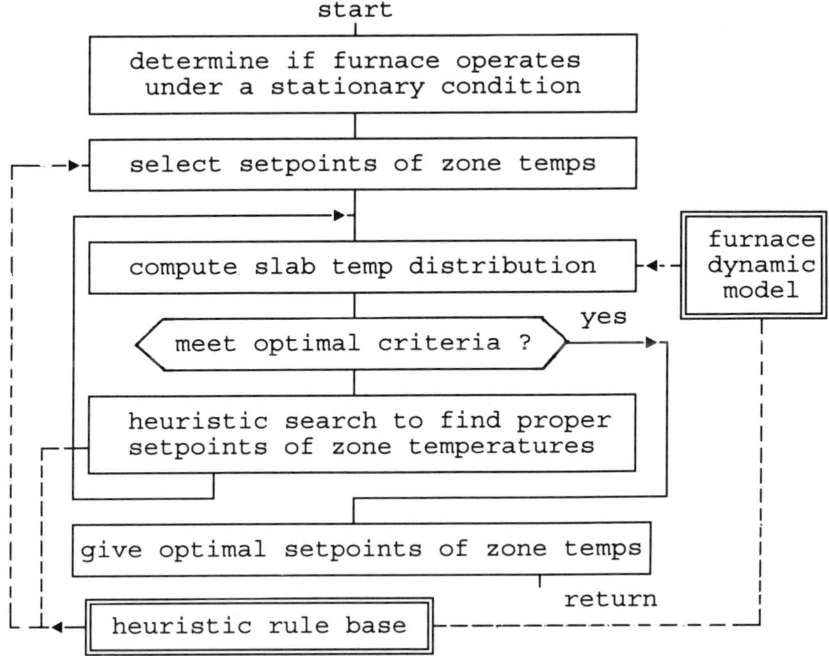

Figure 6.6 Flowchart of dynamic model based expert optimization control for reheating furnace

Obviously, rules of this kind rather rough and vague. Consequently, the major part of the production rules are extracted from the dynamic model based simulation results as shown in Figure 6.7. In addition, the applications of energy conservation can also produce some rules. For instance, the actions of optimization should move the center of heat inputs far from the exhaust gas outlet; in other words, fuel consumption can be reduced with a lower temperature level at the preheating zone.

Industrial application results

The system has been implemented in a production scale slab mill with three parallel pushing type continuous reheating furnaces. The global system architecture is shown in Figure 6.8, from which we can see that a setpoint compensator is used to make the correction based on the difference between the aim and actual observed slab surface temperatures at the furnace outlet. Figure 6.9 shows a comparison of the fuel consumptions between two furnaces running under almost the same conditions, but one was only furnished with the level one control, and the other was controlled by the expert optimization control as shown in Figure 6.8. Significant energy and over heating steel losses can be saved through applying a real-time optimization control system for reheating furnaces.

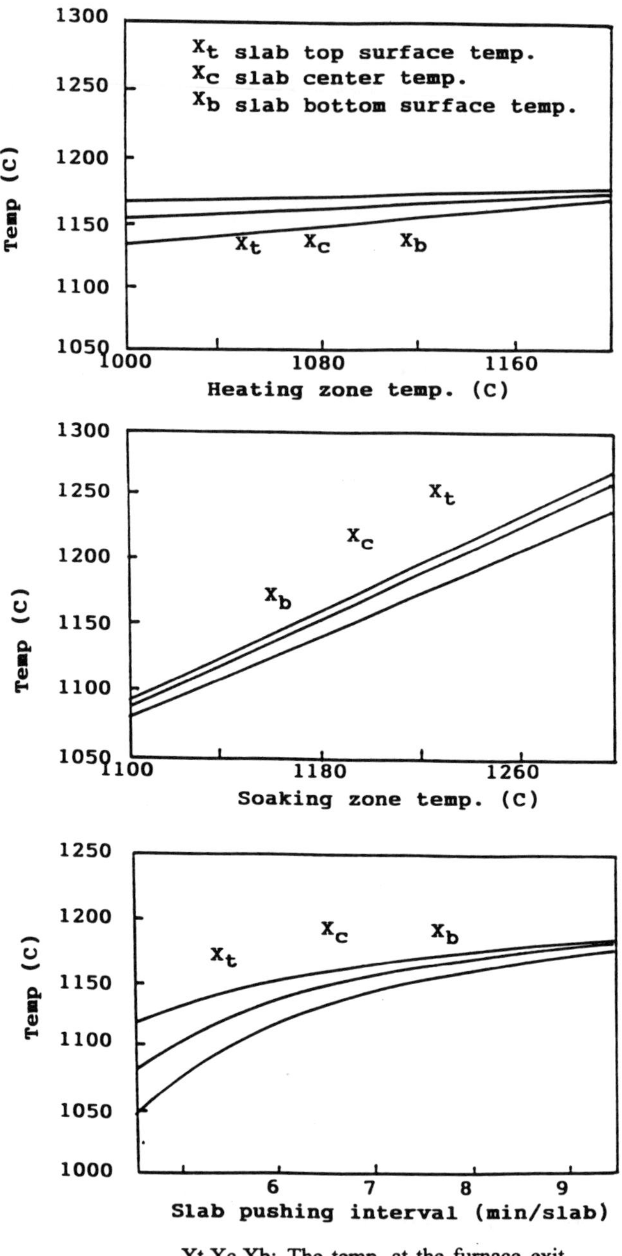

Xt,Xc,Xb: The temp. at the furnace exit

Figure 6.7 Numerical relations between operating conditions and slab temperature profile at the furnace exit

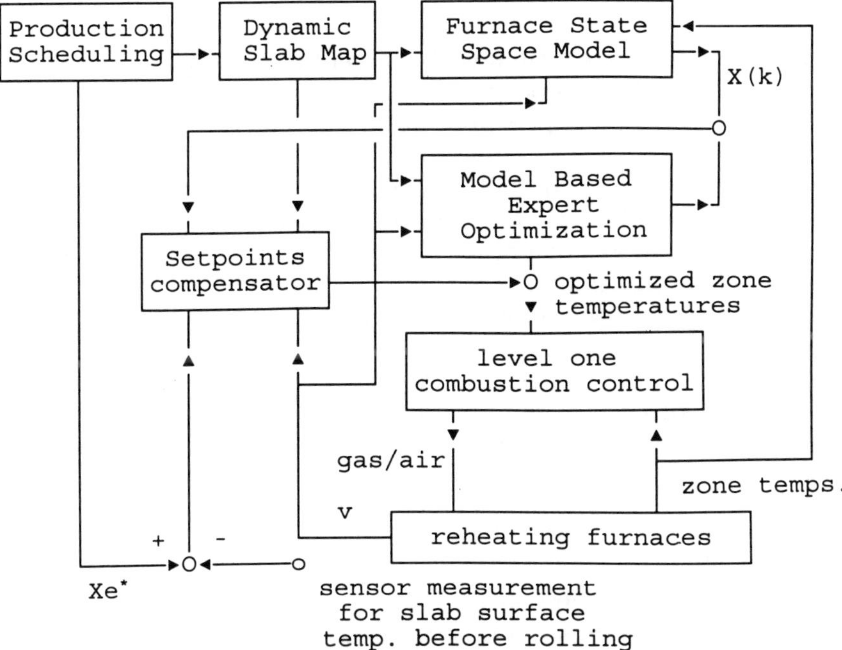

Figure 6.8 Diagram of real-time optimization control system for a reheating furnace

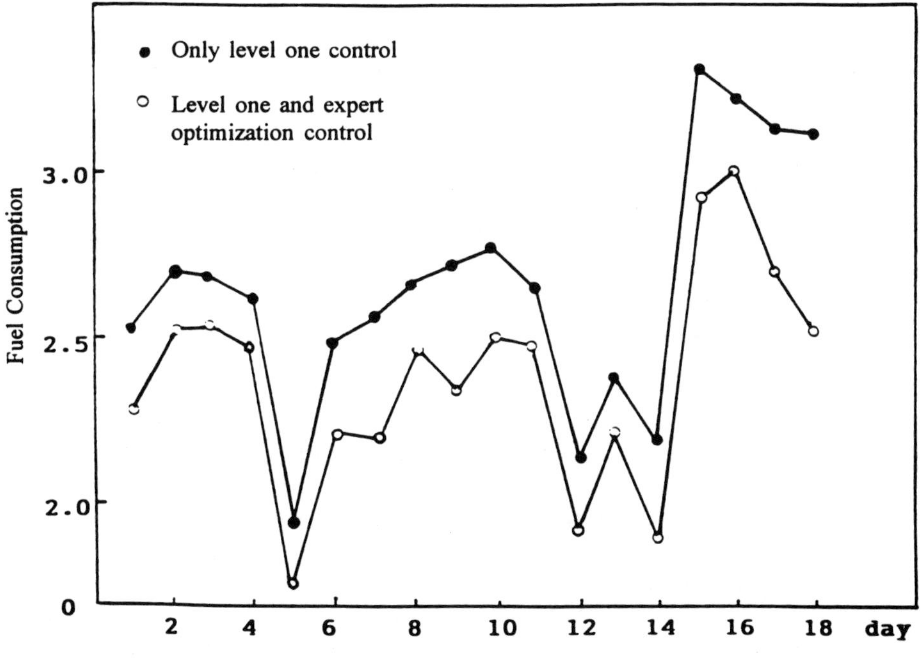

Figure 6.9 Comparison of fuel consumptions

6.4.3 Comments

A combination of an expert system with a dynamic math-model may provide an approximate optimization solution much more efficients. The following points help us to design and apply an expert optimization in industrial control:

1. Expert optimization can also be carried out by combining a rule based expert system with operations research algorithms. Wang and Lu (1990) investigated an algorithm of knowledge-based successive linear programming for linearly constrained optimization. The results show that when introducing the human knowledge to math-programming, the efficiency of problem solving can be considerably improved. The combination of expert systems and operations research methods has been an attractive direction in solving complex optimization problems; the detailed benefits were described by O'Keefe (1985) and O'Keefe *et al.* (1986).
2. Expert optimization techniques may not provide an exact optimization solution; in contrast, they only provide a 'satisfactory' or an 'approximately optimal' solution.
3. Finally, it should be emphasized that 'local optima' have been a critical problem in many industrial optimizations, if there exists multiple optimal points for the process, since the solution of a nonlinear programming algorithm, e.g. the gradient descent search, is obtained with a series of iterative calculations i.e. the future solution is dependent on a comparison between current and previous solutions under the direction of the gradient descent search. Obviously, the selection of the initial point will significantly affect convergence of the solution; in other words, it is quite possible to enter a local optimal region. However, since the solution of an expert optimization is driven by human expertise knowledge, heuristic search and deep reasoning, the search domain can be much wider than that of the gradient descent search. As a result, with some knowledge concerning a feasible global optimization region, the local optimum problem can sometimes be avoided. On the other hand, decomposition in the global search space will also help to get a global solution. This means that a global search space can be decomposed into a number of sub-spaces; then the heuristic search can be performed in a decomposed manner, i.e. the system starts to search with each sub-space and then further seeks a global solution.

During the last few decades, many efforts have been made to explore more efficient optimization strategies, in addition to expert optimization. Both '*genetic algorithms*' (or, in more general by, so-called '*evolutionary computation*') and '*Hopfield network and simulated annealing*' have been recognized as attractive potential directions in developing novel industrial optimization strategies. However, in the limited space of this book, we will give only a general introduction to these two topics in sections 6.6 and 6.7, respectively. More detailed descriptions of these two topics have been given in many publications, for instance the book on simulated annealing and Boltzmann machines written by Aarts and Korst (1989), that on generic algorithms written by Michalewicz (1992) and Sucek, *et al.* (1992), and a special issue of *IEEE Trans. Neural Networks* on evolutionary computation (IEEE, 1994).

6.5 EXPERT SYSTEM FOR PRODUCTION SCHEDULING

6.5.1 Introduction

Computer aided intelligent production scheduling has become an attractive R&D subject to simulate human behavior in the production scheduling activities. Fox *et al.* (1983) proposed the well-known 'Intelligent Scheduling and Information System (ISIS)', which provides a basic principle and direction for development of an intelligent scheduling tool. In Fox's proposal, a general scheduling problem can be governed by the goals of organization, physical constraints, causal constraints, availability and preference, and the approaches to hierarchical constraint-guided search in a solution space. However, the key problem of applying an expert system in production scheduling is how to make a system have learning capability, namely knowledge acquisition and upgrade. The emphasis of this section is to describe the development of an intelligent scheduling tool with object oriented relational model and the applications of deep reasoning and rule learning. A working example of an intelligent production scheduling system for a multi-stage and multi-product (MSMP) process is also addressed (Lu *et al.*, 1995).

6.5.2 Problem statement

The layout of an MSMP process is illustrated in Figure 6.10. The major characteristics of the process can be summarized as follows:

- the production system consists of a number of stages, each with one or more parallel production facilities, and a number of buffer storage areas with a limited number of tanks being located between two neighboring stages;

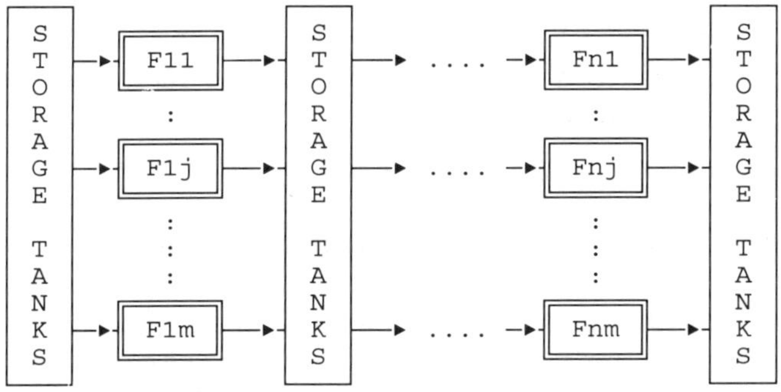

Figure 6.10 Layout of MSMP production process

- the production facilities can produce relevant products under the particular production environment which includes the raw material resources available, production constraints and other related conditions;
- since the volume of the buffer storage is quite limited, the desired amount of products can only be produced through switching the production schemes;
- the undesirable process during the transition phase from one production scheme to another will increase the extra operating costs and production time, due to the additional energy and material consumptions. Hence the minimum switching time becomes a major criterion in production scheduling.

The example constraints involving the production planning, safety, storage spaces, quality, etc., can be summarized as follows:

- actual total weight of all products must be equal to or greater than the planned total weight;
- weight of *each* product should be equal to or greater than the planned weight; however, this is only a soft constraint;
- storage tank levels must be within the desired 'min–max' ranges;
- minimal continuous production time for each facility should be equal to or greater than that desired;
- time-sequential switch order for changing production schemes should follow a set of specified rules;
- willingness and preference of human scheduler, etc.

The constraints are divided into two categories, hard and soft constraints, which can be described by both quantitative and qualitative forms. The scheduling problem can be described as follows: based on a production plan and initial production conditions, the task of scheduling is to provide a time sequence of production schemes, which requires 'minimum' or 'less' production scheme switching times and satisfies all hard constraints. When making the schedule, there are a huge number of potential candidates to be selected; in other words, the 'combinatorial explosion' may occur in problem solving. In this study, an intelligent scheduling system was developed to solve such problems, which exist in many semi-continuous production processes.

6.5.3 Strategies and structure of development tool

Like many expert system tools, this tool also consists of an interface for system development, generator and editor of the scheduling scheme, a unit of evaluation, and an interface with database and knowledge base. Here we focus on two major topics: knowledge representation and scheduling strategy.

Knowledge representation

A hybrid knowledge representation with the forms of the frames and production rules is designed to present the production background, such as production structure, facilities, mass flows, intermediate and finished products and their properties, major production parameters, etc. An example frame structure is given as follows:

Product Name:	*Prod1, . . . , Prodm*
Facility Name:	*Fac1, . . . , Facn*
Storage Area Name:	*Sto1, . . . , Ston*
Process Name:	*Pro1, . . . , Pron*
Production Rule Set 1:	*Rule_S11 , . . . , Rule_Sn1*
⋮	⋮
Production Rule Set n:	*Rule_Sn1, . . . , Rule_Snn*
Constraint Set 1:	*Cons_S11, . . . , Cons_Sn1*
⋮	⋮
Constraint Set n:	*Cons_Sn1, . . . , Cons_Snn.*

Similarly, each facility, tank and product can also be represented by the relevant frame, which describes its structure, parameter properties, and their relationships.

The production rule and constraint sets in the frame are represented by 'if–then' rules and 'min–max' or other limitations. In addition, some computational algorithms describing basic mass and energy balance are embedded into the frame associated with the rule base. A hierarchical frame structure is used to describe the production background. However, since there are too many production systems with different structures, to effectively complete reasoning, the application of an object oriented relational knowledge model with highly structured knowledge and databases is introduced.

A hierarchical, multi-dimensional information model consisting of a number of two-dimensional relational tables is used to describe the *'object'* and its *'attribute'*. Both object and attribute, and their relationships, can be effectively applied to describing the complex structure in the production scheduling problem. The object is an abstract of the system environment and its principles. In this particular application, the object consists of a real facility, or so-called 'entity'.

The 'attribute' represents some common behaviors in an object; based on different information resources, the attribute can be divided into the following three categories:

- index attribute: describing the various entities in an object;
- describing attribute: the characteristics of an object;
- reference attribute: the fact used to connect the entities in the different objects.

The 'relationship' is an abstract of internal relations between objects; in other words, the relationship represents the corresponding relations between the entities in the different objects. The application of the object oriented relational knowledge

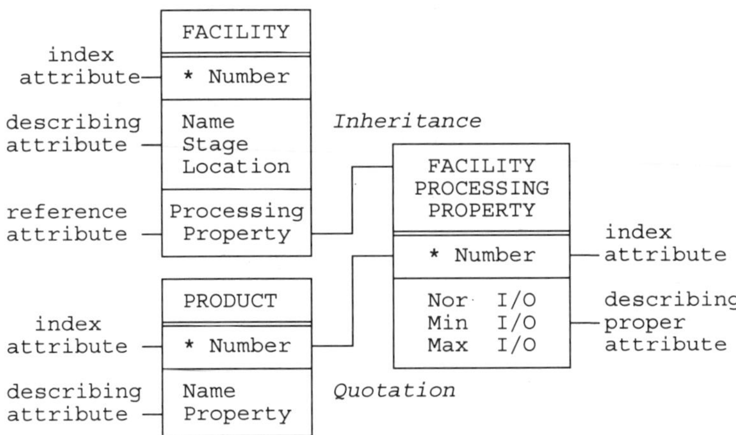

Figure 6.11 Structure of object oriented knowledge model

model in representing the production background is illustrated in Figure 6.11. It can be seen that relationships between the production facilities, products and their properties can then be readily operated and managed in a knowledge base.

Application of object oriented knowledge model in production scheduling

The knowledge of the scheduling background can be directly represented by the relevant object and attribute through system structure decomposition and restructure.

The definition of object

OBJECT: *Object Name*
COMMENT: *Object Description*
ATTRIBUTE: *Number of Attributes*
Attr. Name 1 | Comment | Type
⋮
Attr. Name N | Comment | Type

The definition of entity

ENTITY: *Entity Name*
COMMENT: *Entity Description*
OBJECT: *Object Name*
Attr. Name 1 | Attr. Value | 1...
⋮
Attr. Name N | Attr. Value | N ...

Consequently, the production scheme can be described by a set of interconnected object entities and the related data; moreover, both rules and math-model

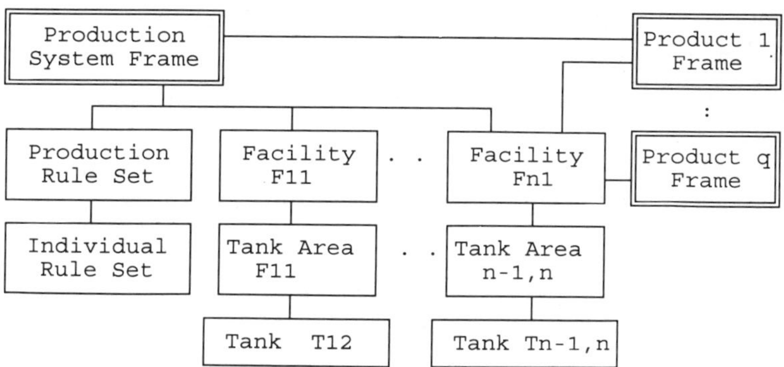

Figure 6.12 A hybrid knowledge frame structure

fragments being used for production scheduling can be considered as the attribute values of the entity. As a result, a hierarchical hybrid knowledge frame can be constructed through the information exchanges, namely 'inheritance' and 'quotation', which are shown in Figure 6.12.

Since there is a significant difference between the various scheduling problems, scheduling knowledge can be implemented through embedding the functions into the tool. The detailed procedures are as follows:

Design Since the decision making in production scheduling is to elegantly use the scheduling knowledge through visiting the interface between the production environment, such as status, data and conditions, and the system knowledge frame, a visiting interface is designed to get the relevant scheduling knowledge based on the production environment, and its major functions are:

- sequentially visiting the entities

  First Entity (Entity, p IDvalue)
  Next Entity (Entity, p IDvalue)
  ⋮

- visiting a particular entity to get and/or put its corresponding attribute values based on the ID values, for instance

  Get Attr (Entity Name, Attribute, ID Value, pAttrValue)
  Put Attr (Entity Name, Attribute, ID Value, pAttrValue)

- adding and/or deleting a particular entity in the object, for instance

  Add Entity (Entity Name, Attr 1, . . . , Attr N)
  Del Entity (Entity Name, IDvalue)

The design of scheduling rules, constraints and computational algorithms usually has the following procedures:

- receive the production state via visiting the knowledge frame interface;

- make logical judgment, reasoning and computation based on both production state and scheduling knowledge;
- fill the results in the knowledge frame of a particular object.

It can be seen that scheduling knowledge is running and applied in an event-driven manner. In fact, scheduling knowledge is represented as independent production state oriented functions.

Register The scheduling knowledge is filled in the particular forms in the intelligent scheduling tool, for example:

```
Structure Function Table {
   Byte Name [Name_Len]; --Knowledge name;
   Void (*Function)( ); -- Function Index;
                        }Appfunt [];
```

Embedded frame The scheduling knowledge should also be filled in the relevant knowledge frame, in order to classify the knowledge into different categories. In general, the scheduling knowledge for MSMP processes can be divided into the following categories:

Generate rules – Generate set of the product candidates to be selected in switching;

Cutting rules – Delete some product candidates to match the desired production constraints;

Cal stage I/O – Calculate the buffer tank capacity available, facility throughput in each production stage, etc., based on production mass balance;

Dispatch rules – Provide the time sequences of the possible production scheme candidates;

Balance rule – Modify the production time and continuous production time related to the preliminary scheduling solution based on various production constraints;

Evaluation rules – Evaluate the performance of the scheduling solution based on both quantitative and qualitative criteria, which include the human preference and other related factors.

Scheduling strategy for MSMP processes can be divided into two phases, namely scheduling for the single facility and scheduling for the entire process. In fact, the iterative searching between these two phases is used to seek a satisfactory solution under the given production constraints. The scheduling strategy used here is similar to the approach to decomposition and coordination in large scale systems (Singh and Titli, 1978).

(1) Scheduling for single facility
The single facility scheduling problem is only a single stage and multiple product one, which can be described by the structural linguistic form in terms

of knowledge based intelligent search strategies. The single facility scheduling only provides the potential scheduling schemes which will be added into the scheduling set for the entire scheduling.

(2) Scheduling for entire process
The applications of the scheduling knowledge in this phase are to establish the preliminary scheduling scheme for individual facility based on check of the satisfaction of I/O constraints for all facilities and modify the production scheme using balance rules. The major function of the balance rules is to make both individual facilities and the entire process satisfy 'min–max' limitations of the intermediate storage, I/O mass flow and other related constraints.

Finally, it should be noted that to make the intelligent scheduling tool more flexible and effective in establishing a scheduling system for various production environments, a modular structure is used in design of the intelligent scheduling tool, which will not be given here.

The 'layered-generate-test', 'deep reasoning' and 'rule learning' as introduced in section 3.9.4 are applied in developing this tool to make the resulting system more elegant and flexible, when the pre-specified knowledge is not applicable in providing a feasible solution. This happens because either the constraints are not realistic or there exist conflicts among the rules. For instance, if one assigns a long time interval as a minimum continuous operating time for a specific production scheme in the constraints, then some intermediate tanks may overflow or empty. However, this kind of problem may be automatically found and solved through deep reasoning and rule learning. Correspondingly, the new knowledge (new rules and/or upgraded rules) can then be automatically reorganized into a particular category of the knowledge base.

6.5.4 Industrial test results

An intelligent scheduling system was developed to carry out computer aided scheduling for lube oil production in a large scale refinery. The entire production system is divided into three stages, namely ketone-benzene de-wax, (furfural) refiner I and refiner II. Each of them has two parallel facilities and there is limited storage between two stages. The criterion of lube oil production scheduling is to minimize the total switch times, which can be written in the following explicit forms:

$$\min J = W_{11}L_{11} + W_{12}L_{12} + W_{21}L_{21} + W_{22}L_{22} + W_{31}L_{31} + W_{32}L_{32} \tag{6.32}$$

subject to the specified production constraints, such as production plan, tank volume limitation and others.

The sample comparison of the production scheme sequences between the human scheduler and expert system are shown in Table 6.2. It can be seen that the expert scheduling provides fewer switch times. The average switch times are reduced by about 10% by using an expert scheduling system.

Table 6.2 Comparison results between human scheduler and expert system

Facility	Scheduler	Production scheme sequence
F11	Human	A42-A52-A51-A41
	Expert system	A42-A52-A51-A41-A51
F12	Human	A31-A21-A33-A21-A22-A31
	Expert system	A31-A21-A22-A31
F21 and F31	Human	A33-A32-A41-A42-A52-A41-A31-A22
	Expert system	A33-A31-A41-A52

6.6 EVOLUTIONARY COMPUTATION

6.6.1 Introduction

It has been a challenge to explore new techniques of finding a global optimization solution through search in either a multi-dimensional Euclidean space or various discrete spaces. In many industrial optimization (control, job-machine assignment, scheduling or planning) problems, the cost functions could be non-differentiable and/or discontinuous with various forms of constraints; as a result, the classical gradient or heuristic search strategies will fail to provide a satisfactory solution. The global random search approaches, such as the simulated annealing methods and evolutionary computation, have been recognized as the attractive future directions in developing industrial optimization systems. This section will briefly introduce the general concepts and algorithms of evolutionary computation.

Evolutionary computation (EC) has become a term that encompasses 'genetic algorithms' (GA), 'evolution strategies' (ES) and 'evolutionary programming' (EP) techniques (Fogel, 1994). They were developed independently and have different characteristics; however, all these techniques rely on the fundamental principles of 'natural evolution'. Before describing the concepts of EC or GA, it will be helpful to learn the natural idea behind these techniques through a rabbit population example given by Michalewicz (1992): *'At any given time there is a population of rabbits. Some of them are faster and smarter than other rabbits. These faster, smarter rabbits are less likely to be eaten by foxes, and therefore more of them survive to do what rabbits will do best: make more rabbits. Of course, some of the slower, dumber rabbits will survive just because they are lucky. The surviving population of rabbits starts breeding. The breeding results in a good mixture of rabbit genetic material: some slow rabbits breed with fast rabbits, some fast with fast, some smart rabbits with dumb rabbits, and so on. And on top of that, nature throws in a 'wild hare' every once in a while by mutating some of the rabbit genetic material. The resulting baby rabbits will (on average) be faster and smarter than the original population because some faster, smarter parents survived the foxes. (It is a good thing that the foxes are undergoing a similar process - otherwise, the rabbits might become too fast and smart for the foxes to catch any of them.)'*. From this example we can see that if a simulated computer program can imitate an optimized genetic process, an optimization solution may be provided through a controlled genetic process from generation to generation.

Some terms used in EC were originally introduced from natural evolution, for instance 'individuals' (or 'genotypes' or 'structures') in a population (e.g. each trial solution) are called 'chromosomes' (or 'strings'); and 'genes' (or 'features', 'characters', 'decoders') are the elements (or units) in chromosomes. The general structure of an evolution program (Michalewicz, 1992) can be described as follows:

```
'begin
   t<-0
   initialize P(t)
   evaluate  P(t)
   while (not termination-condition) do
   begin
     t->t+1
    select P(t) from P(t - 1)
    recombine P(t)
    evaluate P(t)
  end
end '
```

The evolution program is a probabilistic algorithm which maintains a population of individuals, $P(t) = x_1(t), \ldots, x_n(t)$, where t denotes the number of iterations. In a classical GA, the original problem should first be converted into a binary string oriented one, then all operations (e.g. crossover) are also made under a binary frame. To make GAs communicate with a real-world environment, information coding (form a digital or an analog signal to a binary string) and decoding (transfer from binary string to real physical values) are needed. In contrast, EP can directly deal with the natural data structure without having such kind of coding and decoding processes. However, they both share an identical natural fundamental of random optimization search. Now we start to present some EC algorithms.

6.6.2 Genetic algorithms

Fraser (1962), Bremermann (1968), Reed *et al.* (1967) and Holland (1975, 1992b) proposed similar algorithms. The common feature of these algorithms is that they simulate a genetic system, and are now defined as 'genetic algorithms'. The general procedures of GAs can be summarized as follows (Fogel, 1994):

(1) The problem can be defined and captured in an objective function, or so-called fitness of any potential solution, commonly used in GA.

(2) A population of candidate solutions is initialized subject to certain constraints. The candidate solutions usually can be represented as a vector X, defined as a 'chromosome' with its elements being described as 'genes'.

(3) Each chromosome, $X_i, i = 1, \ldots, P$, in the population is decoded into a form appropriate for evaluation and is then assigned a fitness score, $S(X_i)$, according to the objective.

Table 6.3 Sample results of reproduction via fitness

No.	String	Fitness	% of total
1	01101	169	14.4
2	11000	576	49.2
3	01000	64	5.5
4	10011	361	30.9
	Total	1170	100.0

(4) Each chromosome is assigned a probability of reproduction, $p_i, i = 1, \ldots, P$, so that its likelihood of being selected is proportional to its fitness relative to the other chromosomes in the population. For example, if the fitness is strictly a positive number to be minimized, then roulette wheel selection can be used, and the sample results are shown in Table 6.3 (Goldberg, 1989).

(5) According to the assigned probabilities of reproduction, $p_i, i = 1, \ldots, P$, a new population of chromosomes can be generated by probabilistically selecting strings from the current population. The selected chromosomes generate 'offspring' via the use of specific genetic operators, such as 'crossover' and 'bit mutation'. Crossover is applied to two chromosomes (parents) and creates two new chromosomes (offspring) by selecting a random position along the coding and splicing the section that appears before the selected position in the first string with the section that appears after the selected position in the second string, and vice versa. The one-point crossover operator is illustrated in Figure 6.13 (Goldberg, 1989). The typical method of recombination in GA is to select two parents and randomly choose a splicing point along the chromosomes. The segments from two parents are exchanged and two new offspring are created. On the other hand, bit mutation simply offers the chance to flip each bit in the coding of a new solution. Typical values for the probabilities of crossover and bit mutation range from 0.6 to 0.95 and from 0.001 to 0.01 respectively.

(6) The process is halted if a suitable solution has been found, or if the available computing time has expired, otherwise the process proceeds to step (3) where the new chromosomes are scored and the cycle is repeated.

```
          Crossover
            point
              /
Parent #1:1101|0111101       Offspring #1:10100111101

                        ⇒

Parent #1:1010|0000100       Offspring #1:11010000100
```

Figure 6.13 Illustration of one-point crossover operator (Reproduced from Fogel, 1994, © 1994 IEEE)

There are many discussions (Goldberg, 1989, Dejong, 1988) concerning the feasibility and reality of applying GAs in solving real-world optimization problems. Michalewicz (1992) indicates that for valued optimization problems, floating-point representations outperform binary representations, because they are more consistent, more precise, and lead to faster execution. The applications of GAs with different data representations in solving a very large scale optimization problem are another topic to move GAs into practice. To avoid losing the best chromosomes in the iterative population generations is an important measure that can ensure asymptotic convergence to a global optimization. The applications of heuristic oriented elitist selection (Grefenstette, 1986) may always retain the best chromosome in the population and guarantee asymptotic convergence (Rudolph, 1994).

6.6.3 Evolutionary algorithms

'Evolutionary algorithms' (EA), also commonly called 'evolution strategies' (ES) or 'evolutionary programming' (EP), were developed by Schwefel (1965), Rechenberg (1973) and Fogel (1962). When applying evolutionary algorithms to real-valued function optimization, the computational procedures can be described as follows (Fogel, 1994):

(1) A problem is defined as finding the real-valued n-dimensional vector $X \in R^n$, which maximizes a fitness function of $F(X)$.

(2) An initial population of the parent vectors, $X_i, i = 1, \ldots, P$, is randomly selected from a feasible range in each dimension, and the distribution of the initial trials is typically uniform.

(3) An offspring vector, $X_i', i = 1, \ldots, P$, is created from each parent X_i by adding a Gaussian random variable with zero mean and preselected standard deviation to each component of X_i.

(4) Selection then determines which of these vectors to maintain by comparing the error $F(X_i)$ and $F(X_i'), i = 1, \ldots, P$. The P vectors that possess the least error become the new parents for the next generation.

(5) The process of generating new trials and selecting those with least error continues until a sufficient solution is reached or the available computation is exhausted.

It can be seen from EA that the resulting change in each trial solution will follow a Gaussian distribution with zero mean difference and some standard deviation. In comparing GA with EP and ES, we will find these methods share a common feature, namely, improving the solution performance generation after generation through the random changes in the population. However, the major differences between GA and EP or ES as analyzed by Fogel (1994) are that *GAs emphasize models of genetic operators as observed in nature, such as crossover, inversion and point mutation, and apply these to abstracted chromosomes. EP and ES emphasize the mutational parent and its offspring, respectively, at the level of the individual or the spaces.* Finally, it should be emphasized that both GAs and EAs as a stochastic optimization oriented direct search method do not require derivative of objective function nor continuity of the search space (Rao, 1978).

6.6.4 Applications of EAs in job assignment problems

Job assignment problems (Koopmans and Bechmann, 1957) may be modeled as a quadratic assignment problem (QAP) or an integer programming problem, stated as follows (Kusiak and Heragu, 1987):

$$\min Z = \sum_{i=1}^{N}\sum_{j=1}^{N} a_{ij}x_{ij} + \sum_{i=1}^{N}\sum_{j=1}^{N}\sum_{k=1}^{N}\sum_{l=1}^{N} f_{ik}c_{jl}x_{ij}x_{kl} \tag{6.33}$$

subject to

$$\sum_{j=1}^{N} x_{ij} = 1, i = 1, \ldots, N$$

$$\sum_{i=1}^{N} x_{ij} = 1, j = 1, \ldots, N$$

$$x_{ij} \in \{0, 1\}$$

where

N = total number of facilities/locations;
a_{ij} = fixed cost of locating facility i at location j;
f_{ik} = flow of material from facility i to facility k;
c_{jl} = cost of transferring a material unit from location j to location l;
x_{ij} = 1 if facility i is at location j, or 0 otherwise.

The concept of flow-dominance introduced by Vollmann and Buffa (1966) serves as the interconnection strength which can be measured by the variation of the values in the flow matrix with its elements, 100 x standard deviation ÷ mean.

The QAP is a very common problem involved in many machine scheduling and job assignment subjects. When N becomes a huge number, it will be difficult to provide an optimal solution with the approaches to operations research and rule based expert systems. Nissen (1994) proposed a combinatorial variant of ES (CES) for solving QAP. The solution representation of CES is a straightforward permutation coding (see Figure 6.14 (a)). *Each position on a solution vector represents a facility location and integer values are assigned as numbered facilities. The cost values are calculated from the given problem matrices and directly taken as fitness information of CES. A predefined number of generations is set as the termination criterion. CES uses a simple population concept. It is basically a $(1, \lambda)$-ES (see Figure 6.14(b)), i.e., in each generation, children are generated from one parent solution by first copying the parent and then randomly swapping integer values on the coding string via mutation. The best child becomes*

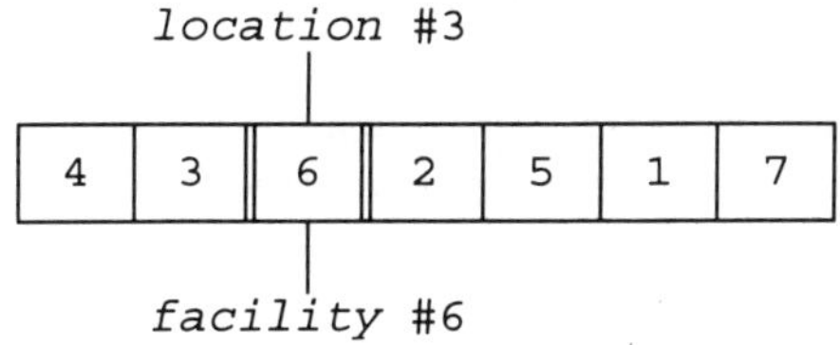

(a) Genetic representation of a solution in CES (N=7)

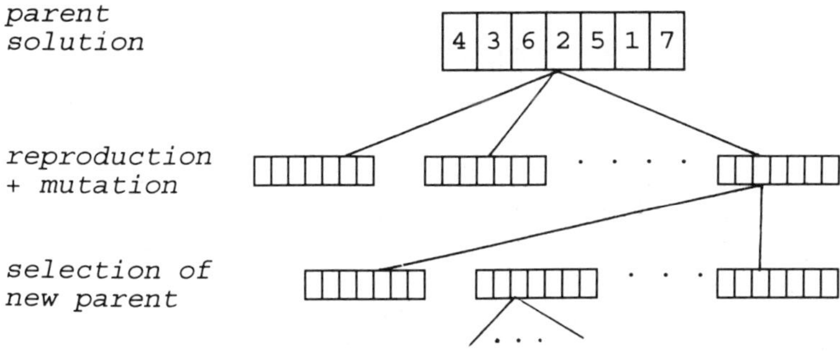

(b) Basic concept of CES

Figure 6.14 Genetic representation and basic concept of CES

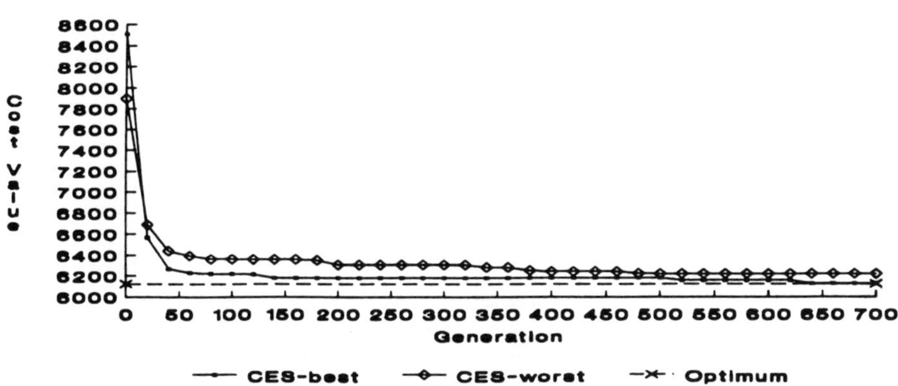

Figure 6.15 Convergence of fitness on NUG 30 (Reproduced from Nissen, 1994, © 1994 IEEE)

the new parent. If its fitness value is no better than the former parent's value, the counter is increased. The counter is reset to zero whenever a CES generation is successful.

To test the applicability of CES, Nissen (1994) provided many simulation results with some benchmark combinatorial problems. The CES was run on the test problem NUG 30 (Nugent and Vollmann, 1968) and the simulation results (with $N = 30$) given by CES are shown in Figure 6.15. It can be seen that after only 37 generations, the fitness of best CES was convergent to its optimum value.

6.6.5 Evolutionary neural network training

The applications of evolutionary computation in neural network training (both topology and parameters) have been an attractive field in the design of a neural network with better generalization. Yao (1993) gave a review of evolutionary artificial neural networks. In Yao's review, the applications of GAs in neural network training can be divided into three categories, namely

- evolution of neural network architectures and connection weights;
- evolution of neural network weights architectures;
- evolution of neural network learning rate and associated parameters.

Many researchers (Whitehead and Choate, 1994; Maniezzo, 1994; Angeline *et al.*, 1994) have applied GAs and ES to various neural network training problems. Here we only highlight an evolutionary neural network training for the radial basis function (RBF) proposed by Whitehead and Choate (1994). The activation Gaussian radial basis function can be described as follows:

$$y_j = \exp\left[-\sum_{k=1}^{n}\frac{(x_k - \mu_{jk})^2}{2\sigma_j^2}\right] \tag{6.34}$$

where $x_k, k = 1, \ldots, n$ denote network inputs and y_j is the activation of output node j. In traditional RBF networks, both width σ_j and center vector $\mu_{jk}, k = 1, \ldots, n$, being a point in the input space are normally fixed, and the neural network training only modifies the interconnection weights. In the evolutionary approach, by

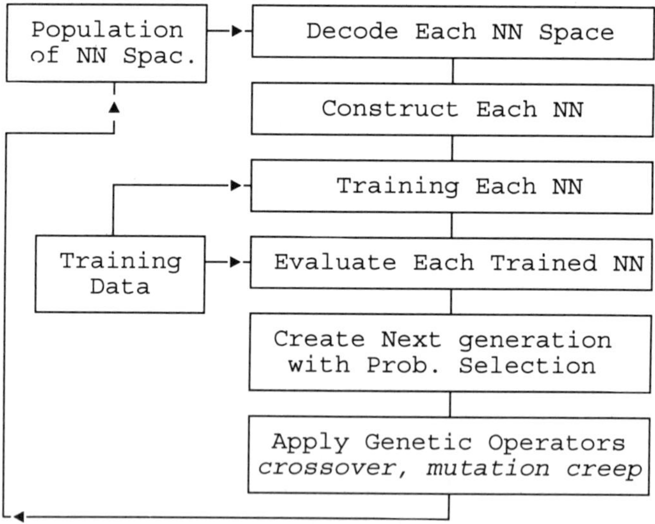

Figure 6.16 Illustration of evolutionary RBF network training (Reproduced from Whitehead and Choate, 1994, © 1994 IEEE)

contrast, the centers and widths are not fixed, but are specified by a genetic encoding that evolves in a population over the generations of the genetic algorithm. The algorithm of one generation of GA is schematically given in Figure 6.16 (Whitehead and Choate, 1994).

6.6.6 Fuzzy rule learning by GAs

How to create fuzzy rules to be applied in fuzzy inference is still a difficult task in the design of a fuzzy model or fuzzy control system. The fuzzy rules can be found through human knowledge and/or neural network unsupervised learning. This section addresses the applications of GAs in finding the fuzzy rules for process modeling and control.

Applications of GA in fuzzy rule learning for enzyme proportion prediction

Shiba *et al.* (1994) proposed the algorithm of GA learning based fuzzy rule extraction, and its application in seeking the optimal enzymes in proportion of rice koji, the enzymes being α-amylase, glucoamylase, acid protease and acid carboxypeptidase.

To apply GAs in fuzzy rule learning, the fuzzy rules can be coded in digital numbers. Here the linguistic statement of preconditions, 'small', 'big' and 'dummy', are coded 0, 1 and 2 respectively, and 'small' and 'big' in the post-proposition are coded 0 and 1 respectively. The codings of the component parts are shown in Figure 6.17, in which GA/AA denotes the standard ratio of glucoamylase to α-amylase,

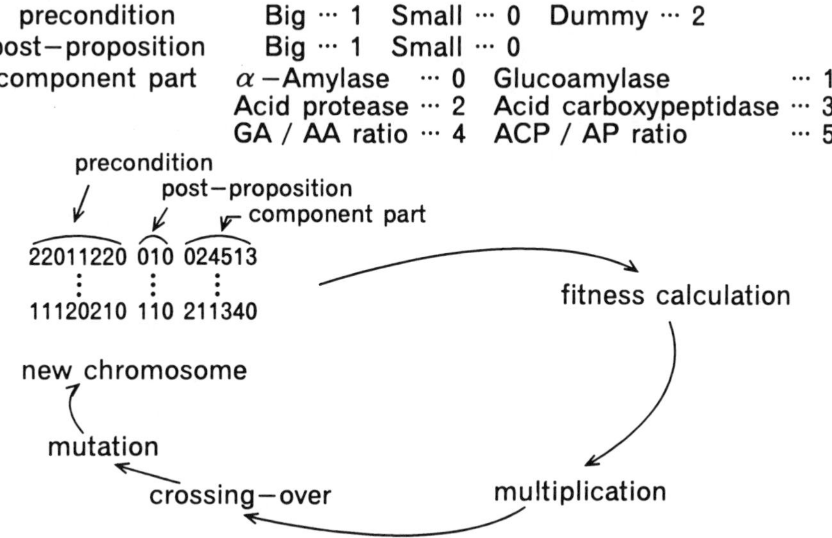

Figure 6.17 Coding of fuzzy rules for applications of GAs

Table 6.4 Comparison of learning and prediction results

	GA-min–max	GA-algebraic product sum	Multiple regression
Amino acid:			
learning	0.079 986	0.057 016	0.080 730
prediction	0.127 872	0.082 004	0.122 412
Total acid:			
learning	0.124 517	0.094 799	0.162 223
prediction	0.150 699	0.126132	0.195 747
Iso-amyl achhol:			
learning	0.159 566	0.089 818	0.159 012
prediction	0.170 932	0.135 249	0.206 130
Ethyl caproate:			
learning	0.126 559	0.085 900	0.139 476
prediction	0.181 531	0.168 338	0.207 887

and ACP/AP is the standard ratio of acid carboxypeptidase to acid protease. Each allocated code was arranged in series as chromosomes. The generation process of the new chromosomes is illustrated in Figure 6.17 (Shiba, *et al.*, 1994). The fitness was defined by the sum of square error between teaching data and resulting fuzzy inference values. Shiba *et al.* (1994) applied GA in both regular min–max product and algebraic-product-sum fuzzy inference; in the latter, a grade of the post-proposition part was the product of all grades of the membership function corresponding to the input values in the rule (algebraic product operation). The details can be found in Shiba *et al.* (1994). The comparison of learning and prediction results between the approaches to min–max product and algebraic-product-sum (both with GA oriented fuzzy rule extraction, at 200 generations) and multi-variable regression are shown in Table 6.4. It can be seen from Table 6.4 that the best results were provided by fuzzy inference of the algebraic-production-sum method with GA rule learning. After 1000 generations, the fuzzy inference with the rules extracted by GA has a small prediction error, and the contour maps of each component are quantitatively shown in Figure 6.18 (Shiba, *et al.*, 1994).

Applications of GA in real-time fuzzy rule learning for a bioreactor

Another GA application of fuzzy rule generation in the control of substrate concentration of fed-batch operation of a cell culture process was developed by Choi *et al* . (1994). The behavior of a bioreactor is very difficult to model by process first principle and/or math-model based system identification. The application of fuzzy logic in bioreactor modeling and control is an opportunity to simplify control system design and implementation. However, knowledge acquisition, determination of fuzzy membership and its real-time tuning are the key issues in the design of

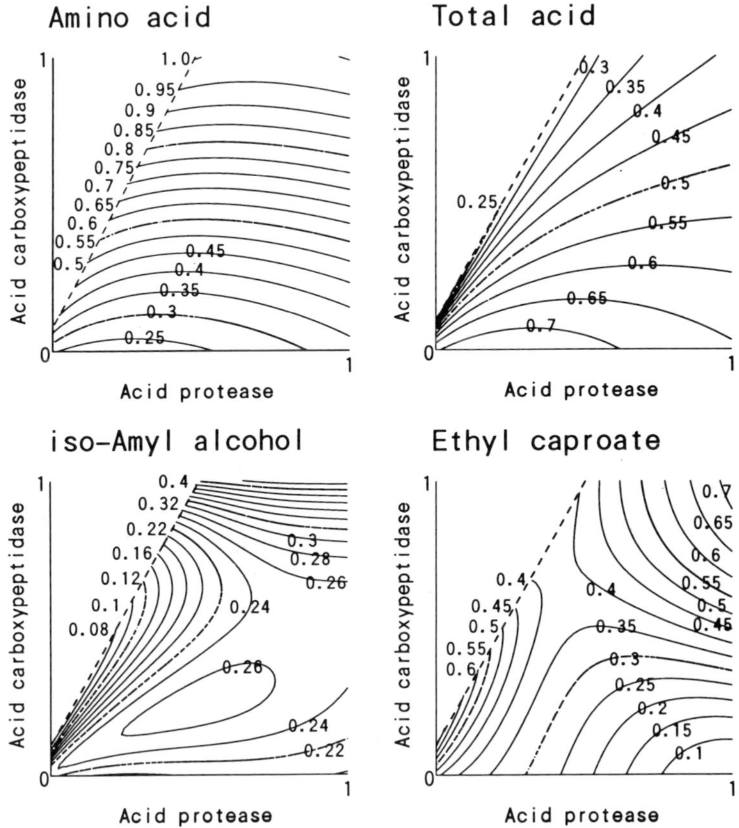

Figure 6.18 Contour maps of each component

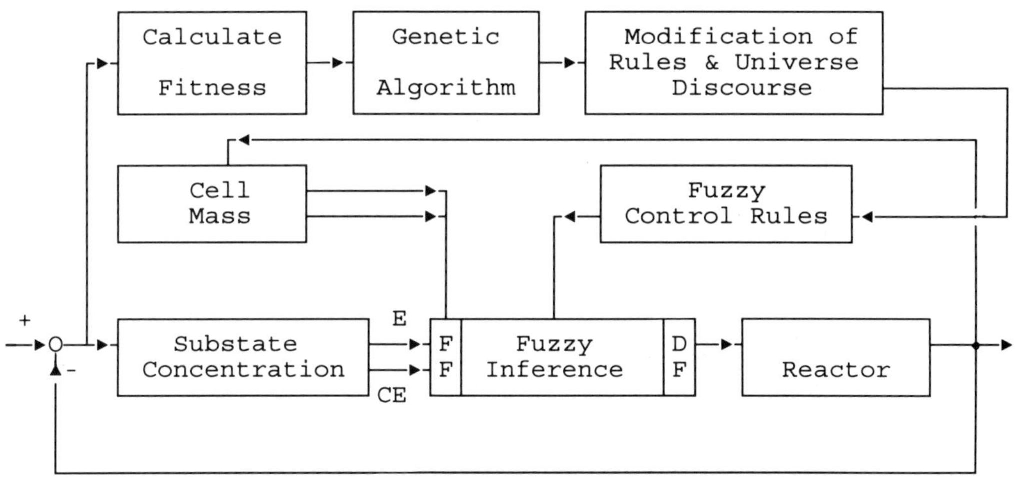

Figure 6.19 Self-organizing fuzzy controller with GA learning

a fuzzy controller for a complex process. Choi and his co-workers (1994) developed a self-organizing fuzzy logic control with GA adaptive learning for a bioreactor process which is shown in Figure 6.19. The fuzzy controller is designed to regulate the glucose concentration at the given set point to maximize the synthesis of berberine in a fed-batch culture of *Thalictrum rugosum* plant cell. The structure of the fuzzy control algorithm shown in Figure 6.19 is similar to regular fuzzy control, but in contrast, both fuzzy rules and universe of discourse are automatically modified by GA optimization search. To apply GA, the fuzzy rules are converted to the binary code string to be used as a chromosome, and a scalar valued fitness function is defined as follows:

$$\text{fitness} = \frac{K}{\sum_{i=1}^{n} |y_{set} - y_i| + \sum_{i=1}^{n} (y_{set} - y_i)^2} \tag{6.35}$$

The simulation results of the closed-loop responses of substrate trajectory in fed-batch culture are given in Figure 6.20, from which we can see that the self-organizing fuzzy control with GA optimization learning provides good dynamic performance for glucose control.

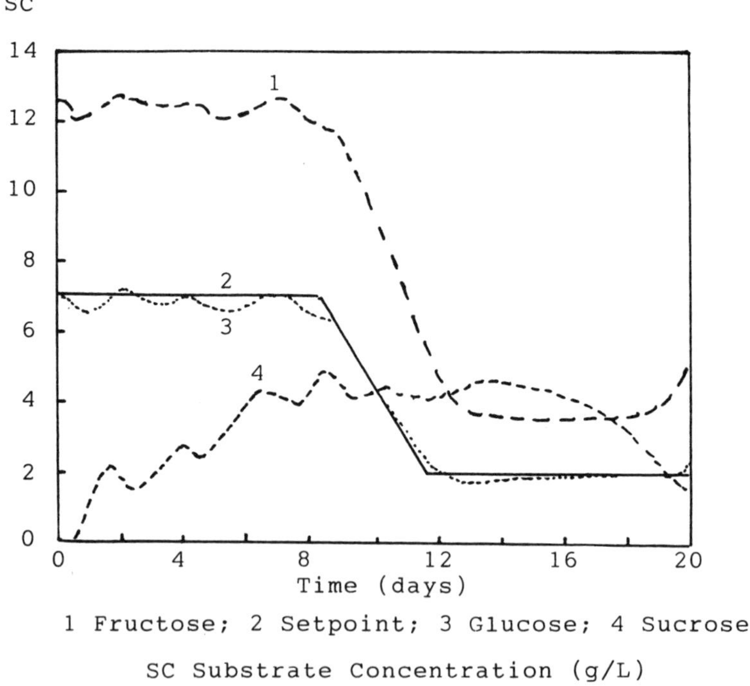

Figure 6.20 Dynamic tracking responses of glucose in fed-batch culture

6.7 HOPFIELD NETWORK AND SIMULATED ANNEALING

6.7.1 Hopfield network for combinatorial optimization problems

Hopfield and Tank (1985) have shown how to solve a traveling salesman problem (TSP) with a single layered analog network (refer to section 2.5) in their pioneering work. The TSP can be defined as given the detailed positions of a number of desired cities on a map, the salesman is to find an optimal tour with a minimal traveling distance that starts and finishes at the same city. Suppose we only have eight cities to be visited; a sample solution can be described by the following matrix:

Hopfield network and simulated annealing

	Visit order							
City	1	2	3	4	5	6	7	8
A	0	0	1	0	0	0	0	0
B	1	0	0	0	0	0	0	0
C	0	0	0	0	1	0	0	0
D	0	1	0	0	0	0	0	0
E	0	0	0	0	0	0	1	0
F	0	0	0	0	0	0	0	1
G	0	0	0	1	0	0	0	0
H	0	0	0	0	0	1	0	0

or simply represented by the tour 'B-D-A-G-C-H-E-F-B'.

The TSP is a typical combinatorial optimization problem, or is so-called 'NP-complete'. Physically, when using a Hopfield network to solve the TSP, the synaptic weights of the network are determined by the distances between the cities visited on the tour, and the optimum solution to the problem is a fixed point of the neurodynamic equations (Haykin, 1994). If there are N cities, then we need N^2 to represent this knowledge. To apply a Hopfield network in the combinatorial optimization problem solving, a Lyapunov function should be established in terms of an objective function subject to the hard constraints (Gee *et al.*, 1993). For a TSP, we have the following hard constraints:

- there should be only one '1' in each row (the salesman can visit each city only once);
- there should be only one '1' in each column (the salesman cannot be in two cities at the same time).

Similar to the common method used in operations research, a constrained optimization problem can be converted to the relevant unconstrained one through introducing constraint–oriented penalty terms in the objective function, such that a

Lyapunov (energy) function can be constructed as follows:

$$E = E^{\text{opt}} + C_1 E_1^{\text{cos}} + C_2 E_2^{\text{cos}} + \cdots \tag{6.36}$$

where

E^{opt} denotes the optimization function to be minimized;
E_i^{cos} represent the penalty functions optimizing the satisfaction of the constraints;
C_i are constant weights related to the respective E_i^{cos}.

It should be emphasized that the most difficult task of applying a Hopfield network in solving a combinatorial optimization problem is to construct a network which well represents the physical problem, or in other words, to establish a mapping between the original optimization problem and the Hopfield model with a proper energy function. Many application-oriented problems have been discussed in the literature (Wilson and Pawley, 1988; Gee, 1993). In fact, as Hopfield optimization networks show only narrow and limited applications, e.g. TSP, much more R&D are needed to move Hopfield networks from academic research to practical industrial applications.

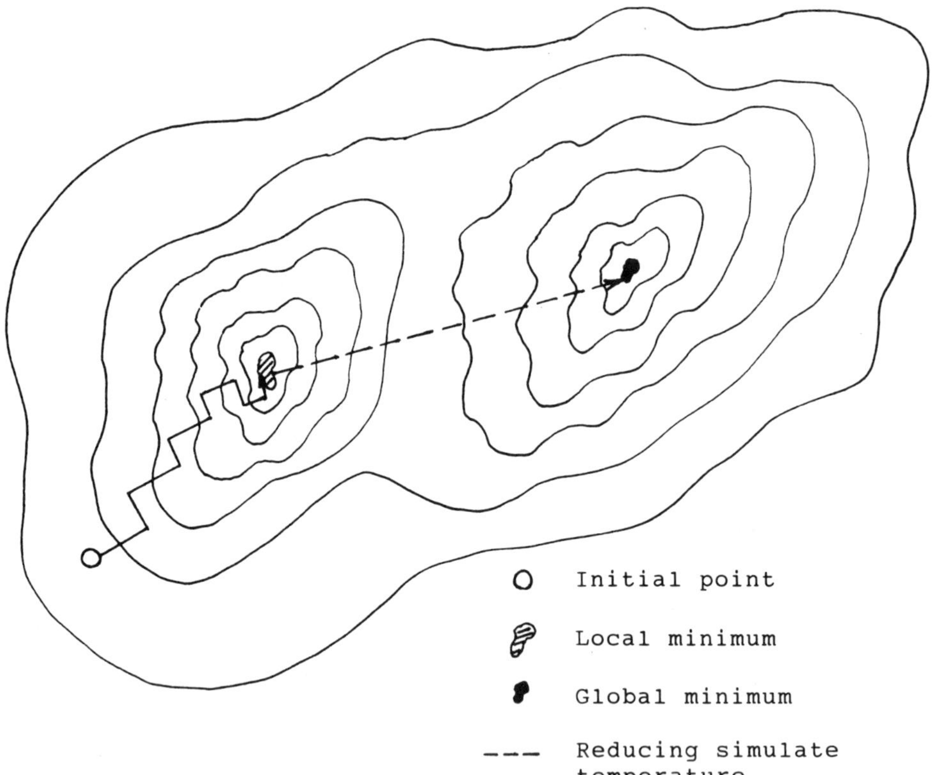

Figure 6.21 Escaping from local minima to global optimal point through a simulated annealing process

6.7.2 Simulated annealing algorithm

To solve a highly nonlinear optimization problem with deterministic search algorithm, the algorithm may suffer from a weakness of gradient descent search, namely, the solution may be 'local minima' instead of 'globally optimum'. Obviously, any gradient descent search-oriented neural network training will also have the local minima problem. 'Simulated annealing' originally developed by Kirkpartrick *et al.* (1983) is an effective measure to explore a global optimization solution. The simulated annealing algorithm is based on the analogy between the behavior of a physical system with many degrees of freedom in thermal equilibrium at a series of finite temperatures, as encountered in statistical physics, and the problem of finding the minimum of a given function depending on many parameters, as in combinatorial optimization (Kirkpartrick *et al.*, 1983; Haykin, 1994). The geometrical illustration of escape from local minima to a global optimal point with the simulated annealing process is shown in Figure 6.21. It can be seen that with decreasing simulated temperature, e.g. T in the sigmoid function $y = 1/(1 + \exp(-x/T))$, the thermal equilibrium will move from an original point to a new one, and finally, when $T = 0$, the thermal equilibrium may reach a global optimum state. Now the problem becomes how to control (or schedule) the annealing speed. For instance, if the cooling is too rapid, there is not enough time to reach a new equilibrium point, which depends on the simulated temperature. Finally, it should be noted that simulated annealing is particularly suited for joint working with a Hopfield network and Boltzmann machine optimization. A more detailed introduction to these issues is beyond the scope of this text.

7

Multivariate statistics and quality control

7.1 INTRODUCTION

7.1.1 SPC and process control

Statistical process control (SPC) has been popularly used in quality control (QC) and productivity improvement for both manufacturing and process industries. Deming's pioneering work (1972, 1982) has revolutionized quality control first in Japan and then in North America (MacGregor, 1988), and now SPC charts have become a standard tool from operator to manager in almost all enterprises. SPC sounds so close to process control, and eventually they both have the same goal, but, in fact, they belong to different groups in an enterprise and use different methodologies to deal with the problems.

The goal of process control is to design a computer control system which performs real-time modeling, prediction, optimization and dynamic control for a controlled environment. In general, *'real-time'* and *'dynamic'* are two key features of any process control system. Consequently, the design and analysis of a process control system usually use some dynamic (deterministic or stochastic), optimal and mathematical optimization approaches. For instance, system stability, robustness and other dynamic performances (e.g. tracking time, overshoot, offset, setting time or an error integration) are the major concerns in evaluating a dynamic control system.

In contrast, statistics oriented SPC is used as an analytical tool of the operating status; in other words, the statistical properties of production parameters, e.g. *'mean'* and *'standard deviation'*, are able to evaluate a production system which works under *'normal'* or *'abnormal'* status. If finding an abnormal status from an SPC chart, human expertise analysis associated with some statistical analysis approaches can be used to explore the factors that result in the problems as detected. It should be emphasized that the traditional SPC chart with one-dimensional parameter can only be applied to off-line problem analysis, but not for the automatic diagnosis. In comparison with process control, SPC deals with the system analysis in *'off-line'* and *'steady-state'* manners with statistical methods.

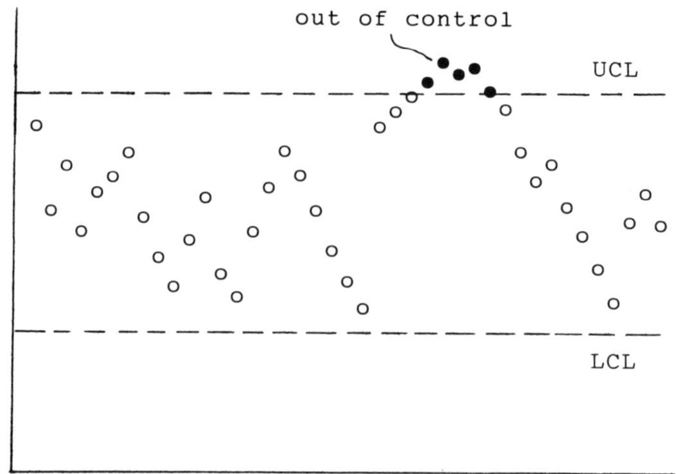

Figure 7.1 A sample SPC (Shewart) chart

The problem of combining SPC with process control has been proposed by MacGregor (1988). In this chapter, in addition to introducing the basic concepts of SPC, we will focus on the topics of on-line quality control with applications of multivariate statistics and intelligent system techniques. The goals of this chapter are to provide both SPC and process control professionals with applications of intelligent system methodologies in designing an on-line quality control system.

7.1.2 Concepts of SPC

In general, the status of a production system may be illustrated by an SPC chart, or so-called Shewart chart, as shown in Figure 7.1. The SPC chart is built with real-production data, e.g. product quality, productivity, or any other production related parameters. Each point in an SPC chart is related to the statistical results obtained from a specific period of time (hour, day, or month, etc.), or a certain number of products and batches. The coordinates of n and y represent time (or sequence) and quality related parameter respectively. If almost all of the points are located within upper and lower action limits, it means that the production system is *'in control'*, otherwise some control actions should be taken to bring the production system back to an 'in control' status. Generally, the action limits are set at plus and minus three standard deviations about the target. The purpose behind the SPC chart for a production system is to indicate an *'out-of-control'* status, when some relevant control actions must be taken to steer the system back to an 'in control' status. However, obviously it is hardly possible to find the control actions just from an SPC chart, because an SPC chart only presents the statistics of the quality space and does not have any associative memories or knowledge concerning the pattern mapping between the system inputs and outputs.

7.1.3 Problem statement

To design and implement an on-line quality control system, we must deal with the following two fundamental problems.

On-line quality prediction

From the previous section, we realize that the traditional SPC technique is of an off-line data oriented quality control; obviously it cannot be applied in on-line quality prediction and control. In fact, on-line quality prediction is similar to system modeling and estimation as we described in Chapter 4. The problem of on-line quality control may be defined as *'quality and/or other production related information predicted on-line with real-time production data and knowledge in terms of a quality prediction model'*. The form of quality prediction model can be designed based on the problem background.

On-line quality control

The task of on-line quality control is to steer the production system from an 'out-of-control' status to 'in-control' status through real-time control actions (e.g. supervised or setpoint signals) taken. In other words, the problem of on-line quality control is to determine the real-time control actions with predicted quality and a quality control model. The quality control model can be viewed as an associative memory or pattern mapping from a desired quality domain and a uncontrollable input vector to a control action vector. In fact, the on-line quality control is a 'recall' process from an associative memory, and thus results in finding the control actions to maintain the quality in a desired area.

To solve these two problems, we can use either traditional multivariate statistical approaches, e.g. multivariate regression, or the combination of multivariate statistics with intelligent system methodologies. We will first introduce the fundamentals of the statistics-oriented multivariate quality control, and then further investigate the applications of intelligent system methodologies in the on-line quality control.

7.2 MULTIVARIATE QUALITY CONTROL

7.2.1 Problem statement

Suppose that a quality prediction model for a MISO linear system can be stated as follows:

$$y = b_1x_1 + b_2x_2 + \cdots + b_rx_r + e \tag{7.1}$$

where y is defined as a quality variable, or so-called 'dependent variable'; $\{x_i, i = 1, \ldots, r\}$ represent the system inputs, or 'independent variables', $b_i, i =$

$1, \ldots, r$ are the model parameters to be identified, and e is the error or *'residual'*. Equation 7.1 can also be represented in vector form:

$$y = X^T B + e \tag{7.2}$$

where $X^T = [x_1, x_2, \ldots, x_r]$, and $B = [b_1, b_2, \ldots, b_r]^T$. Let there be k observations (samples) for both y and X, then 7.2 can be written as follows:

$$\begin{aligned} y_1 &= b_1 x_{11} + b_2 x_{12} + \cdots + b_r x_{1r} + e_1 \\ y_2 &= b_1 x_{21} + b_2 x_{22} + \cdots + b_r x_{2r} + e_2 \\ &\vdots \\ y_k &= b_1 x_{k1} + b_2 x_{k2} + \cdots + b_r x_{kr} + e_k \end{aligned} \tag{7.3}$$

or in the relevant matrix form

$$Y_{k\times 1} = X_{k\times r} B_{r\times 1} + E_{k\times 1}. \tag{7.4}$$

In real applications, if the number of samples is greater than that of independent variables, an exact solution of B is thus not available, and one can solve B only by minimizing the residual vector E. The most popular approach to solving this problem is the *'least-squares method'* (Draper and Smith, 1981), and the least-squares solution is

$$B = (X^T X)^{-1} X^T Y. \tag{7.5}$$

From 7.5 we can see that the parameter vector B with respect to minimal residual can be simply solved in terms of the observed data boxes X and Y. The results can be extended to the systems with multiple dependent variables, i.e. MIMO system, then the multiple linear regression should be used to solve the parameter matrix $B_{n\times r}$ with n dependent variables. However, for some systems the inverse of $X^T X$ may not exist, due to the problems of zero determinant, singularity, collinearity, etc. To solve these problems, we will further investigate principal component analysis and partial least-squares methods, and their applications in quality prediction and control.

7.2.2 Principal component analysis

General concepts

How to deal with raw data has been a key problem both in *'multivariate statistics'* (e.g. regression modeling) and in *'statistical pattern recognition'* (e.g. feature extraction). For example, both standard least-squares and pattern recognition methods may fail to provide a model with satisfactory precision due to the following factors:

- the input (or output) variables are interrelated, for instance if, $x_1 = a, x_2 = b, x_3 = c, x_4 = (a+b)/c$ and $x_5 = ac$, it can be seen that only three of them are independent. As a result, the matrix X^TX has a large condition number or is singular;
- there are a large number of inputs (or outputs), but, in fact, only part of the inputs has significant variance.

Principal component analysis (PCA) was originally developed by Pearson (1901) and has been used as a standard multivariate statistical method in many text books (e.g. Anderson, 1984; Wold *et al.*, 1987). PCA may be applied in data processing (e.g. *'data compression'* or *'dimensionality deduction'*) for either data matrix X or Y. The function of PCA is simply to perform a linear projection from the original (r-dimensional) data space to a lower (a-dimensional) new space, i.e., $a < r$, such that all variables in the new data space are independent and contain the main portion of information in the original data block. In fact, the new data space is established through selecting the principal components or factors that are all independent or orthogonal, and contain almost all (e.g. 95%) of the information from the original data block. Principal component transformation of the data matrix X can be mathematically represented as

$$X = T_1P_1^T + T_2P_2^T + \cdots + T_aP_a^T + E \tag{7.6}$$

where $\{T_i, i = 1, \ldots, a\} \in R^{k\times 1}$ and $\{P_i^T, i = 1, \ldots, a\} \in R^{1\times r}$ are defined as *'scores'* and *'loadings'* vectors respectively, and a denotes the number of principal

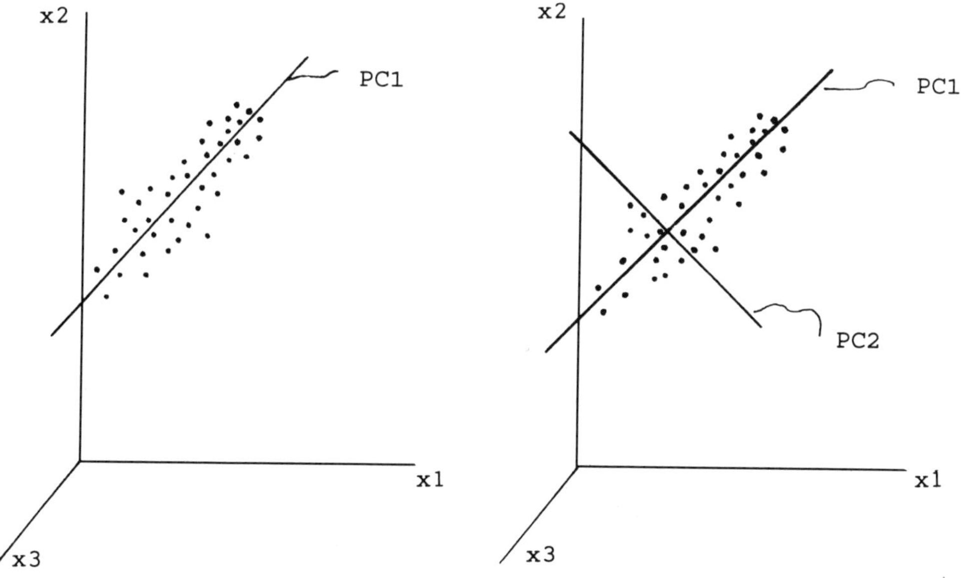

Figure 7.2 Geometric illustration of PCA in a two-dimensional space

components or factors. Each of the matrices $\{T_i P_i^T, i = 1, \ldots, a\}$ has rank one. Equation 7.6 can also be written in matrix form:

$$X = TP^T \tag{7.7}$$

where $T \in R^{k \times a}$ and $P^T \in R^{a \times r}$.

Actually, PCA is a method to explain the variance of data matrix X, namely, the first principal component (or vector) describes the direction of the greatest variability in data matrix X, and the second, principal component, having an orthogonal direction with the first one in the hyper-space represents the second greatest variability, until all principal components are solved. As a result, the residual is small enough in a practical sense. A geometric explanation of PCA is illustrated in Figure 7.2. Geometrically, *the principal components define the plane of greatest variability, and the loadings associated with these principal components define the location of the planes in terms of the original variables, and each observation is located on this planes via its scores. The score is the distance from the origin of the plane along each principal component and is calculated as the product of the loading vector and observation* (Kresta *et al., 1991*).

Computational algorithms

There are a number of approaches to calculating the principal components from a single data matrix (Jackson, 1980); however, here we only introduce an iterative algorithm for PCA computation, which will be further used in PLS calculation. This algorithm does not calculate all principal components at once; instead, the iterative algorithm starts with the calculation of the first pair $\{T_1, P_1^T\}$ in terms of data matrix X, then calculates $\{T_2, P_2^T\}$ with the residual $E_1 = X - T_1 P_1^T$. Finally, the iteration will stop until the residual is small enough to meet the problem under study. The algorithm can be summarized as follows (Geladi and Kowalski, 1986):

(1) let $X \epsilon R^{k \times r}$ be the normalized data matrix obtained with data preprocessing, e.g. 'mean-centering' or 'mean-centering and variance-scaling' from the original data matrix;

(2) set the iteration index $i = 1$, and define $T_i = X_j$ (any column vector X_j in X);

(3) calculate P_i^T with $P_i^T = T_i^T X / T_i^T T_i$;

(4) normalize P_i^T to length 1: $P_i^T = P_i^T / \|P_i^T\|$;

(5) calculate $T_i : T_i = XP_i / P_i^T P_i$;

(6) compare the T_i used in step (3) with that obtained in step (5). If they are not the same, then go to step (3), otherwise stop iteration, and further calculate the residual, $E_i = X_i - T_i P_i^T$. If E_i is less than a specified ϵ, then stop principal component calculation, otherwise $i = i + 1$, set $X_i = E_{i-1}$ and return to step (2).

The number of principal components is equal to i at the end of convergence, being defined as a, and then the scores and loadings for the principal components can be represented as $T = [T_1 \dots T_a]$ and $P = [P_1 \dots P_a]$ respectively. The results of PCA can be used to represent a data matrix X as its score matrix T:

$$T_{k\times a} = X_{k\times r} P_{r\times a} \tag{7.8}$$

where all column vectors in matrix T are independent and the dimension a is usually less than r. The applications of 7.8 in system modeling, data processing and quality control (e.g. regression or math-model free approaches) may result in data compression and decorrelation. PCA can also be used for the system output data matrix Y. Finally, the main results of PCA in regression are as follows:

- a data matrix X may be represented by its score matrix, which has fewer dimensions of column; in other words, the dependent variables and small eigenvalues in the scores are omitted;
- when applying PCA in principal component regression, there are no matrix inversion problems, since it is well conditioned.

7.2.3 Partial least-squares regression

The *'partial least-squares'* or so-called *'project to latent structure'* (PLS) has been an important subject in multivariate statistics and data/information processing in the design of an intelligent system. The pioneering work in PLS was developed by Wold *et al.* (1987) in the late sixties and then used in chemical applications (Kowalski *et al.*, 1982). In contrast to PCA, PLS deals with the development of a system regression model describing numerical relations between multi-inputs and multi-outputs in terms of both I/O data matrices. However, it should be noted that a PLS regression model is established based on the regression between the scores corresponding to I/O data blocks X and Y. As we have shown in the previous section, that the outer

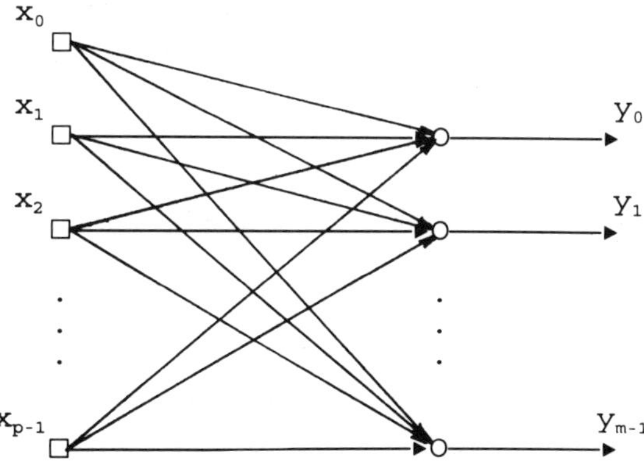

Figure 7.3 Architecture of a feedforward network for PCA (Reproduced from Haykin, 1994, © 1994 IEEE)

relation between X and its scores can be described as

$$X = TP^T + E = \sum_{i=1}^{a} T_i P_i^T + E. \tag{7.9}$$

The outer relation for the Y block can be built in the same way:

$$Y = UQ^T + F = \sum_{i=1}^{a} U_i Q_i^T + F \tag{7.10}$$

where U, Q and F denote the scores, loadings and residual for Y respectively. The a represents the number of principal components.

As a result of PLS regression, the interlinear regression model between the scores U and T can be built much more simply and precisely than using the traditional regression methods directly based on X and Y data blocks. The iterative PLS algorithm can be summarized as follows (Geladi and Kowalski, 1986):

Data block normalization:

(1) let $X \in R^{k \times r}$ and $Y \in R^{k \times n}$ be the normalized data matrices for the original input and output data blocks obtained with data preprocessing, e.g. *'mean-centering'* or *'mean-centering* and *variance-scaling'* from the original data matrix; set factor (component) index $i = 1$;

(2) for each component define $U_i = Y_j$ (any column vector Y_j in Y);

In the X data block:

(3) calculate W_i^T with $W_i^T = U_i^T X_i / U_i^T U_i$;

(4) normalize W_i^T to length 1: $W_i^T = W_i^T / \|W_i^T\|$;

(5) calculate $T_i : T_i = X_i W_i / W_i^T W_i$;

In the Y data block:

(6) calculate Q_i^T with $Q_i^T = T_i^T Y_i / T_i^T T_i$;

(7) normalize Q_i^T to length 1: $Q_i^T = Q_i^T / \|Q_i^T\|$;

(8) calculate U_i: $U_i = Y_i Q_i / Q_i^T Q_i$;

Check convergence:

(9) compare the T_i used in step (3) with that obtained in step (5); if they are not the same then go to step (3), otherwise stop iteration and go to step (10);

Calculate the X loadings and rescale the scores and weights accordingly:

(10) calculate X loadings: $P_i^T = T_i^T X_i / T_i^T T_i$;

(11) normalize P_i^T to length 1: $P_i^T = P_i^T / \|P_i^T\|$;

(12) update T with $T_i = T_i \|P_i^T\|$;

(13) update W with $W_i = W_i \|P_i^T\|$;

Calculate the regression coefficients B_i for the inner relation:

(14) $B_i = U_i^T T_i / T_i^T T_i$;

Calculate the residuals for factor i:

(15) calculate the residuals E and F for the blocks X and F respectively, namely $E_i = X_i - T_i P_i^T$ and $F_i = F_i - B_i T_i Q_i^T$;

(16) let $i = i + 1$ and set $X_i = E_{i-1}$ and $Y_i = F_{i-1}$ then return to step (2) to calculate next principal component, until all principal components are calculated.

It should be noted that during the calculation of individual principal components, P_i^T, Q_i^T, W_i and B_i should be saved for later application, e.g. modeling and prediction, and T_i and U_i are saved for diagnosis and/or classification purposes. The determination of the number of components is an important issue in working on PLS. In general, for a linear model, the required number of components is equal to the rank of the data matrix X; however, for a nonlinear model, the extra components are needed to describe the nonlinearity. On the other hand, since the data matrices are never noise-free, some smaller components describe only noise. How to determine the number of components for the problem under consideration has been investigated. In general, the following two methods can be used in practice (Geladi and Kowalski, 1986):

- the number of components can be determined based on the norm of F_i, $\|F_i\|$; in other words, the iteration can be stopped when $\|F_i\|$; is less than a predefined threshold;
- the iteration for the component calculation can also be stopped when the error between the model (PLS regression) predicted outputs and the actual output data becomes small enough;
- 'cross-validation' is another measure to determine the number of components. If the test results of the PLS model for a set of non-training data are satisfied, the iteration can be ended, otherwise the iteration is still needed to find more components.

In practice, a combination of different methods would be preferable. Actually, this problem is similar to the structure learning in neural or fuzzy systems as introduced in Chapter 3.

7.3 PCA USING SELF-ORGANIZED LEARNING

PCA can also be explained by eigenvalues and eigenvectors; as noted by Oja (1982), *to perform dimensionality reduction on some input data, we compute the eigenvalues and eigenvectors of the correlation matrix of the input data vector, and then project the data orthogonally onto the subspace spanned by the eigenvectors belonging to the largest eigenvalues.* In the present section, we only address the design of a self-organized neural network that performs principal component analysis of arbitrary size on the input vector (Sanger, 1989; and Haykin, 1994).

Consider a single-layered, computational feedforward network as shown in Figure 7.3 (Haykin, 1994). Structurally, suppose that each neuron in the output layer of the network is linear, and the network has fewer outputs than inputs (i.e. $m < p$). The network outputs at time n can be calculated as a linear combiner:

$$y_i(n) = \sum_{i=0}^{p-1} W_{ji}(n)x_i(n), j = 0, 1, \ldots, m-1. \tag{7.11}$$

The synaptic weight is modified with a generalized form of Hebbian learning as follows:

$$\Delta W_{ji}(n) = \eta\,[y_j(n)x_i(n) - y_j(n)\sum_{k=0}^{j} W_{ki}(n)y_k(n)], \quad i = 0, \ldots, p-1 \quad j = 0, \ldots, m-1 \tag{7.12}$$

$$\Delta W_{ji}(n) = \eta\,y_j(n)(x_i'(n) - W_{ji}(n)y_j(n)), \quad i = 0, \ldots, p-1 \quad j = 0, \ldots, m-1 \tag{7.13}$$

where η denotes the learning rate, and $x_i'(n)$ is a modified version of the ith element of the input vector $X(n)$, that can be represented as a function of the index j:

$$x_i'(n) = x_i(n) - \sum_{k=0}^{j-1} W_{ki}(n)y_k(n) \tag{7.14}$$

The principal components can be solved by the generalized Hebbian algorithm based on the input observations as follows (refer to Sanger, 1989; and Haykin, 1994):

(1) For the first neuron in Figure 7.3, we have

$$j = 0, X'(n) = X(n) \tag{7.15}$$

In this case, there is only a single neuron, and the first principle component (i.e. the largest eigenvalue and associated eigenvector) of the input vector $X(n)$ can be calculated from the correlation matrix $E[X(n)X^T(n)]$ of the input vector $X(n)$ based on the definition of the principal components.

(2) For the second neuron, let

$$j = 1, X'(n) = X(n) - W_0(n)y_0(n) \tag{7.16}$$

and since the first neuron has already converged to the first principal component, the second neuron works based on its input vector $X'(n)$ provided by 7.16 in which the first eigenvector of the correlation matrix has been removed. As a result, the second principal component (i.e. the second largest eigenvalue and associated eigenvector) of the original data vector $X(n)$ can be solved.

(3) Similarly, for the third neuron, we have

$$j = 2, X'(n) = X(n) - W_0(n)y_0(n) - W_1(n)y_1(n) \tag{7.17}$$

Since the first and second neurons have converged to the relevant principal components, the third principal will be solved in terms of $X'(n)$ provided by 7.17. When we follow the procedures as noted above, the eigenvalues will be decreasing as we add more principal components. The approaches to determining the number of principal components introduced in the previous section can also be used here. The stability or convergence is controlled by the learning rate following a stability theorem (refer to Sanger, 1989).

7.4 NONLINEAR PLS MODELING USING NEURAL NETWORKS

The applications of the PLS regression method make it possible to solve linear regression problems with high dimensionality and collinearity in data matrices. However, the PLS linear regression method cannot be applied in solving many real-world problems in industry, since they are nonlinear, time-varying and/or uncertain. This section will address the system architecture which combines PCA/PLS with intelligent (neural, fuzzy, pattern recognition, etc.) system methodologies.

Based on the universal approximation property of multilayered neural networks, Qin and McAvoy (1992) proposed an architecture and algorithm of neural network oriented PLS (NNPLS) for nonlinear system modeling. The general scheme of NNPLS is shown in Figure 7.4 (Qin and McAuoy, 1992). In the NNPLS algorithm, a multilayered feed-forward neural network replaces the inner linear regression model, and keeps the original outer transform. The NNPLS algorithm can be summarized as follows:

(1) set the index of principal components, $i = 1$, and let X_i and Y_i be the normalized data matrices for both input and output data blocks;

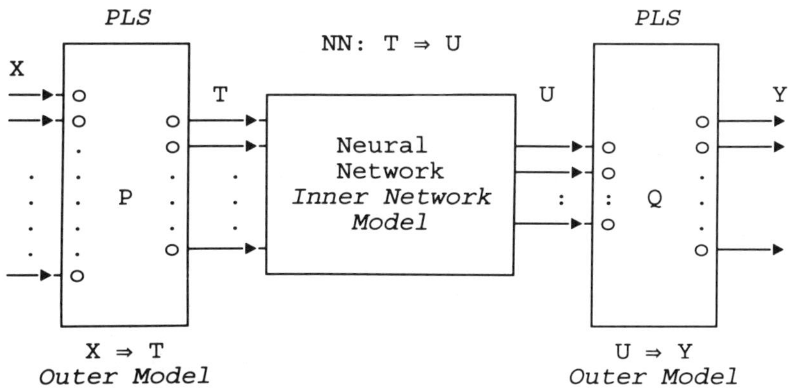

Figure 7.4 Scheme of neural network-oriented PLS

(2) for each component define $U_i = Y_j$ (any column vector Y_j in Y);

(3) in matrix X

- calculate W_i^T with $W_i^T = U_i^T X_i / U_i^T U_i$;
- normalize W_i to length 1;
- calculate T_i: $T_i = X_i W_i$;

(4) in matrix Y,

- calculate Q_i^T with $Q_i^T = T_i^T Y_i / T_i^T T_i$;
- normalize Q_i^T to length 1;
- calculate U_i : $U_i = Y_i Q_i$;

(5) return to step (3) until the iteration converges;

(6) calculate the X loadings: $P_i^T = T_i^T X_i / T_i^T T_i$;

(7) find the *'inner model'*: the training neural network with the *conjugate gradient* method, minimizing the error $J_i = \|U_i - NN(T_i)\|^2$;

(8) calculate the residuals E and F for blocks X and Y respectively, namely $E_i = X_i - T_i P_i^T$ and $F_i = F_i - U_i' Q_i^T$, where $U_i' = NN(T_i)$;

(9) let $i = i + 1$ and set $X_i = E_{i-1}$ and $Y_i = F_{i-1}$ then return to step (2) to calculate the next principal component, until all principal components are calculated.

The number of principal components can also be determined by the methods noted in the previous section. In fact, the inner model (in step (7)) can also be implemented by a fuzzy model $U_i' = FL(T_i)$, or a pattern recognition model $U_i' = PR(T_i)$. Finally, it should be pointed out that the NNPLS algorithm can also be extended to its on-line form, in which the I/O data blocks are upgraded with the time and/or the changes of the system environment. As a result, the 'time-oriented' or 'event-driven' scores, loadings and neural model $NN(*)$ are used to adapt to the changes of system environment.

7.5 INDUSTRIAL APPLICATIONS OF PCA AND PLS

The PCA/PLS multivariate statistics can be widely used in various fields, such as data and information processing, feature extraction, dimension deduction, quality control and industrial control as well. This section will focus on the PCA/PLS applications in the quality prediction and control area.

7.5.1 Dimension deduction and data compression using PCA

From the above discussions, we realize that PCA only deals with a single data matrix, i.e. system inputs (process variables) or system outputs (quality related variables). The task of PCA is to transfer (or project) the original data space onto a relevant new space, such that the dimensionality of the space is reduced, but the information existing in the original data space is still kept, and all variables (principal components) in the new space are independent, or so-called 'orthogonal'. In many real-world industrial processes, it is quite common to measure hundreds of on-line process variables and tens of off-line product quality variables. However, since they are usually highly correlated through mass/energy exchanges and/or chains of chemical reactions, consequently the real (independent) dimension in process or quality variable space is much lower than the number of measurements. For example, we can describe a chemical reactor with concentrations $(A, B, \ldots)$, reaction temperature, pressure, etc.; however, since there exist strong correlations between those process variables, the number of independent dimensions is much lower than the number of measurement parameters.

Theoretically, this kind of dimension deduction can also be performed by process principal analysis. However, since many industrial processes are too complicated to be dealt with based on explicit principles, PCA becomes a practical and effective way to extract features directly from data. On the other hand, the applications of PCA can also avoid the situation of *'data rich'* but *'information poor'*, since PCA does not only reduce the data dimension, but also makes the extracted principal components contain greater variance. It should be kept in mind that after linear projection, $P : X \to T$, the new data space then loses the physical meaning in the original data space.

7.5.2 Applications of NNPLS in system modeling and prediction

From the above discussions, we realize that PLS can provide much better results than regular linear regression methods. Moreover, NNPLS can be applied in high dimensional, nonlinear system modeling. Figure 7.5 shows a scheme of NNPLS with both *'training'* and *'recall'* phases. From Figure 7.15 we can see that the

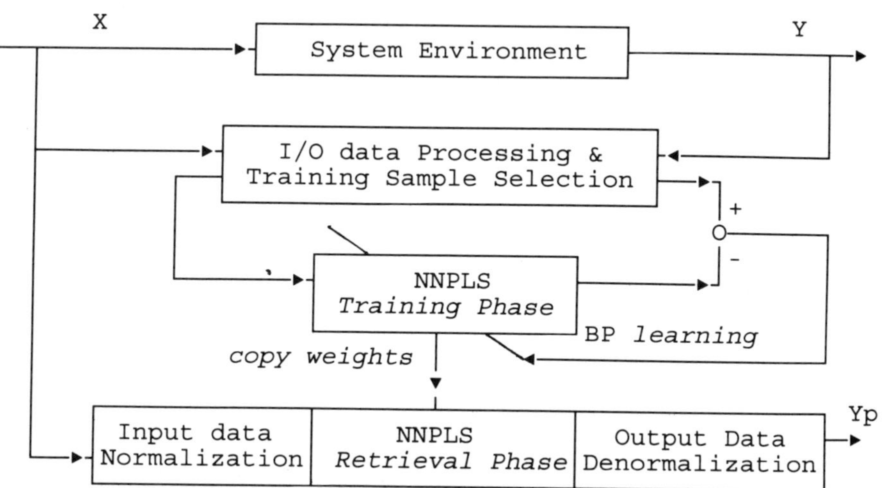

Figure 7.5 Architecture of NNPLS for system modeling and prediction

convergent loadings P and Q, and the inner network model $NN(*)$ governed by the frozen weights, are saved and copied to a program for the operation of recall. As a result, the predicted output can be provided through the following forward path in terms of the observations of system input vector X:

$$X \xrightarrow{P} T \xrightarrow{NN(*)} U \xrightarrow{Q} Y \tag{7.18}$$

In comparison with the approaches to regular neural network modeling, with NNPLS, *a MIMO nonlinear system modeling task is decomposed into linear outer relations and simple nonlinear inner relations, which are performed by a number of SISO networks* (Qin and McAvoy, 1992). As a result, the improvement in over-parameterized, local-minimum, and training time problems may be made through applying NNPLS in nonlinear system modeling.

Similar to the regular neural networks, the performance of generalization should be a major concern in order to put the resulting neural network with PLS into practice. Besides the factors of generalization which we investigated in section 3.4.2, the number of principal components (or factors) is an additional important factor to affect the performance of generalization. In general, the optimal number of principal components may be determined by cross-validation. Figures 7.6 and 7.7 (Qin and McAvoy, 1992) show the cross-validation results (for both training and test

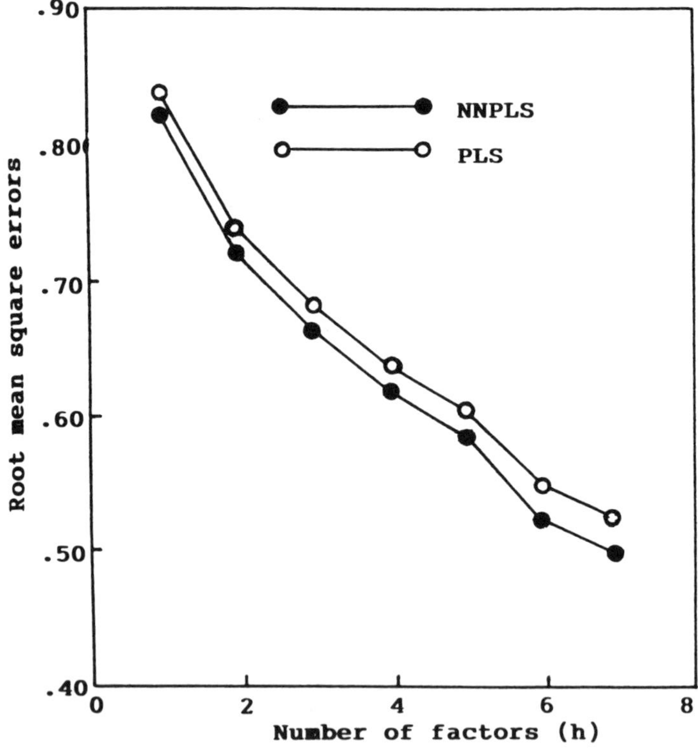

Figure 7.6 Training errors by NNPLS and PLS for the cosmetic data

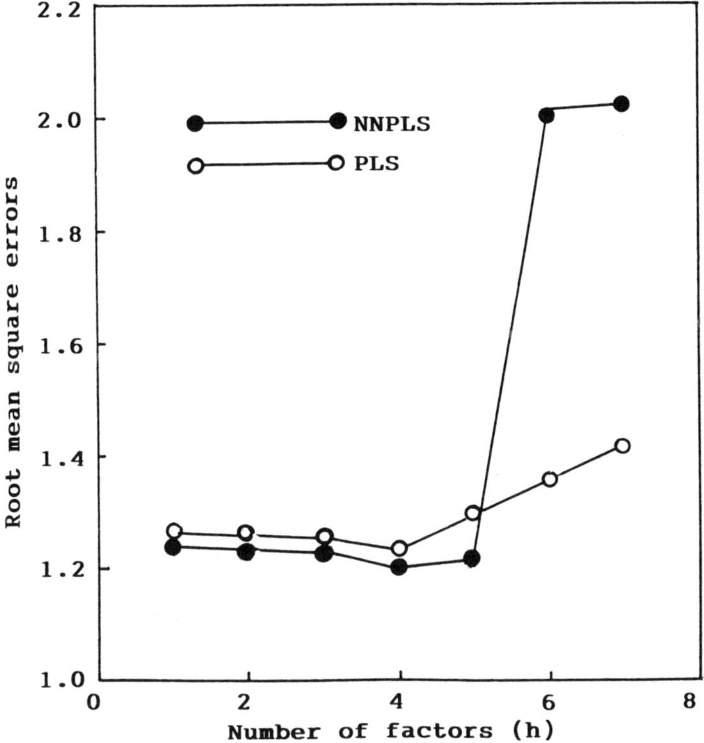

Figure 7.7 Prediction errors by NNPLS and PLS models for the cosmetic data

data sets) for the cosmetic data. It can be seen that NNPLS generally provides lower errors for both training and test sets. However, it is not true that the more principal components, the lower the predicted errors. For this particular example, we can see that the optimized number of principal components equals four. However, if NNPLS works under a variable environment, NNPLS model should be periodically upgraded through learning under a upgraded environment.

In the quality control area, NNPLS can be used to predict product quality and productivity variables with the process variables. The predicted results can then serve as off-line and/or on-line monitoring information. Traditionally, an SPC chart is used to detect process status, 'in-control' or 'out-of-control'. However, the decisions and actions for quality control and monitoring are usually taken by off-line analysis. In contrast, the applications of NNPLS may not only provide real-time quality prediction, but also result in decision making for process monitoring. The architecture and strategy of real-time quality prediction and control will be discussed in section 7.6.

7.5.3 Statistical process control by PCA

As we discussed in the previous section, it requires both process and quality data blocks to make quality control. However, in many industrial processes, either only

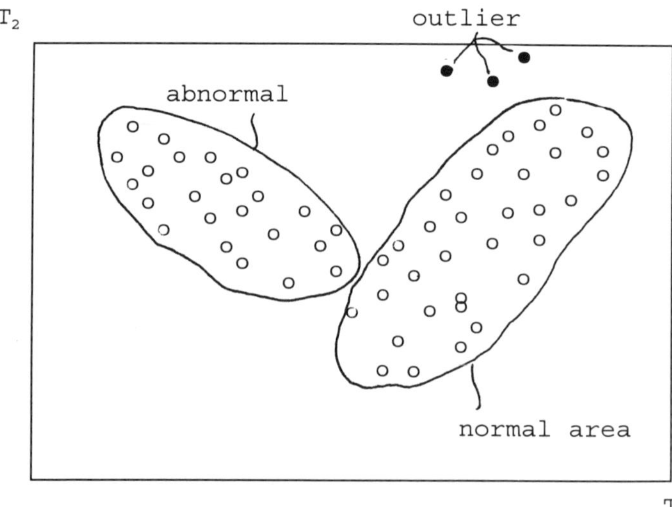

Figure 7.8 Sample plot with two principal components for process analysis

a single data block (e.g. process variables) is available, or the quality data are observed in an infrequent time base; then the applications of PCA can help us to detect and monitor an upset or 'out-of-control' process status. Any operating condition governed by r-dimensional process variables can be represented as an r-dimensional vector, $X = [x_1, \ldots, x_r]^T$ in a hyperspace. However, it will be very difficult to carry out process analysis based on the observations of X, if

- there are correlations between process variables, and/or
- the vector X has very high dimensions.

An important benefit of PCA is to perform an orthogonal projection from the original data matrix X to the relevant score matrix T, which is orthogonal and lower dimensional. Geometrically, with more than three principal components, it will be impossible to plot those values as a physical plot. However, if the first two or three components account for a large proportion of the total variation, PCA enables us to plot the data (scores) in two or three dimensions. Figure 7.8 shows a two-dimensional process variable plot with the first two components, T_1 and T_2. Since T_1 and T_2 are independent and retain the major portion of the variation, it is easier to evaluate the 'normal' and 'abnormal' status of the process with data grouping. On the other hand, the 'outlier' may also be detected by using a PCA oriented chart. The application of pattern recognition provides the opportunity to represent multi-dimensional vector space. For instance, if the number of components is equal to $p\,(p > 3)$, each score vector (pattern) can be represented as a point in a p-dimensional hyperspace. The process status can then be classified into a number of clusters in terms of the Euclidean distance between the score patterns. The multi-dimensional 'mean', 'lower-limit' and 'upper-limit' vectors under score coordinates can also be represented in the same space; then the distances between the individual score patterns and the vectors of 'mean', 'upper-limit' and 'lower-limit' can be further represented in the same plot. It should be emphasized that a weighting vector describing the priority of process variables may be introduced when calculating the Euclidean distance. On the other hand, the approach to self-organized learning (supervised or unsupervised) for pattern classification may also be applied in quality control and monitoring.

However, whatever kind of pattern classification method is used, the orthogonal data transformation by PCA will be beneficial in obtaining more precise and reasonable numerical results in an efficient way. Finally, it should be kept in mind that since PCA only works with a single data block, some knowledge based inference is needed to provide further analysis for diagnosis.

7.6 ON-LINE QUALITY CONTROL

We know that the present activities of SPC and quality assurance groups in enterprises are to make quality analysis in terms of historical production data and infrequent off-line (laboratory) observations. MacGregor (1988) encouraged reducing the gap between SPC and process control through developing on-line SPC or QC. How to design an on-line quality control system is still an open problem; however, the latest R&D in intelligent systems, machine learning, and particularly the combination of multivariate statistics and intelligent system methodologies, provide great opportunities to develop on-line quality control. An on-line quality control system generally consists of the following functions:

- real-time quality prediction associated with an inference model in terms of on-line observations of the process variables and infrequent laboratory measurements;

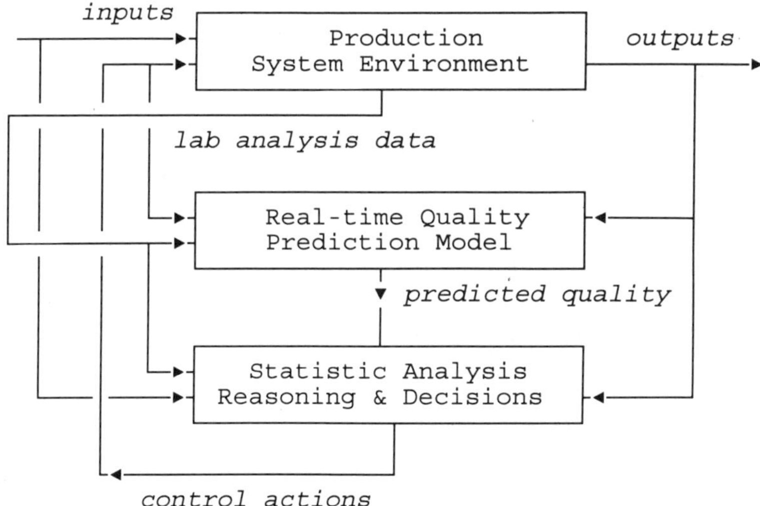

Figure 7.9 Architecture of on-line QC system

- periodic statistical analysis for quality distribution, such as mean, standard deviation and other criteria;
- decision making for control actions (e.g. setpoints of the dedicated control loops) with a quality control (monitoring) model in terms of the aim quality and process variables.

A generic architecture of an on-line QC is illustrated in Figure 7.9. The quality prediction (quantitative, qualitative or multivalued) can be performed by a combination of multivariate statistics and intelligent system methodologies, such as NNPLS, fuzzy/PLS, pattern recognition/PLS, and rule based expert system, etc. As a result, the real-time quality information described by analog, discrete or binary number can be estimated by a pre-trained prediction model (or associative memory) with real-time observations of the process variables. To adapt to the changes of production environment, the infrequent laboratory measurements are used as the feedback signals for self-learning, for instance in upgrading neural network weights or fuzzy membership. In contrast to process control, in an on-line QC system, control actions are not instantly taken based on the predicted quality information; instead, the control actions are taken only if the quality distribution analysis has been made. Generally, a moving window with specific size of time or number of events can be designed to make periodic statistical analyses for quality and/or production status. The quality control model in Figure 7.9 can be performed in many different ways; however, from a system control viewpoint, this control mode functionally belongs to feedback/feedforward control. In fact, quality control can also be viewed as an associative memory or rule reasoning which carries out the pattern mapping from quality (aimed and predicted) and production environment information (e.g. uncontrollable inputs) to control actions (manipulated variables). In general, neural network, fuzzy logic and rule based expert system associated with multivariate statistics can be applied in building such a quality control model.

7.7 CONCLUDING REMARKS

'Which one is better for system functional modeling and prediction? Statistics or pattern mapping oriented associative memory?' has been one of the key problems in both industrial and academic societies. In many cases they have the same purpose but use different methods. From above discussions, we may realize that a hybrid structure with an elegant combination of multivariate statistics and intelligent system methodologies, for instance *'neural network–PCA/PLS'* or *'fuzzy logic–PCA/PLS'*, will be mutually beneficial to both methods.

8
Fault detection and diagnosis

8.1 INTRODUCTION

Fault detection and diagnosis have been an important part of an integrated computer control system to maintain production running under a normal and stable status. Since many complex and/or large scale industrial processes are highly nonlinear, time-varying and seriously interconnected, with uncertainties, it becomes a very difficult task to detect and diagnose abnormal production conditions with real-time observations of process variables. Traditionally, a math-model based state and parameter estimation can be directly used in fault detection and diagnosis. However, due to the complexity of many industrial processes, sometimes it is difficult to develop a realistic mathematical model for real-time applications in fault detection and diagnosis. With the recent R&D in intelligent systems and machine learning technologies, the applications of intelligent systems in process fault detection and diagnosis have become an attractive subject in the control industry. In this chapter, we will start with a general description of the problems in fault detection and diagnosis, and then investigate the applications of some intelligent system methodologies in fault detection and diagnosis with a number of working examples.

8.2 MATH-MODEL BASED FAULT DETECTION AND DIAGNOSIS

The concepts and background of math-model based fault detection and diagnosis are apparent for practicing engineers. For example, the leakage of a natural gas pipe can be detected by a simple fluid dynamic model with observations of gas flow rate and/or pressure at a number of limited locations. However, for a complicated natural gas network system with a huge number of users and unsteady-state operating conditions, the detection and diagnosis of the leakage become much more difficult, and then a complicated network mathematical model is needed to perform the leakage detection and diagnosis.

Suppose we have a nonlinear state space model with the following state and output equations:

$$X_{k+1} = F(X_k, U_k, \theta) \tag{8.1}$$

$$Y_k = CX_k \tag{8.2}$$

where $U \epsilon R^r$, $Y \epsilon R^m$ and $X \epsilon R^n$ denote system input, output and state vectors $F(*)$ denotes a nonlinear function C is a constant matrix describing the relation between the state and output vectors, and θ presents a parameter vector. Generally speaking, the status of the fault, e.g. *'fault size'* and *'fault location'*, can be estimated by either *'parameter estimation'* or *'state estimation'*. For instance, the reduction of heat conductivity of the boiler pipes mostly means sludge formed in the boiler pipes' internal and/or external surfaces. Because some model parameters are functions of heat conductivity, the estimation of heat conductivity-related model parameters will help us to detect such kinds of faults in a boiler. On the other hand, application on state estimation can also provide internal information on a process which is related to a process fault. For instance, the real-time estimation of the carbon accumulation

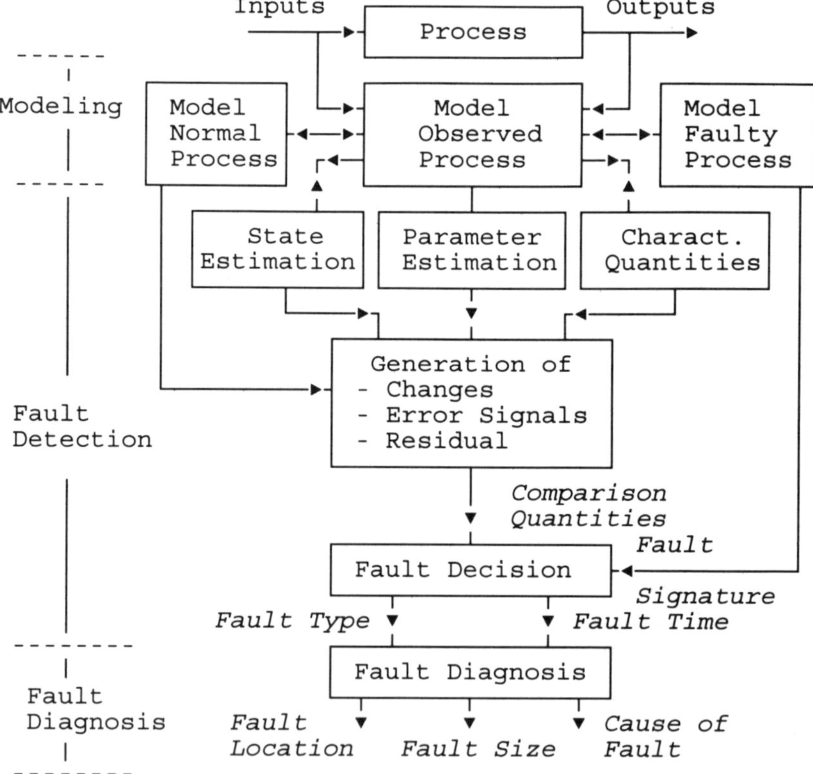

Figure 8.1 System configuration of math-model based fault detection and diagnosis system

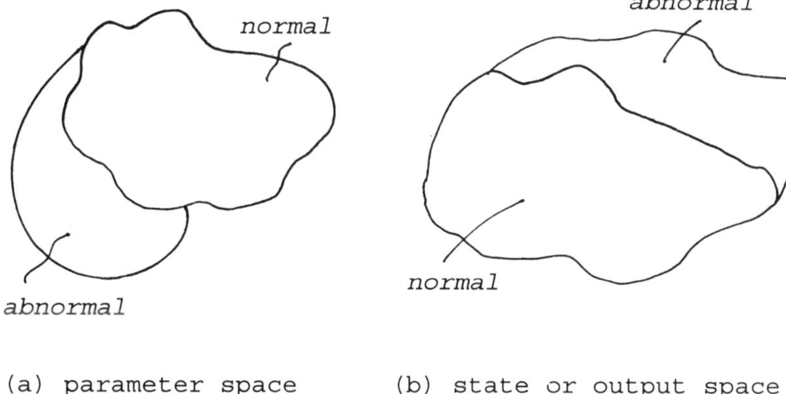

Figure 8.2 Geometrical illustration of process faults in parameter and/or state hyperspace

on the surface of the catalyst in a fluidized catalytic cracking unit (FCCU) can detect abnormal operating conditions of a reactor–regenerator system in FCCU.

A general architecture of a math-model based fault detection and diagnosis system is shown in Figure 8.1 (Isermann, 1984). The system provides the real-time (model parameter and/or state variable) information which is related to process faults in terms of sensory I/O measurements. As shown in Figure 8.2, geometrically the parameter and/or state domains in their hyperspaces can be divided into a number of regions which represent 'normal' and 'abnormal' statuses, or describe the fault size with its corresponding regions.

Many math-model based real-time parameter and state estimation algorithms, e.g. recursive least-squares and extended Kalman filtering, have been developed. However, it is still quite difficult to apply those methods in some complex industrial systems with high dimensional, nonlinear, time varying and uncertain behavior. On the other hand, applications of expert systems, neural networks and fuzzy logic have provided great opportunities to deal with fault detection and diagnosis problems. As a result, a combination of human knowledge and process principle with production data becomes a strategic concern in developing a knowledge based fault detection and diagnosis system.

8.3 GENERAL VIEW OF MATH-MODEL FREE FAULT DETECTION AND DIAGNOSIS

As we mentioned in the previous section, the process status, 'normal' or 'abnormal', can be classified in either a parameter or a state variable hyperspace in terms of a math-model based estimation algorithm. This chapter will focus only on the design of math-model free fault detection and diagnosis for industrial applications. Structurally, the most popular strategies of math-model free fault detection and diagnosis can be summarized as follows:

(1) Rule based expert system

The expert system has been recognized as one of the most effective tools in performing fault detection and diagnosis in industry. Particularly when a system has a set of linguistic (logic or fuzzy logic) 'if-then' statements, a knowledge base with inference mechanism can be developed to detect and diagnose the faults. Generally, reasoning or deep reasoning is carried out in terms of process 'symptoms' and/or 'conditions' observed from the sensor measurements and some simple first principle model segments as well. However, for many sophisticated modern industrial processes, if the fault related knowledge is not available or clear enough, it will be difficult to develop an expert system for this purpose. With the rapid progress on information and data processing technology, the applications of neural network, fuzzy logic and pattern recognition have opened a new avenue in carrying out fault detection and diagnosis.

(2) Associative memory (Pattern mapping)

As we have learned from section 8.2, the fault status can be represented and recognized in a fault feature space. From the fundamental concept of 'pattern mapping', an 'associative memory' for fault detection and diagnosis can be established through supervised or unsupervised learning. Figures 8.3 and 8.4 show the geometrical illustrations of associative memories for fault detection and diagnosis respectively. When the associative memory is established, from any operating point in the input (independent) vector space, $U = [u_1, \ldots, u_r]^T$, we can determine a corresponding point in the fault related output (dependent) vector space, $Y = [y_1, \ldots, y_n]^T$, through a pattern mapping or so-called 'retrieval' process. Similarly, another associative memory from the fault variable space, $Y = [y_1, \ldots, y_n]^T$, and

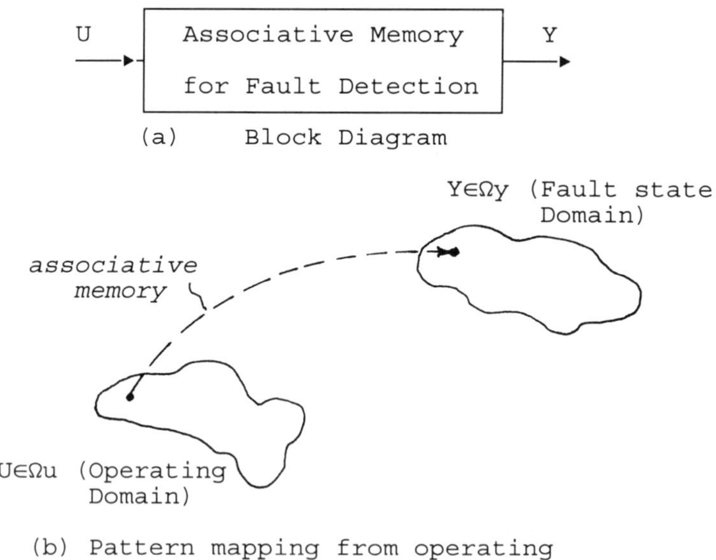

Figure 8.3 Geometrical illustration of associative memory for fault detection

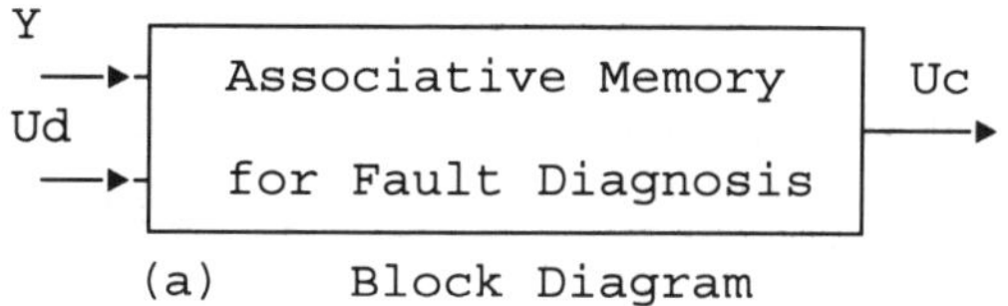

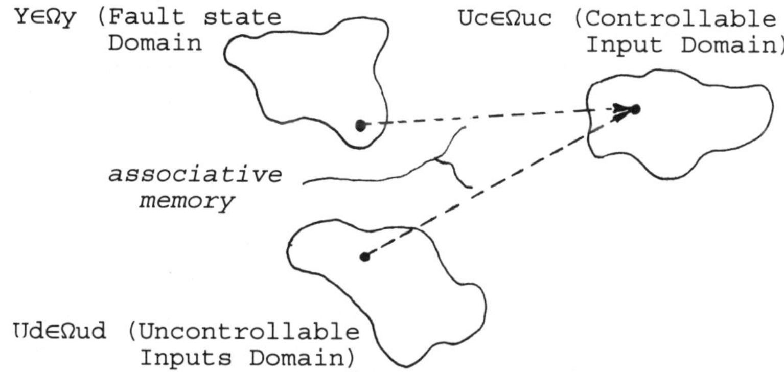

Figure 8.4 Geometrical illustration of associative memory for fault diagnosis

uncontrollable input space, $U_d = [u_{d1}, \ldots, u_{dr}]^T$, to the control action space, $U_c = [u_{c1}, \ldots, u_{cr}]^T$, can also be established through learning. To simplify the problem, the pattern mapping in multi-dimensional spaces may be simplified as in the single-dimensional one with the data coding, i.e. the space can be classified into a number of subspaces with a corresponding discrete index. The associative memories can be performed by pattern recognition, neural network or fuzzy logic models through learning from examples.

Finally, it should be pointed out that the associative memories can also be established through combining a pattern based mode (unstructured knowledge) with a rule based mode (structured knowledge). As a result, some hybrid form of associative memories may be developed. This chapter will present only the applications of neural networks and fuzzy logic in fault detection and diagnosis.

8.4 APPLICATIONS OF NEURAL NETWORKS IN FAULT DETECTION AND DIAGNOSIS

8.4.1 General architectures

A neural network being applied for fault detection and diagnosis may also be developed through 'learning', i.e. 'learning from examples', even if one does not know the 'if-then' kind of linguistic rules or process principle in detail. The design of neural network architectures and learning algorithms depends on the forms of learning examples. Let us consider the following different forms of learning samples:

(1) If the training samples are represented by a set of numerical inputs and fault related outputs, then a neural network can be designed in the following ways:

(a) Keep the original I/O data mode and establish a neural network through BP-learning. As a result, the fault related outputs can be predicted by recall of the neural associative memory in terms of the process inputs. The fault can then be detected and identified directly from fault related variables. This is a quite straightforward method to deal with fault detection.

(b) The output data space is classified into a number of fault related domains with coding. One can then use a few discrete numbers to describe the fault statuses, e.g. 'normal', 'abnormal', etc., The neural network can be designed with only a single output neuron with analog value, that describes the fault level. Or, on the other hand, the neural network can be designed with multiple binary output neurons. The position of the exciting neuron represents the fault status; however, it should be noted that the coding of the training samples should dictate that at any time only one neuron is allowed to be exciting.

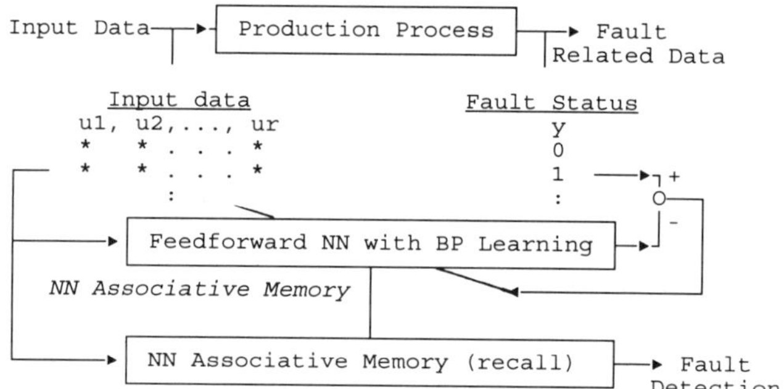

(a) Associative memory from direct supervised learning

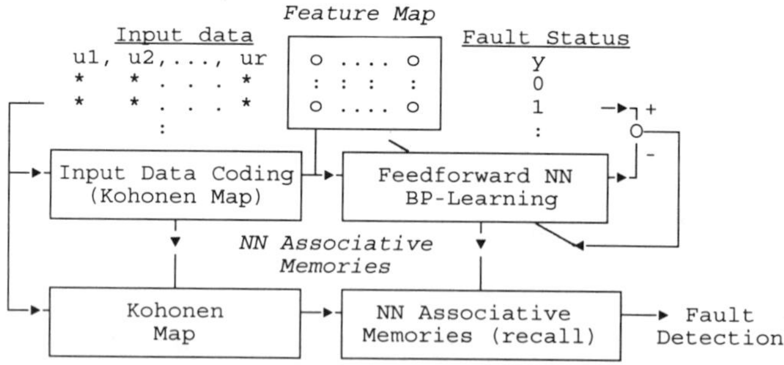

(b) Supervised learning with self-organizing (Kohonen) coding

Figure 8.5 General architecture of neural network oriented fault detection and diagnosis system

(c) Both I/O data spaces are classified into a number of domains respectively and then the I/O data can be coded into the relevant binary numbers. A neural network with the binary input and output neurons can be designed.

(2) If the training samples are represented by a set of numerical inputs and fault related linguistic outputs respectively, then a neural network can be designed with coding for either only linguistic variables or both input and output (linguistic) variables. Nevertheless, whatever kind of coding methods are applied, the key task is to establish a neural network with an associative memory which can classify the input space into a number of fault related domains. The learning algorithms designed mainly depend on the informative form and content in the training samples; generally, 'BP-learning', 'reinforcement learning' or 'self-organized learning' can be used.

Moreover, the fault diagnosis or decision making for control actions is a more complicated problem, particularly for a high dimensional, nonlinear and/or uncertain system. To solve this problem, we first divide the input space into both U_c and U_d subspaces, then an associative memory from the desired operating condition, $Y \in \Omega_y$ and U_d to U_c can be established through pattern learning or search on Y space with respect to U_c. However, if Ω_y is a fixed domain, then the problem can be simplified as a pattern mapping directly from U_d to U_c. A general architecture of neural network oriented fault detection and diagnosis system is illustrated in Figure 8.5.

8.4.2 Neural network based mold breakout prediction for steel continuous caster

In the continuous casting process, the breakout of molten steel spilling through a rupture in the solidifying shell is a serious factor that has to force a prolonged production shutdown and significant production loss. Principally, when a breakout occurs due to shell rupture, the hot energy gradually propagates in the mold width and casting directions. Since the molten steel directly contacts the mold at the shell rupture, the temperature of the mold near the location of the shell rupture will increase and be observed by the thermocouples in that neighbourhood.

Conventionally, the breakout is detected by both top and bottom temperature patterns observed from two sets of thermocouples which are embedded near the top and bottom of the mold respectively as shown in Figure 8.6(a). A typical time-series temperature pattern indicating the breakout is given in Figure 8.6(b). The decisions of the mold breakout are made by a set of simple rules associated with human analysis in terms of the rate of temperature changes, the relative position of the upper and lower temperature time-series trajectories, casting speed, propagation time, etc. However, with the variable and sophisticated operating conditions, and measurement noise, the traditional method may provide delayed and/or wrong prediction.

To improve the prediction accuracy, a neural network system shown in Figure 8.7 (Hammerstrom, 1993) for breakout prediction has been developed and successfully

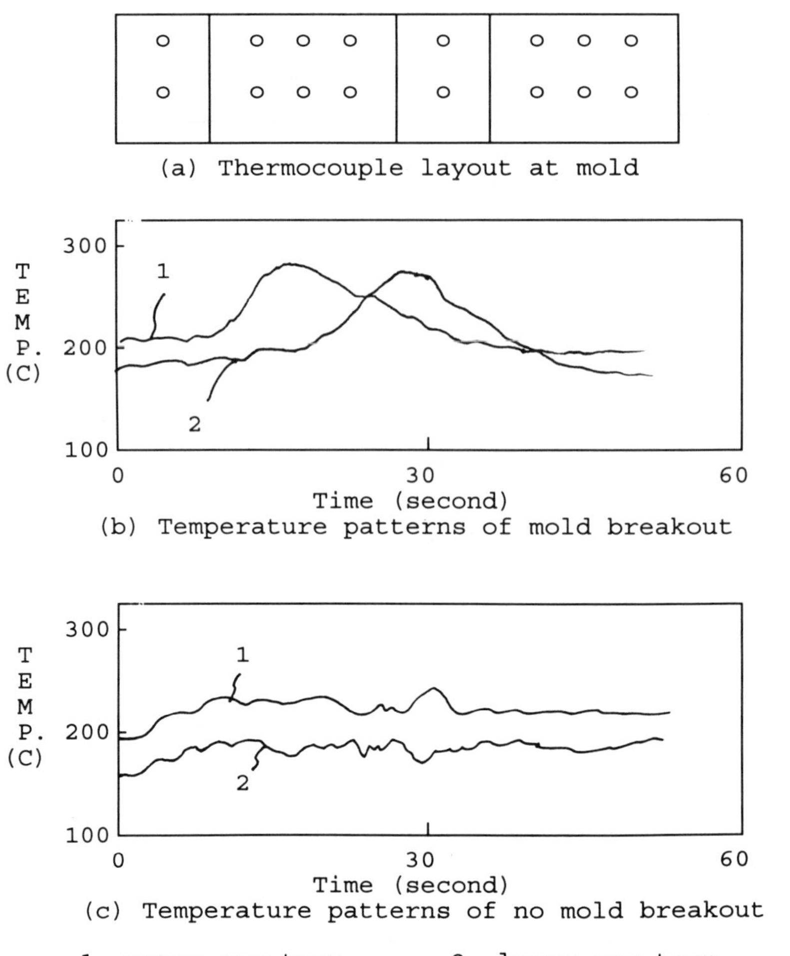

Figure 8.6 A typical example of temperature pattern showing mold breakout

applied in Nippon Steel Co. (Kominami *et al.*, 1991). The system consists of both 'time-series' and 'spatial' networks. The inputs of each time-series network are the trajectory of the observed temperatures (normalized from the means), e.g. $T(k), k = k, \ldots, k - N$, with a moving window. The shift window, N, depends on the dynamic response of the temperatures, and can usually be selected at around 30 seconds. The outputs of the time-series networks are the certainties which describe the probability of breakout, ranged within 0.0 to 1.0. Obviously, it is not good enough to detect the breakout only based on the dynamic responses of the individual temperatures.

As we can see from Figure 8.7, a number of spatial networks are connected with the time-series networks. The function of the spatial networks is to detect the breakout by comparing the temperature patterns from adjacent pairs of sensors at the upper-row and lower-row. The outputs of the time-series network serve as the

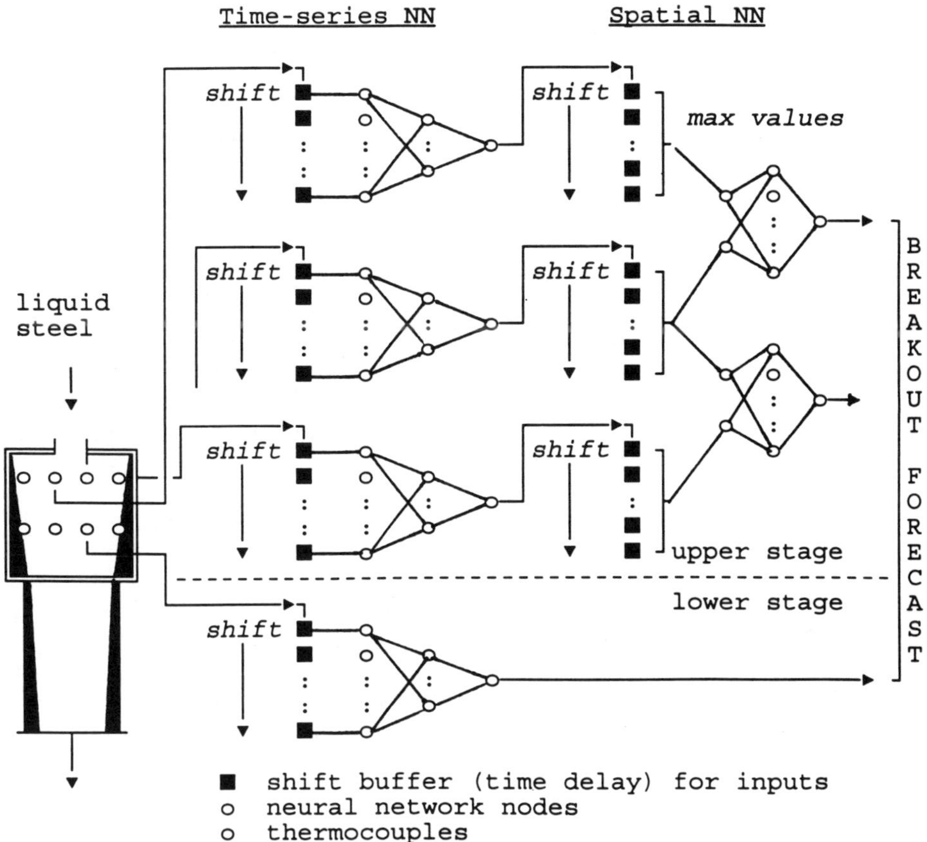

Figure 8.7 Architecture of neural network oriented mold breakout detection for caster

inputs of the spatial network through time shift. As a result, the outputs of the spatial network are able to detect the breakout based on the information gathered from adjacent pairs of thermocouples. The number of spatial networks depends on the number of adjacent pairs. Finally, the breakout can be predicted in terms of comparison between the outputs of spatial networks and the output of the corresponding time-series network at the lower-row.

Both time-series and spatial networks are performed with multi- layers, sigmoid activate function and BP-learning. They are trained separately with selected training samples ('breakout', 'over-detected' and 'non-breakout' patterns). During the beginning of on-line tests, there were many over-predicted alarms; however, after introducing many non-breakout patterns in the training data set, the prediction performance significantly improved. Based on Nippon Steel's report (Kominami *et al.*, 1991), the system has been running on-line with almost 100% prediction accuracy. The system can predict breakout 3 to 14 seconds earlier than the conventional method.

8.5 FUZZY MODEL BASED FAULT DETECTION AND DIAGNOSIS

8.5.1 General architecture

Generally, a fuzzy model based fault detection and diagnosis system can be developed by combining fuzzy linguistic rules with non-fuzzy numerical data. However, on the other hand, if the fuzzy rules are not available, as we introduced in Chapter 3, a fuzzy associative memory can also be extracted through self-organized learning. A generic architecture illustrated in Figure 8.8 shows that the structure of a fuzzy inference system is established by fuzzy linguistic 'if-then' rules, and then the off-line and/or on-line data are used to further determine fuzzy membership function by fuzzy membership function learning. The design of the fuzzy reference set, inference mechanism, and signal coding and decoding policies are dependent on the problem background. The following example will provide us with the concepts and design procedures of a fault detection and diagnosis system with fuzzy modeling.

8.5.2 Process monitoring for a low pressure chemical vapor deposition

In this section, we present a fuzzy inference system for prediction of average grain size of polysilicon deposition films developed by Chen and Spanos (1992). This

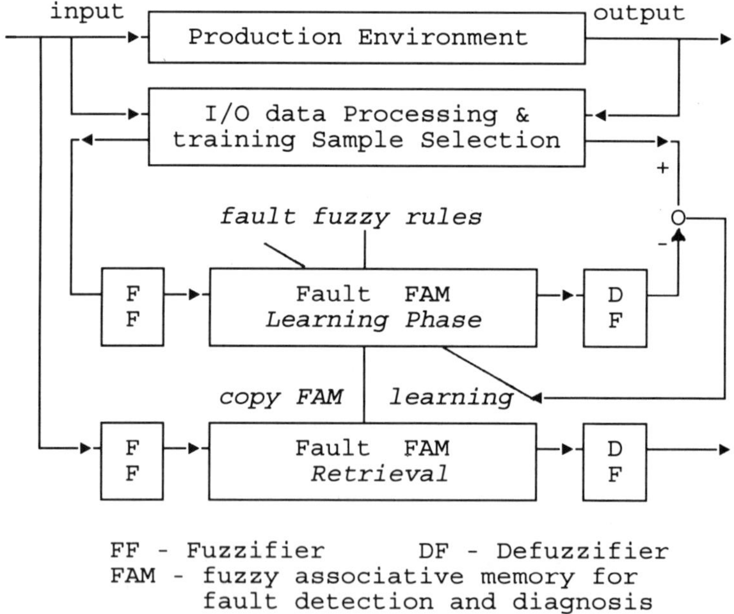

Figure 8.8 Architecture of fuzzy model based fault detection and diagnosis

work gives us a successful application of fuzzy inference in qualitative modeling and process monitoring for semiconductor manufacturing equipment.

Process background

A glass tube reactor (deposition furnace) with three zone coil heaters, heat baffle, aluminum cantilever rod and wafers used for polysilicon low pressure chemical vapor deposition (LPCVD) is illustrated in Figure 8.9 (Chen and Spanos, 1992). The goal of LPCVD operation is to maintain product quality related parameters, such as grain size, surface roughness and grain orientation, and step coverage of a deposited film within a desired domain through monitoring process setting, deposition and annealing temperatures, doping profile, silane flow rate, pressure, deposition time, etc. However, since there are many qualitative aspects in this process, it is difficult to develop a mathematical model to describe the numerical relations between operating conditions and polysilicon film quality with process principle and/or math-model based estimation techniques. A fuzzy inference system with qualitative modeling has been developed to predict the average grain size of polysilicon deposition film in terms of the operating conditions, e.g. deposition temperature, T_d and annealing temperature, T_a. The following sections will describe the fuzzy rules, self-learning for fuzzy inference and the simulation results.

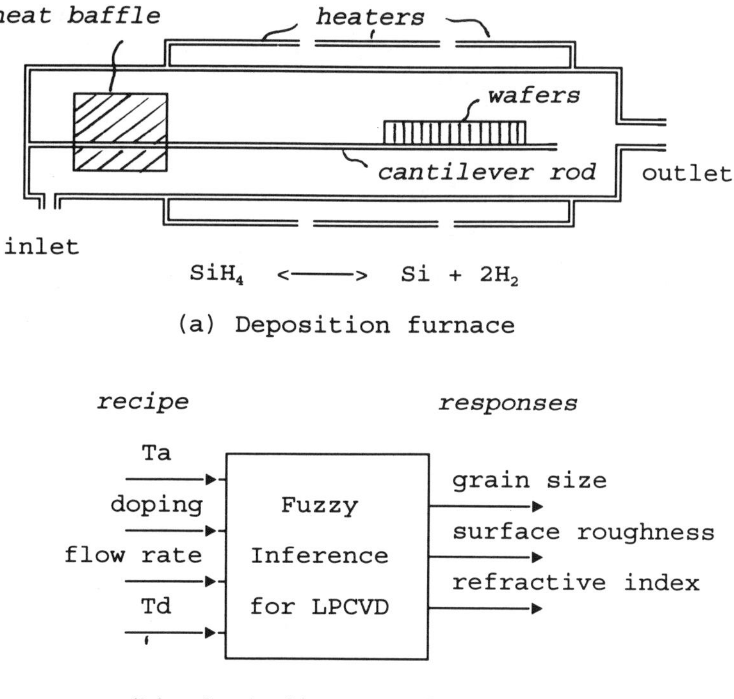

Figure 8.9 Illustration of a deposition furnace (From Chen and Spanos, 1992, © 1992 IEEE)

Fuzzy rules

The average grain size (S) of the deposited polysilicon film is an important quality that will affect the performance of semiconductor devices. The average grain size (between 50 and 300 nm) can be represented by a linguistic variable with 'small', 'medium' and 'large' fuzzy sets through 'fuzzification' in terms of a defined fuzzy membership function. Similarly, both the T_d and T_a can also be transferred to the fuzzy (linguistic) variables with 'low', 'medium' and 'high' sets. The linguistic fuzzy rules describing the qualitative relations between the major operating conditions (T_d: deposition temperature and T_a: annealing temperature) and the average grain size can be summarized as follows:

— if T_d = 'low' and T_a = 'high', then S = 'large';
— if T_d = 'med' and T_a = 'med', then S = 'med';
— if T_d = 'high' and T_a = 'low', then S = 'small'.

Fuzzy inference and self-learning

If the membership functions are defined as shown in Figure 8.10, then the fuzzy rules can be represented as Rule $\{i, j\}$: 'If x_1 is A_{1i} and x_2 is A_{2i}, Then y is W_{ij}' where $\{i, j\}(i = 0, \ldots, m; j = 0, \ldots, n)$ denote rule numbers, and A_{1i} and A_{2j} denote the linguistic values of x_1 and x_2, which are represented by membership values, m_{1i}, and m_{2j}. Similarly w_{ij} denotes the linguistic value of y with membership m_{yij}. The membership function of x_1 can be described as follows:

$$m_{1i} = \begin{cases} \dfrac{x_1 - a_{1(i-1)}}{a_{1i} - a_{1(i-1)}} & a_{1(i-1)} < x_1 \leq a_{1i} \\ \dfrac{a_{1(i+1)} - x_1}{a_{1(i+1)} - a_{1i}} & a_{1i} < x_1 \leq a_{1(i+1)} \\ 0 & \text{otherwise} \end{cases} \tag{8.3}$$

and the membership function for x_2, m_{2j}, is the same as 8.3.

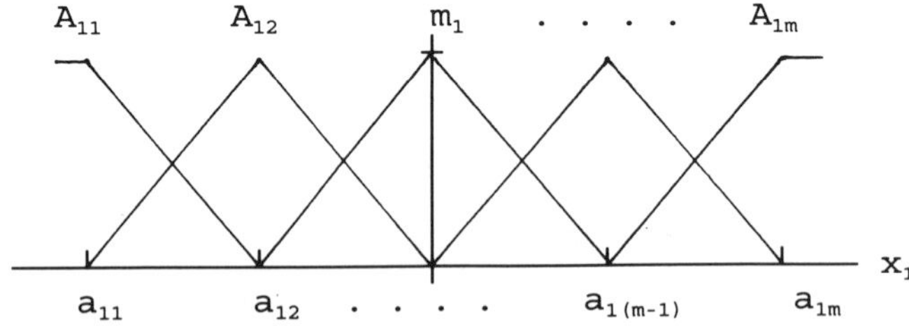

Figure 8.10 Membership function for x_1 (From Chen and Spanos, 1992, © 1992 IEEE)

The output y can thus be derived by the weighted average:

$$y = \sum_{i,j} m_{yij} W_{ij} = \sum_{i,j} (m_{1i} m_{2i} W_{ij}) \tag{8.4}$$

A self-learning algorithm with descent optimization (Nomura, *et al.*, 1992) was used in this development. The problem is to minimize the following 'cost function' with the optimized values of (a_{1i}, a_{2j}, w_{ij}) based on the training data set $\{x_{1k}, x_{2k}, y_{kr}\}$:

$$E = \frac{1}{2}\sum_{k}(y_k - y_k^r)^2 = \frac{1}{2}\sum_{k}\left(\sum_{i,j} m_{1i} m_{2j} W_{ij} - y_k^r\right)^2 \tag{8.5}$$

The problem can be further represented as

$$\frac{\partial E}{\partial W_{ij}} = \sum_{k=1}^{q}(y_k - y_k^r)\cdot(m_{1i}\cdot m_{2j}) \tag{8.6}$$

$$\frac{\partial E}{\partial a_{1i}} = \sum_{j=1}^{n}\sum_{k=1}^{q}(y_k - y_k^r)\left(\sum_{p=i-1}^{i+1} u_p \cdot W_{ij}\right) \tag{8.7}$$

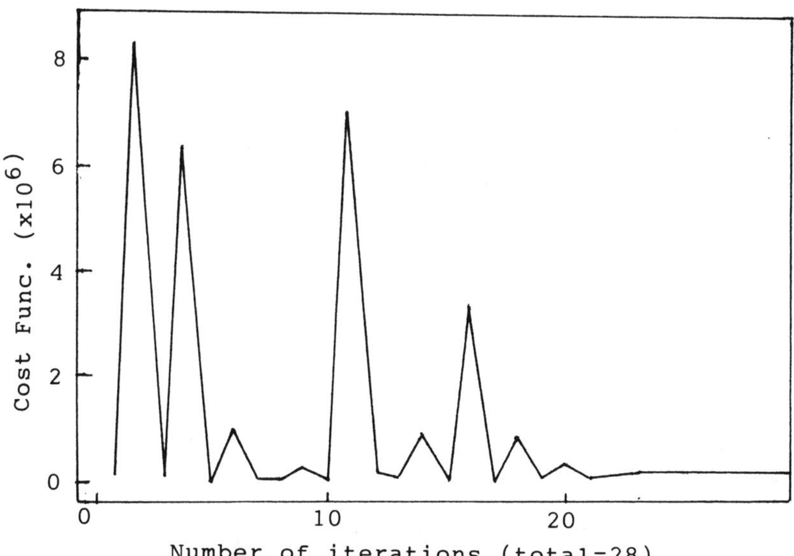

Figure 8.11 Cost function versus number of training iterations (From Chen and Spanos, 1992, © 1992 IEEE)

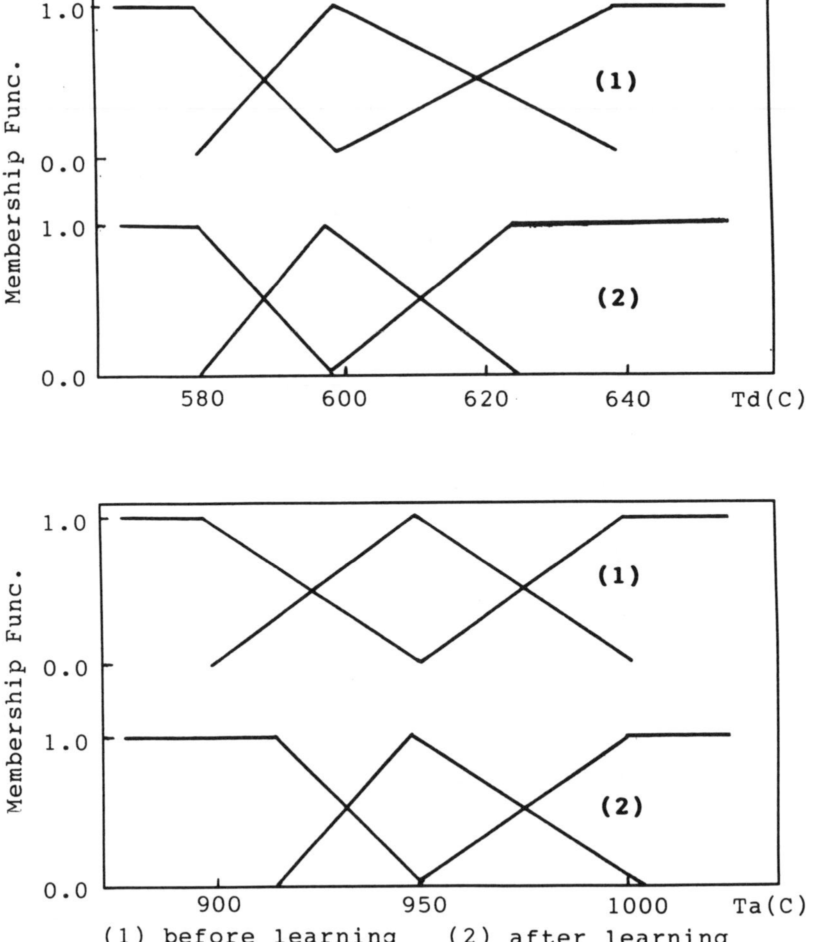

Figure 8.12 Membership function before and after training (from Chen and Spanos, 1992, © 1992 IEEE)

where

$$u_i = \begin{cases} \dfrac{-m_{1i} \cdot m_{2j}}{a_{1i} - a_{1(i-1)}} & a_{1(i-1)} < x_1 \leq a_{1i} \\ \dfrac{m_{1i} \cdot m_{2j}}{a_{1(i+1)} - a_{1i}} & a_{1i} < x_1 \leq a_{1(i+1)} \\ 0 & \text{otherwise} \end{cases} \tag{8.8}$$

$$u_{i-1} = \begin{cases} \dfrac{1 - m_{1(i-1)} \cdot m_{2j}}{a_{1i} - a_{1(i-1)}} & a_{1(i-1)} < x_1 \leq a_{1i} \\ 0 & \text{otherwise} \end{cases} \tag{8.9}$$

$$u_{i+1} = \begin{cases} \dfrac{(m_{1(i-1)} - 1) \cdot m_{2j}}{a_{1(i+1)} - a_{1i}} & a_{1i} < x_1 \leq a_{1(i+1)} \\ 0 & \text{otherwise} \end{cases} \tag{8.10}$$

A similar processing procedure can be applied for $\partial E/\partial a_{2j}$. The constraints are represented as follows:

$$\begin{aligned} a_{11} \leq a_{12} \leq \cdots \leq a_{1m} \\ a_{21} \leq a_{22} \leq \cdots \leq a_{2n} \end{aligned} \tag{8.11}$$

Simulation results

The cost function versus the number of training iterations is shown in Figure 8.11. It can be seen that after 28 iterations the cost function was reduced to 1.18×10^4. The membership functions before and after training for deposition and annealing temperatures are given in Figure 8.12.

As noted by Chen and Spanos (1992), due to the large number of process variables and their complex interconnections in a semiconductor environment, the pertinent expert knowledge is mostly qualitative and incomplete. A computer-aided integrated system for a future automated fabrication factory can be facilitated by the fuzzy logic based qualitative models which can capture qualitative human knowledge into a computer software simulation system. This fuzzy inference system can be self-learning to accommodate the updated expert knowledge.

Applications of expert systems and pattern recognition for fault detection and diagnosis are given in the literature (Moore, 1991; Qian and Lu, 1989; Ye and Lu, 1986; Pau, 1983).

Appendix

LIST OF APPLICATION EXAMPLES

1. Optimal estimation of steel ingot temperature profile using Kalman filter (section 4.2).
2. Product distribution prediction of fluidized catalytic cracking unit using fuzzy model based estimation (section 4.5).
3. Gasoline vapor pressure prediction at a stabilization tower using fuzzy pattern recognition estimator (section 4.6).
4. Application of neural network in establishing a common gas P–V–T state model (section 4.7).
5. Application of neural network in continuous optical strip centering system modeling (section 4.7).
6. A neural adaptive controller with direct feedback (section 5.3).
7. An integrated neural system for coating weight prediction and control (section 5.4).
8. Application of expert control in fluidized catalytic cracking unit (section 5.6).
9. Fuzzy expert optimization control for fluidized catalytic cracking unit (section 6.3).
10. Dynamic model based optimization control for steel reheating furnaces (section 6.4).
11. Expert system for lube oil production scheduling (section 6.5).
12. Application of evolutionary algorithms in job assignment problem (section 6.6).
13. Application of GA in real-time fuzzy rule learning for a bioreactor (Section 6.6).
14. Emzyme proportion prediction by fuzzy GA fearning (Section 6.6).
15. Application of neural network partial least-squares for cosmetic data modeling (section 7.5).
16. Mold breakout prediction of continuous steel caster using neural networks (section 8.4).
17. Prediction of average grain size of polysilicon deposition films using fuzzy inference (section 8.5).

Bibliography

Aarts, E. and Korst, J. (1989) *Simulated Annealing and Boltzmann Machines: A Stochastic Approach to Combinatorial Optimization and Neural Computing*, John Wiley & Sons, New York.

Alexander, S. T. (1986) *Adaptive Signal Processing: Theory and Applications*, Springer-Verlag, New York.

Amari, S.-I. (1990) Mathematical foundations of neurocomputing, *Proc. IEEE*, **78**(9), 1443–1463.

Amit, D. J. *et al.* (1985) Spin-glass models of neural networks, *Physical Review* A, **32**(2), 1007–1018.

Anderson, T. W. (1984) *Introduction to Multivariate Statistical Analysis*, 2nd edn, John Wiley & Sons, New York.

Angeline, P. J. *et al.* (1994) An evolutionary algorithm that constructs recurrent neural networks, *IEEE Trans. Neural Networks*, **5**(1).

Astrom, K. J. (1983) Theory and applications of adaptive control – a survey, *Automatica*, **19** (471).

Astrom, K. J. and Wittenmark, B. (1989) *Adaptive Control*, Addison-Wesley, Reading, MA.

Astrom, K. J. *et al.* (1977) Theory and applications of self-tuning regulator, *Automatica*, **13**, 457.

Astrom, K. J. *et al* (1986) Expert control, *Automatica*, **22**(3).

Astrom, K. J. *et al.* (1991) *Progress report – Facing challenge of computer science in industrial applications of control: a joint IEEE CSS – IFAC Project*, IEEE and IFAC.

Azimi-Sadjadi, M. R. *et al.* (1993) Recursive dynamic node creation in multilayer neural networks, IEEE *Trans. Neural Networks*, **4**.

Ball, C. H. (1965) Data analysis in social sciences: what about the detail?, *Proc. FJCC*, pp. 533–560, Spartan Books, Washington, DC.

Barto, A. G. (1992) Reinforcement learning and adaptive critic methods, in White, D. A. and Sofge, D. A. (eds), *Handbook of Intelligent Control*, Van Nostrand-Reinhold, New York, pp. 469–491.

Barto, A. G. and Anderson, C.W. (1985) Structural learning in connectionist system, *Proc. 7th Annual Conf. Cognitive Sci. Society.*

Barto, A. G., Sutton, S. R. and Anderson, C. W. (1983) Neuronlike adaptive elements that can solve difficult learning control problems, *IEEE Trans. on System, Man & Cybernetics*, SMC-13, pp. 834–846.

Bavarian, B. (1988) Introduction to neural networks to intelligent control, *IEEE Control System Magazine*, April.

Bellman, R. (1957) *Dynamic Programming*, Princeton University Press, Princeton, NJ.

Bellman, R. and Dreyfus, S.E. (1962) *Applied Dynamic Programming*, Princeton University Press, Princeton, NJ.

Benveniste, A. and Astrom, K. J. (1993) Meeting challenge of computer science in the industrial applications of control: An introductory discussion to the special issue, *IEEE Trans. on Automatic Control*, **38**(7).

Bhat, N. and McAvoy, T. J. (1989) Use of dynamic modeling and control of chemical process systems, *Proc. American Control Conf.*

Box, G. E. P. and Jenkins, G. M. (1976) *Time Series Analysis, Forecasting and Control*, Holden Day, San Francisco, CA.

Bremermann, H. J. (1968) Numerical optimization procedures derived from biological evolution processes, in Oestreicher, H. L. and Moore, D. R. (eds), *Cybernetic Problems in Bionics*, Gordon & Breach, New York, pp. 543–562.

Brice, J. C. *et al.* (1983) Improve FCC productivity with better control, *Hydro-Carbon Processing*, **5**.

Brogen, W. L. (1985) *Modern Control Theory*, Prentice-Hall, Englewood Cliffs, NJ.

Bromley, J. A. (1981) FCC control: a structure analysis approach, *IEC Des. Dev* **20**.

Buckley, J. J. (1992) Universal fuzzy controller, *Automatica*, **28**(6).

Buckley, J. J. (1990) Managing uncertainty in a fuzzy expert system, in Gaines, B. and Boose, J. (eds) *Machine Learning and Uncertain Reasoning*, Academic Press, New York.

Cannon, R. H. (1967) *Dynamics of Physical Systems*, McGraw-Hill, New York.

Cendrowska, J. (1988) PRISM: an algorithm for inducing modular rules, *Intel. Man-Machine Studies*, **27**.

Chen, R. L. and Spanos, C. J. (1992) Self-learning fuzzy modeling of semiconductor processing equipment, in Marks, R. J. (ed.), *Fuzzy Logic Technology and Applications*, IEEE Press, New York.

Choi, J. W. *et al.* (1994) Application of genetic algorithm to fuzzy rule generation in bed-batch culture of *Thalictrum rugosum* plant cell", *Proc. Asian Control Conf.*, July, Tokyo.

Cook, P. A. (1986) *Nonlinear Dynamic Systems*, Prentice-Hall, London.

Cox, E. (1993) Fuzzy fundamentals, A *Spectrum*, Oct.

Dejong, K. (1988) Learning with genetic algorithms, *Machine Learning*, **3**, 121–138.

Deming, W. E. (1972) Report to management, *Qual. Prog*, **5**.

Deming, W. E. (1982) Quality, productivity, and competitive position", MIT CASE.

Draper, N. and Smith, H. (1981) *Applied Regression Analysis*, John Wiley & Sons, New York.

Dubois, D. and Prade, H. (1980) *Fuzzy Sets and Systems: Theory and Applications*, Academic Press, New York.

Dutta, A. (1985) Reasoning with imprecise knowledge in expert systems", *Information Science*, **37**, 3–24.

Farley, A. M. (1983) A probabilistic model for uncertain problem solving, *IEEE Trans. Systems, Man, & Cybernet.*, **13**, 568–579.

Fogel, D. B. (1962) Autonomous automata, *Industrial Research*, **4**, 14–19.

Fogel, D. B. (1994) An introduction to simulated evolutionary optimization, *IEEE Trans. Neural Networks*, **5**(1).

Fox, M. S. *et al.* (1983) ISIS: a constraint-directed reasoning approach to job scheduling: system summary, GMU-RE-83-8.

Franks, R. G. E. (1972) *Modeling and Simulation in Chemical Engineering*, John Wiley & Sons, New York.

Fraser, A. S. (1962) Simulation of genetic systems, *J. Theor. Biol.* **2**, 329–346.

Fu, K. S. (1970) Learning control systems – review and outlook, *IEEE Trans. Automatic Control*, AC-15.

Fu, L. M. and Fu, L. C. (1990) Mapping rule based system into neural architecture, *Knowledge-Based Systems*, **3**, 48.

Fujioka, H. R. and Tanaka, H. (1993) Neural networks that learn from fuzzy if-then rules, *IEEE Trans Fuzzy Systems*, **1**(2).

Gaines, B. R. and Boose, J. H. (eds), *Machine Learning and Uncertain Reasoning*, Academic Press, London.

Gee, A. H. (1993) *Problem Solving with Optimizing Networks*, PhD Dissertation, University of Cambridge, UK.

Gee, A. H., Aiyer, S. V. B. and Prager, R. (1993) An analytical framework for optimizing neural networks, *Neural Networks*, **6**, 79–97.

Geladi, P. and Kowalski, B. R. (1986) Partial least squares: a tutorial, *Analyt. Chim. Acta*, **185**, 1–17.

Gelb, A. *et al.* (1974) *Applied Optimal Estimation*, MIT Press, Cambridge, MA.

Glorioso, R. M. and Osorio, F. C. C. (1980) Engineering intelligent systems: concepts, theory and applications, *Digital Equipment Co. Report*, Bedford, MA.

Goldberg, D. E. (1989) *Generic Algorithms in Search Optimization and Machine Learning*, Addison-Wesley, New York.

Graebe, S. F. (1993) Robust and adaptive control for an unknown plant: A benchmark of new format, *Preprints of 12th IFAC World Congress*, Sydney.

Gray, R. M. (1984) Vector quantization, *IEEE ASSP Magazine*, **1**.

Greenspan, H. K. *et al.* (1992) Combined neural network and rule based framework for probabilistic pattern recognition and discovery, in Moody, J. E. *et al.* (eds), *Neural Information Processing Systems*, Morgan Kaufmann, San Mateo, CA.

Grefenstette, J. J. (1986) Optimization of control parameters for genetic algorithms, *IEEE Trans. System, Man & Cybernetics*, **16**(1).

Grossberg, S. (1987) Competitive learning: from interactive activation to adaptive resonance, *Cognitive Science*, **11**, 23–63.

Gu, T. (1982) An optimal upturn performance index classifier – OUPIC, *Scientia Sinica (Series A)*, **25**, 565–570.

Gupta, M. M. (ed.) (1986) *Adaptive Methods for Control Systems Design*, IEEE Press, New York.

Gupta, M. M. *et al.* (eds) (1977) *Fuzzy Automata and Decision Processes*, North–Holland, New York.

Hamilton, J. C. (1973) An experimental evaluation of Kalman filtering", *AIChE J.*, **19**, 901–909.

Hammerstrom, D. (1993) Neural network at work, *IEEE Spectrum*, June.

Haykin, S. (1994) *Neural Networks: A Comprehensive Foundation*, IEEE Press, New York.

Hebb, D. O. (1949) *The Organization Behavior*, John Wiley & Sons, New York.

Higashi, M. and Klir, G. J. (1984) Identification of fuzzy relation systems, *IEEE Trans. SMC*, **(14)**, 2.

Hinton, G. E. and Anderson, J. A. (eds) (1981) *Parallel Models of Associative Memory*, Hilladale, NJ.

Hinton, G. E. and Sejnowski, T. J. (1986) Learning and relearning in Boltzmann Machine, in *Parallel Distributed Processing (PDP): Vol. 1 Foundation*, MIT Press, Cambridge, MA.

Hirota, K. and Pedrycz, W. (1994) OR/AND neuron in modeling fuzzy set connectives, *IEEE on Fuzzy Systems*, **2**(2).

Hiroyoshi, N. *et al.* (1992) A learning method of fuzzy inference rules by decsent method, *Proc. of IEEE Intel. Conf. on Fuzzy Systems*, San Diego, CA.

Holland, J. H (1975) *Adaptation in Natural and Artificial Systems*, University of Michigan Press, Ann Arbor, MI.

Holland, J. H. (1992a) *Adaptation in Natural and Artificial Systems*, MIT Press, Cambridge, MA.

Holland, J. H. (1992b) Genetic algorithms, *Scientific American*, July, pp. 66–72.

Holmblad, L. P. and Osterguard, J. J. (1982) Control of a cement kiln by fuzzy logic, in Gupta, M. M. & Sanchez, E. (eds), *Fuzzy Information and Decision Process*, North–Holland, New York.

Hopfield, J. J. (1982) Neural network and physical systems with emergent collective computational abilities, *Proc. Nat. Acad. Sci. USA*, **79**, 2554–2558.

Hopfield, J. J. and Tank, T. W. (1985) Neural computation decision in optimization problems, *Biological Cybernetics*, **52**, 141–152.

Hornik, K. *et al.* (1989) Multilayer feedforward neural networks are universal approximations, *Neural networks*, **2**(5).

Hu, M. and Lu, Y. Z. (1988) Hybrid intelligent controller, *Information and Control*, (in Chinese), **17**(4).

Hunt, K. J. *et al.* (1992) Neural network for control systems – A Survey, *Automatica*, **28**(6).

Ichikawa, Y. and Sawa, T. (1992) Neural network applications for direct feedback control, *IEEE Trans. on Neural Networks*, **3**(2).

IEEE (1987) Challenges to control: a collective view, *IEEE Trans. Auto. Control*, **AC-32**, 275–285.

IEEE (1994) Special issue on evolutionary computation, *IEEE Trans. on Neural Networks*, **5**(2).

IFAC (1993) *Preprints of IFAC 1993 World Congress*, Vol. 3, Sydney.

Isermann, R. (1982) Parameter adaptive control algorithms – A tutorial, *Automatica*, **18**, Sept.

Isermann, R. (1984) Process fault detection based on modeling and estimation – A survey, *Automatica*, **20**, 387–404.

Ishibuchi, H., Fujioka, R. and Tanaka, H. (1993) Neural networks that learn from fuzzy if-then rules, *IEEE Trans Fuzzy Systems*, **1**(2).

Jackson, J. E. (1980) Principal components and factor analysis: Part I – Principal components, *J. Quality Technology*, **12**(4).

Kailath, T. (1981) *Lectures on Wiener and Kalman Filtering*, Springer Verlag, New York.

Kalman, R. E. and Bucy, R. S. (1960) A new approach to linear filtering and prediction problems, *Trans. ASME, J. Basic Engineering*, **82**, 35–46.

Keller, J. M., Yager, R. R. and Tahani, H, (1992) Neural network implementation fuzzy logic, *Fuzzy Sets and Systems*, **45**, 1–12.

Kirkpartrick, S. *et al.* (1983) Optimization by simulated annealing, *Science*, **220**, 671–680.

Kohonen, T. (1988) *Self-Organization and Associative Memory*, Springer-Verlag, Berlin.

Kominami, H. *et al.* (1991) Neural network system for breakout prediction in continuous casting process, *Nippon Steel Report*, No. 49, April.

Kong, S. G. and Kosko, B. (1992) Adaptive fuzzy systems for backing up a truck-and-trailer, *IEEE Trans. on Neural Networks*, **3**(2).

Koopmans, T. C. and Bechmann, M. J. (1957) Assignment problems and the location of economic activities, *Econometrica*, **25**, 53–76.

Kosko, B. (1988) Bidirectional associative memories, *IEEE Trans. on SMC*, SMC-18, Jan.

Kosko, B. (1992) *Neural networks: A Dynamic Systems Approach to Machine Learning*, Prentice-Hall, Engelwood cliffs, N J.

Kowalski, R. *et al.* (1982) Chemical systems under indirect observation", in Joreskog, K. and Wold, H. (eds), *Systems under Indirect Observation*, North-Holland, Amsterdam, pp. 191–209.

Kresta, J. V. *et al.* (1991) Multivariate statistical monitoring of process operating performance, *Can. J. Chem. Eng.*, **69**, Feb.

Kung, S. Y. and Hwang, J. N. (1989) Neural network architectures for robotic applications, *IEEE Trans. on Robotics and Automation*, **5**, 641–656.

Kuschewski, J. G., Hui, S. and Zak, S. H. (1993) Application of feedforward neural networks to dynamical system identification and control, *IEEE Trans on Control System Tech*, **1**(1).

Kusiak, A. and Heragu, S. S. (1987) The facility layout problem, *EJOR*, **29**, 229–251.

Lacher, R. C., Hruska, S. I. and Kuncicky, D. C. (1992) Back propagation learning in expert networks, *IEEE Trans. Neural Networks*, **3**(1).

Latour, P. R. (1979) Online computer optimization 1: what is it and where to do it, *Hydrocarbon Processing*, June–July.

Lee, C. C. (1990) Fuzzy logic in control systems: fuzzy logic controller, Part II, *IEEE Trans. Systems, Man & Cybernetics*, **20**.

Lee, S. and Kil, B. (1989) Bidirectional continuous associator based on Gaussian potential function network, *Proc. IEEE/INNS Int. Joint Conf. Neural Networks*, **1**, 45–54.

Leonard, J. and Kramer, M. A. (1990) Improvement of the back propagation algorithm for training neural networks, *Computers Chem. Engng.*, **14**(3).

Li, B. S. and Liu, Z. J. (1980) Application of fuzzy set theory to identification of system models (in chinese), *Information and Control*, **9**(3).

Lin, J. N. and Lu, Y. Z. (1988) Decomposition algorithm of dynamic programming for large scale time – delay systems, *Int. J. Systems Sciences*, **19**(3).

Lippmann, R. J. (1989) Pattern classification with neural networks, *IEEE Communications Magazine*, **27**.

Lippmann, R. P. (1987) An introduction to computing with neural nets, *IEEE ASSP Magazine*, April, 4–22.

Ljung, L. (1987) *System Identification*, Prentice-Hall, Englewood Cliffs, NJ.

Lu, Y. Z. (1992) The new generation of advance process control, *Control Engineering*, March.

Lu, Y. Z. (1994) Applications of neural networks in the metal industry, *3rd AISE Advanced Modeling and Control Seminar*, Cleveland, OH.

Lu, Y. Z. (1994a) "A neural network learns from rules, unpublished note.

Lu, Y. Z. and He, M. (1991) An expert control system for the fluidized catalytic cracking unit in refinery, in Kompass, E. J. *et al.* (eds), *Expert Systems Applications in Advanced Control*, Purdue Research Foundation and Control Engineering, W. Lafayette; IN.

Lu, Y. Z. and Markward, S. W. (1995) Development and applications of an integrated neural system for HDCL, to appear at *Proc. 1995 ACC*, Seattle, WA.

Lu, Y. Z. and Williams, T. J. (1983) *Modeling, Estimation and Control of the Soaking Pit*, ISA Publishers, Research Park, NC.

Lu, Y. Z. and Williams, T. J. (1983a) Computer control strategies of optimal state feedback methods for the control of steel mill soaking pits, *ISS Trans.*, **2**, 35–43.

Lu, Y. Z. and Williams, T. J. (1983b) Energy savings and productivity increases with computers – a case study of steel ingot handling process, *Int. J. Computers in Industry*, **4** (1).

Lu, Y. Z., Cheng, G. S. and Manoff, M. (1992) Universal process control using artificial neural networks, *US Patent*, 5,159,660.

Lu, Y. Z., He, M. and Xu, C. W. (1995) Fuzzy modeling and expert optimization control for industrial processes, submitted to *IEEE Trans. Control System Technology*.

Lu, Y. Z. *et al.* (1995) "An intelligent scheduling tool for multi-stage and multi-product process, to be published.

Luenberger, D. G. (1964) Observing the state of a linear system, *IEEE Trans. on Milo Electronics*, MIL–**8**, April.

Luenberger, D. G. (1971) An introduction to observers, *IEEE Trans on AC*, **AC-16**(6).

Luenberger, D. G. (1984) *Linear and Nonlinear Programming*, Addison–Wesley, Reading, MA.

MacGregor, J. F. (1988) On-line statistical process control, *Chem. Engng. Prog.*, Oct.

MacQueen, J. (1967) Some methods for classification and analysis of multivariate observations, in LeCam, L. M. and Neyman, J. (eds), *Proc. 5th Berkeley Symposium on Mathematics, Statistics and Probability*, pp. 281–297.

Mamdani, E. H. (1977) Application of fuzzy logic to approximate reasoning using linguistic synthesis, *IEEE Trans. Computers*, **C–26**(12).

Mamdani, E. H. and Assilian, J. J. (1974) Application of fuzzy algorithms for control of a dynamic plant, *Proc. Inst. Elec. Eng.*, **121**, 1585–1588.

Maniezzo, V. (1994) Genetic evolutionary of the topology and weight distribution of neural networks, *IEEE Trans. Neural Networks*, **5**(1).

Markward, S. W. and Lu, Y. Z. (1995) Integrated neural system for coating weight prediction and control of HDCL, to be presented at *95 AISE Spring Convention*, Salt Lake City, UT.

Michalewicz, Z. (1992) *Genetic Algorithms + Data Structures = Evolution Programs*, Springer-Verlag, Berlin.

Moody, J. and Yarvin, N. (1992) Networks with learned unit response functions, in Moody, J. and Yarvin, N. (eds), *Advances in Neural Information Processing Systems*, Morgan Kaufmann, San Mateo, OA.

Moore, R. L. (1991) G2: a software platform for intelligent process control", Proc. *IEEE Symposium on Intelligent Control*, Arlington, VA.

Morari, M. and Zafiriou, E. (1989) *Robust Process Control*, Prentice-Hall, Englewood Cliffs, NJ.

Narazaki, H. and Raleson, L. (1992) A Connectionist Approach for Rule-Based Inference Using an Improved Relaxation Method, *IEEE Trans. on Neural Network*, **3**(5).

Narendra, K and Parthasarathy, K. (1990) Identification and control of dynamical systems using neural networks, *IEEE Trans. Neural Networks*, **1**(1).

Narendra, K. and Thathachar, M. (1989) *Learning Automata: An Introduction*, Prentice-Hall, Englewood Cliffs, NJ.

Nilsson, N. J. (1980) *Principle of Artificial Intelligence*, Springer-Verlag, New York, NY.

Nissen, V. (1994) Solving the quadratic assignment problem with clues from nature, *IEEE Trans. on Neural Networks*, **5**(1).

Nomura, H. *et al.* (1992) A learning method of fuzzy inference rules by descent method, *Proc. IEEE Conf. on Fuzzy Systems*, San Diego, CA.

Nugent, E. N. and Vollmann, T.E. (1968) An experimental comparison of techniques for the assignment of facilities to locations, *Operations Research*, **16**, 150–173.

Ogata, K. (1990) *Modern Control Engineering*, Prentice-Hall, Englewood Cliffs, NJ.

Oja, E. (1982) A simplified neuron model as a principal component analyzer, *J. Mathematical Biology*, **15**, 267–273.

O'Keefe, R. M. (1985) Expert systems and operational research – mutual benefits, *J. Operational Research Society*, **36**, 125–129.

O'Keefe, R. M. *et al.* (1986) Experiences with using expert systems and O.R., *J. Operational Research Society*, **7**, 657–688.

Orfanidis, S. J. (1990) Gram–Schmidt neural nets, *Neural Computation*, **2**, 116–129.

Park, D. C. *et al.* (1991) An adaptive trained neural network", *IEEE Trans. Neural Networks*, **2**(3).

Parker, D. (1982) Learning logic, *Stanford University, Dept. of Elec. Eng., Invention Report*, 584–64, Oct.

Pau, L. F. (1983) Application of pattern recognition to fault detection and diagnosis, AUTOTESCON'83, 389.

Pearson, K. (1901) On lines and planes of closest fit to systems of points in space, *Philos. Mag.* **2**, 559–572.

Pedrycz, W. (1983) Numerical and application aspects of fuzzy relational equations, *Fuzzy Sets & Systems*. **11**, 1–18.

Pedrycz, W. (1984) An identification algorithm in fuzzy relation systems, *Fuzzy Sets & Systems*, **13**, 153–167.

Pedrycz, W. and Rocha, A. F. (1993) Fuzzy-set based models of neural and Knowledge-based networks, *IEEE Trans. Fuzzy Systems*, **1**(4).

Pontryagin, L. S. *et al.* (1962) *The Mathematical Theory of Optimal Processes*, Wiley Interscience, New York.

Psaltis, D. *et al.* (1988) A multilayered neural network controller, *IEEE Control System Magazine*, April.

Qi, X. and Palmieri, F. (1994) Theoretical analysis of evolutionary algorithms with an infinite population size in continuous space, *IEEE Trans. Neural Networks*, **5**(1).

Qian, D. Q. (1987) *Knowledge Representation, Problem Solving and Applications in Dynamic Systems*, Doctoral Thesis, Zhejiang University, Hangzhou, China.

Qian, D. Q. and Lu, Y. Z. (1986) Expert system based fault diagnosis in fluidized catalytic cracking unit, *Proc. World Congress of Chem. Eng.*, Tokyo.

Qian, Q. D. and Lu, Y. Z. (1989) A strategy of problem solving in a fuzzy reasoning network, *Fuzzy Sets and Systems*, **33**, 137–154.

Qin, S. J. (1994) Auto-tuned fuzzy logic control, *Proc. 1994 ACC Conf.*, Baltimore, MD.

Qin, S. J. and Borders, G. (1994) A multi-region fuzzy logic controller for nonlinear process control, *IEEE Trans. Neural Networks*, **2**(1).

Qin, S. J. and McAvoy, T. J. (1992a) Process control through neural computing, Proc. *ITERKAMA Congress*, Düsseldorf, Germany, Oct. 5–9.

Qin, S. J. and McAvoy, T. J. (1992a) Nonlinear PLS modeling using neural networks, *Computers in Chem. Engng.*, **16**(4).

Qin, S. J. and McAvoy, T. J. (1992b) A data-base process modeling approach and its applications, *Preprints 3rd IFAC Dycord Symposium*, College Park, MD.

Qin, S. J. and Rajagopal, B. (1993) Combining statistics and expert systems with neural networks for empirical process modeling, *Preprint ISA/93*, paper 93-411.

Qin, S. J. *et al.* (1992) Comparison of four neural learning methods for dynamic systems identification, *IEEE Trans. Neural Networks*, **3**(1).

Quinlan, J. R. (1986) Introduction to decision trees, *Machine Learning*, **1**, 81–106.

Rao, S. S. (1978) *Optimization Theory and Applications*, John Wiley & Sons, New York.

Rechenberg, I. (1973) *Evolution Strategie: Optimierung Technischer Systeme nach Prinzipien der Biologischen Evolution*, Fromman–Holzboog Verlag, Stuttgart.

Reed, J. *et al.* (1967) Simulation of biological evolution and machine learning, *J. Theor. Biol.*, **17**, 319–342.

Rosenblatt, F. (1958) The perception: a probabilistic model for information storage and organization in the Brain, *Psychological Reviews*, **65**, 386–408.

Rudolph, G. (1994) Convergence properties of canonical genetic algorithms, *IEEE Trans. Neural Networks*, **5**(1).

Rumelhart, D. E. *et al.* (1986) Lerning Representations by Back-Propagations Errors, *Nature* (London), **323**, 533–536.

Rumelhart, D. E. and McClelland, J. L. (eds). (1988) *Parallel Distributed Processing: Vol. 1 Foundation*, MIT Press, Cambridge, MA.

Sanger, T. D. (1989) Optimal unsupervised learning in a single-layer linear feedforward neural networks, *Neural Networks*, **12**, 459–473.

Saridis, N. G. (1983) Intelligent robotic control, *IEEE Trans. Automatic Control,* **AC–28**(5).

Sastry, P. S., *et al.* (1994) Memory neural networks for identification and control of dynamic systems, *IEEE Trans. on Neural Networks,* **5**(2).

Schwefel, H. P. (1965) *Kybernetische Evolution als Strategie der Experimentellen Forschung in der Strömungstechnik,* Diploma Thesis, Technical University of Berlin.

Shannon, C. E. and Weaver, W. (1962) *The Mathematical Theory of Communication,* University of Illinois Press, Urbana, IL.

Shaw, I. S. and Kruger, J. J. (1992) New fuzzy learning model with recursive estimation for dynamic systems, *Fuzzy Sets and Systems,* **48**, 217–229.

Shiba, H. *et al.* (1994) The searching of the optimal enzyme proportion on the solid-state fermentation process by fuzzy inference learned by genetic algorithm, *Proc. Asian Control Conf.,* July, Tokyo.

Siler, W., Buckley, J. and Tucker, D. (1987) Functional requirements for a fuzzy expert system shell, in Zadeh, L. and Sanchez, E. (eds) *Artificial Intelligence: Applications of Quantitative Reasoning,* Pergamon Press, Oxford.

Simpson, P. K. (1991) Foundations of neural networks, in Sanchersinencio, E. and Lau, C. (eds), *Artificial Neural Networks: Paradigm, Application* and *Hardware Implementations,* IEEE Press, New York.

Simpson, P. K. (1992) Fuzzy min-max neural networks – Part 1: Classification, *IEEE Trans. Neural Networks,* **3**(5).

Simpson, P. K. (1993) Fuzzy min-max neural networks – Part 2: Clustering, *IEEE Trans. Fuzzy Systems,* **1**(1).

Singh, M. G. and Titli, A. (1978) *Systems: Decomposition, Optimization and Control,* Pergamon Press, Oxford.

Soucek, B. and IRIS Group (eds.) (1992) *Dynamic, Genetic, and Chaotic Programming,* John Wiley & Sons, Inc., New York.

Steel, L. (1985) Second generation of expert systems, *FGCS,* **1**(4).

Sugeno, M. and Yasukawa, T. (1993) A fuzzy-logic-based approach to qualitative modeling, *IEEE Trans. Fuzzy Systems,* **1**(1).

Sutton, R. S. *et al.* (1991) Reinforcement learning is direct adaptive optimal control, *Proc. of ACC,* Boston, MA.

Takagi, H. (1990) Fusion technology of fuzzy theory and neural networks – Survey and future direction, *1st Intel. Conf. Fuzzy Logic & Neural Networks* (IIZUKA '90), July.

Takagi, H. (1994) Application of neural networks and fuzzy logic to consumer products, in Marks II, R. J. (ed.), *Fuzzy Logic Technology and Applications,* IEEE Press, New York.

Takagi, H. *et al.* (1992) Neural networks designed on approximate reasoning architecture and their applications", *IEEE Trans. Neural Networks,* **3**(5).

Tank, D. W. and Hopfield, J. J. (1986) Simple newal optimization networks: an A/D convertor signal decision circuit and a linear programming circuit, *IEEE Trans. on Circuits & Systems,* **CAS–33**, 533–541.

Togai, M. and Watanabe, H. (1986) "Expert system on a chip: an engine for real-time approximate reasoning, *IEEE Expert,* **1**(3).

Tong, R. M. (1978) Synthesis of fuzzy models for industrial processes, *Int. J. Gen. Syst.* **4**, 143–162.

Tong, R. M. (1979) The construction and evaluation of fuzzy models, in Gupta, M. M. *et al.* (eds), *Advances in Fuzzy Set Theory and Applications,* North-Holland, New York.

Tsetlin, M. L. (1962) On the behavior of finite automate in random media, *Automatic and Remote Control,* **22**, 1210–1219.

Tsetlin, M. L. (1973) *Automation Theory and Modeling of Biological Systems*, Academic Press, New York.

Tsypkin, Ya. Z. (1971) *Adaptive and Learning in Automatic & Systems*, Academic Press, New York.

Vollmann, T. E. and Buffa, E. S. (1966) The facility layout problem in perspective, *Management Science*, **12**(10).

Wang, L. X. (1992) Fuzzy systems are universal approximators, *Proc. IEEE Int. Conf. Fuzzy Systems*, San Diego, CA.

Wang, L. X. (1993) Stable adaptive fuzzy control of nonlinear systems, *IEEE Trans. Fuzzy Systems*, **1**(2).

Wang, S. and Lu, Y. Z. (1990) Knowledge-based successive linear programming for linearly constraints optimization, in Gaines, B. R. and Boose, J. H. (eds), *Machine Learning and Uncertain Reasoning*, Academic Press, London.

Weiss, G. (1990) Artificial Neural Learning, Report FKI-127-90, Technical University of Munich.

Werbos, P. J. (1974) *Beyond Regression: New Tools for Prediction and Analysis in the Behavioral Sciences*, Thesis in Applied Mathematics, Harvard University.

White, H. (1989) Some asymptotic resutls for leaving in single Hidden–Layer feedforward network models, *J. of the American Statistical Society*, **84**, 1003–1013.

Whitaker, H. P., Yamron, P. J. and Kezer, A. (1958) Design of model reference control systems for aircraft, *Report R-164*. Instrumentation Lab., MIT, Cambridge, MA.

Whitehead, B. A. and Choate, T. D. (1994) Evolving space-filling curves to distribute radial basis functions over an input space, *IEEE Trans. Neural Networks*, **5**(1).

Widrow, B. and Hoff, M. E., Jr (1960) Adaptive switching circuits, *IRE WESCON Convention Record*, 96–104.

Widrow, B. and Lehr, M. A. (1990) 30 Years of adaptive neural networks: Perceptron, Madaline, and Backpropagation, *Proc. IEEE*, **78**(9), 1415–1442.

Widrow, B. and Smith, F. W. (1964) Pattern recognition control system, in, Tou, T. J. and Wilcox, R. H. (eds), *Computer and Information Science*, Cleaver Hume Press.

Widrow, B. and Stearns, S. D. (1985) *Adaptive Signal Processing*, Prentice-Hall, Englewood, Cliffs, NJ.

Widrow, B. and Winter, R. (1988) Neural nets for adaptive filtering and adaptive pattern recognition, *IEEE Trans. Computers*, March, 25–39.

Williams, R. J. (1989) Connectionist learning through gradient descent, in Lee, Y. C. (ed.), *Evolution, Learning and Cognition*, World Scientific, Singapore.

Williams, R. J. and Peng, J. (1989) Reinforcement learning algorithms as function optimizers, *Proc. 1989 Intl. Conf. on Neural Networks*, **2**, 89–95.

Williams, T. J. (ed.) (1984) Tasks of functional specification of the steel plant hierarchical control systems, *PLAIC Report No. 98*, Purdue University, W. Lafayette, IN.

Willis, M. J., *et al.* (1992) Artificial neural networks in process estimation and control, *Automatica*, **28**(6).

Wilson, G. V. and Pawley, G. S. (1988) On the stability of the traveling salesman problem algorithm of Hopfield and Tank, *Biological Cybernetics*, **58**, 63–70.

Witten, I. H. and MacDonald, B. A. (1990) Using concept learning for knowledge acquisition, in Gaines, B. R. and Boose, J. H. (eds) *Machine Learning and Uncertain Reasoning*, Academic Press, New York.

Wold, S., Esbensen, K. and Geladi, P. (1987) Principal component analysis, *Chemometrics and Intell. Lab. Systems*, **2**, 37–52.

Wu, T. J. and Lu, Y. Z. (1987) A feasible interaction prediction approach to large scale systems optimal controls, *Large Scale Systems*, **12**, 35-46.

Wu, Q. H., Hogg, B. W. and Irwin, G. W. (1992) A neural network regulator for turbogenerators, *IEEE Trans. Neural Networks*, **3**(1).

Xu, C. W. and Lu, Y. Z. (1987) Fuzzy model identification and self-learning for dynamic Systems, *IEEE Trans. Systems, Man and Cybernetics*, **SMC–17**(4).

Yager, R. R. (1983) Some relationships between possibility and certainty, *Fuzzy Sets & Systems*. **11**, 151–156.

Yang, Y. Y. and Lu, Y. Z. (1986) Development of a computer control model for slab reheating furnaces, *Computers in Industry*, **7**, 145–154.

Yang, Y. Y. and Lu, Y. Z. (1988) Dynamic model based optimization control for reheating furnaces, *Computers in Industry*, **10**, 11–20.

Yao, X. (1993) A review of evolutionary artificial neural networks, *Intel. J. Intelligent Systems*, **8**(4).

Ye, N. and Lu, Y. Z. (1989) Pattern recognition based fault detection systems in industrial processes, *Proc. World Congress of Chem. Eng.*, Tokyo.

Zadeh, L. A. (1965) Fuzzy sets, *Information and Control*, **8**, 338–353.

Zadeh, L. A. (1968) Probability measures of fuzzy events, *J. Mathematical Analysis and Applications*, **23**, 421–427.

Zadeh, L. A. (1973) outline of a new approach to the analysis of complex systems and decision processes, *IEEE Trans. SMC*, **3**(1).

Zadeh, L. A. (1988) Fuzzy logic, Computer, April.

Zheng, J. D. and Lu, Y. Z. (1988) Novel aggregation algorithm for large scale systems control, *Int. J. Control*, **48**(5).

Zhou, L, Ne, N. and Lu, Y. Z. (1989) Modeling and control for nonlinear time-delay system via pattern recognition, *Proc. 2nd IFAC Workshop of AI in Real Time Control.*

Zhou, T. and Kimura, H. (1993) A contribution to Sydney benchmark problem, *Preprints 12th IFAC World Congress*, Sydney.

Zimmermann, H. J. (1987) *Fuzzy Sets, Decision Making and Expert Systems*, Kluwer Academic, Boston, MA.

Zwick, R. and Wallsten, T. S. (1990) Combining stochastic uncertainty and linguistic inexactness: theory and experimental evaluation of four fuzzy probability models, in Gaines, B. and Boose, J. (Eds) *Machine Learning and Uncertain Reasoning*, Academic Press, New York.

Author Index

Subject Index